T0138198

Science in the Archives

SCIENCE IN THE ARCHIVES

Pasts, Presents, Futures

Edited by Lorraine Daston

The University of Chicago Press
Chicago and London

The University of Chicago Press, Chicago 60637
The University of Chicago Press, Ltd., London
© 2017 by The University of Chicago
All rights reserved. No part of this book may be used or reproduced in any manner
whatsoever without written permission, except in the case of brief quotations in critical
articles and reviews. For more information, contact the University of Chicago Press, 1427
E. 60th St., Chicago, IL 60637.
Published 2017.

26 25 24 23 22 21 20 19 18 17   1 2 3 4 5

ISBN-13: 978-0-226-43222-9 (cloth)
ISBN-13: 978-0-226-43236-6 (paper)
ISBN-13: 978-0-226-43253-3 (e-book)
DOI: 10.7208/chicago/9780226432533.001.0001

Library of Congress Cataloging-in-Publication Data
Names: Daston, Lorraine, 1951– editor.
Title: Science in the archives: pasts, presents, futures / edited by Lorraine Daston.
Description: Chicago; London: The University of Chicago Press, 2017. | Includes
    bibliographical references and index.
Identifiers: LCCN 2016028698 | ISBN 9780226432229 (cloth: alk. paper) | ISBN
    9780226432366 (pbk.: alk. paper) | ISBN 9780226432533 (e-book)
Subjects: LCSH: Scientific archives. | Scientific archives—History. | Science—History.
Classification: LCC Q224 .S35 2017 | DDC 026/.5—dc23 LC record available at https://
    lccn.loc.gov/2016028698

CONTENTS

This book is the fruit of the Working Group "Archives of the Sciences" convened at the Max Planck Institute for the History of Science, Berlin (MPIWG) in the summers of 2013 and 2014. At those meetings the Working Group presented drafts, prepared commentaries, and discussed for hours on end. After each meeting we revised our chapters in light of specific comments and criticism but also with an eye toward an emergent whole: the aim of our discussions was not only to improve each chapter but also to develop a shared framework for thinking about how the sciences choose to remember past findings and plan future research.

The ubiquity and longevity of our topic, scientific archives, heightened the challenge of forging a common way of thinking about these remarkable transgenerational enterprises. Astronomy and medicine, anthropology and genetics, philosophy and geology, history and meteorology—all create and conserve precious records in the most diverse media, from papyrus to punch cards to the electronic pulses of the digital age. The timescales of these archives range from decades to millennia. Without any pretension to comprehensive coverage of scientific archives, this volume nonetheless attempts to do justice to the scope and scale of the phenomenon by deliberately breaching the divisions among ancient, early modern, and modern periods, as well as between the natural and human sciences. The composition of the Working Group reflected this ambition and brought together scholars whose paths might not have crossed in the ordinary scheme of things. This book could only have come about through a collective effort, which in turn taxed the stamina, goodwill, and good humor of all involved.

As in the case of other Working Group volumes, this one benefited greatly from the support of the MPIWG and its staff. We are especially grateful to Regina Held, Tanja Neuendorf, and the entire staff of the MPIWG library. Without Josephine Fenger's careful attention to every aspect of our book-in-the-making, from bibliography to images, it might not have become a book at all. As always, it was a pleasure and a privilege to work with Karen Merikangas Darling and her staff at the University of Chicago Press. We thank them, one and all, most heartily.

Third Nature

Lorraine Daston

Scientific empiricism converts first nature into second nature. Under the carefully controlled conditions of the laboratory but also in the selective observations of the field, the teeming, tangled complexity of nature as given is slowed down or speeded up, winnowed or enriched, measured and modeled, probed by instruments, and translated into graphics. Indigestible first nature becomes intelligible second nature, and the scientific work of hypothesizing, testing, explaining, and predicting can begin.[1] But once second nature slips from science present into science past, collective empiricism requires a third nature: the repository of those findings of second nature selected to endure. These are the archives of the sciences.

We are in the midst of an archival moment, simultaneously overwhelmed by the sheer amount of available information ("drowning in data") and obsessed with its fragility ("the page you are looking for no longer exists"). New digital media conjure up Borgesian dreams of libraries with unlimited shelf space and nightmares of irretrievable loss,[2] by both intention (e.g., cyber sabotage) and inattention (e.g., incompatible software or obsolete hardware).[3] Juxtaposed bright and bleak visions of archival futures at moments of media shift are not new; the transition from manuscript to print in early modern Europe kindled similar hopes and fears.[4] But the speed, the scale, and the specificities of the latest media revolution matter: more people are manipulating more information in more ways, and all at a tempo that baffles "what next?" predictions. These are the moments of expanded possibility that stimulate the archival imagination in both prospective and retrospective directions. What will the archives of the future look like? How can we safeguard the continuity of extant archives (and the disciplines that rely upon them) when all is in flux? The carefully crafted links that connect past, present, and future are once again being reforged, and the archives of the sciences (and the humanities) are in the thick of things.

In both the natural and human sciences,[5] archives of the most diverse forms—the herbaria of botanists, the observational records of astronomers, the data banks of geneticists, the fossil compendia of geologists, the microfiches of anthropologists, the digital silos of meteorologists, and of course the libraries of tablets, papyrus rolls, parchment manuscripts, printed books, and ebooks stretching from Antiquity to now—make cumulative, collective knowledge possible. Their contents are the repository of what a discipline considers worth knowing and preserving; their practices, including storage, classification, and retrieval, are the precondition for (and often the essence of) research. Whether it is a botanist consulting the type specimen of a plant species in the Linnaean herbarium now housed in a London strong room, a historian of Antiquity consulting the paper squeeze of a Latin inscription, a physician consulting the description of a remarkable case published in a medical journal a few years or a few centuries ago, or a sociologist consulting the digital traces amassed by a lifelogger, empiricism is as much about forays into the disciplinary archives as it is about laboratory experiments, observatory vigils, or field expeditions.

Yet in contrast to laboratories, observatories, or even the more amorphous field, archives are mostly invisible in accounts of the sites and practices of science. The one glaring exception is the discipline of history, and it is the exception that proves (indeed, enforces) the rule. Since the mid-nineteenth century, historians have baptized themselves in the dust of archival research: wheedling entry into collections usually off limits to the public, deciphering strange scripts and bad hands, piecing together a narrative from scraps and fragments, and reading gimlet-eyed against the grain, alert to how context and perspective have warped whatever documents survive—these are the rites of passage through which most professional historians have passed since universities all over the world imitated Leopold von Ranke's famous Berlin seminar of the mid-nineteenth century.[6] So complete and exclusive has the identification of archive with the discipline of history become that any other kind of archival research is assumed to be ipso facto historical in nature, and any archive to be of the sort protypically investigated by historians: a fixed place with a curated, often official collection consisting mostly of old unpublished papers. Against this background, the Protein Data Bank or Ptolemy's *Handy Tables* hardly look like archives.

Nor do practices such as data mining or doxography seem archival. Not only does archival research dominate the imagination of the historians; the historians' archives dominate our collective imagination of all archival research.

This is one reason why the archives and archival practices of other sciences have largely escaped notice, much less a cross-disciplinary conceptualization comparable to that undertaken for the laboratory or the field in recent studies in the history, philosophy, and sociology of science. Another related reason is that not all scientific archives are about memory of the past. Some certainly are, as Geoffrey Bowker has shown in his seminal study of scientific "memory regimes, which articulate technologies and practices into relatively historical constant sets of memory practices that permit both the creation of a continuous, usable past and the transmission sub rosa of information, stories, and practices from our wild, discontinuous, ever-changing past."[7] But others are as much about the future as about the past, laying up stores (whether of astronomical observations or snapshots of daily life in the mid-twentieth century) for generations to come. And still others are firmly rooted in the here-and-now, using archives to generate and test hypotheses about everything from gene expression to consumer buying patterns. Finally, the dizzying variety of scientific archives, both in their material form (cuneiform tablets, fossils, microfiche, books, electronic impulses, DNA sequences) and their associated practices (reading, excerpting, transcribing, collecting, collating, standardizing, analyzing, cleaning, visualizing, stabilizing, digitizing, searching, forgetting), obscures deeper affinities and continuities of function and practice.

The aim of this book is to reveal these affinities and continuities among the sciences of the archives across many disciplines and centuries. Examples range from astronomy to medicine, paleontology to philology, history to genetics. Not all of the examples come from established disciplines: some are disciplines in the making, such as data management; others are archival practices shared by numerous disciplines, such as data mining; still others are potential sciences inspired by novel practices, such as the knowledge of self and others generated by digital data gathering. The time span reaches backward to the astronomical diaries of ancient Mesopotamia and forward to today's Google PageRank algorithms. Hard-headed economics and tough-minded politics are ubiquitous: archives are often expensive; access to them is rarely uncontroversial. But wild-eyed fantasies also shimmer through the arcana of data: dreams of immortality, gigantism, and omniscience. Often, ambitious scientific archive projects feed upon utopian visions of community: communities in which data is freely shared rather than hoarded, large-scale empirical investigations are coordinated rather than competitive, and future generations of scientists are grateful for rather than indifferent to the stores laid up for them by their diligent predecessors.

Because archive practices cut across the sciences and connect the sciences to bureaucracies, markets, and media technologies, the "science" in the title of this book is understood inclusively. There is no way of disentangling, for example, record keeping in ancient astronomy from that of early Mesopotamian bureaucracies (Hsia, chapter 1) or the scholarly techniques of compiling a biblical concordance from those of semi-automatically indexing the abstracts of scientific journals (Rosenberg, chapter 11) or data mining algorithms used to detect patterns in scientific data from those seeking associations between who buys what products in the supermarket (Jones, chapter 12). The kinds of comprehensive archives of the everyday once envisioned by social scientists to record everything from dreams to playing darts in the pub are now echoed in the minute-by-minute registration of personal data by life-loggers—and the latter may well become fodder for the former in the not-so-distant future (Lemov, chapter 10). When Big Science, both the word and the thing, was invented in the nineteenth century, it was the classical philologists who were in the van, admiringly and enviously imitated by the astronomers and other natural scientists (Daston, chapter 6). To impose the modern, restrictive sense of the English word "science" on other times and places obviously risks anachronism, but the point is deeper than that. Archive practices run both broad and deep, criss-crossing the boundaries among the divisions of knowledge as well as those between knowledge and practice and surviving revolutions in media and thought. Science in the archives is necessarily science writ large.

Do these common themes really have a common referent? No one doubts the existence and significance of scientific archives, especially not in our own data-dazzled epoch, but can they usefully be studied together and compared across disciplines and periods? What is to be gained by such a comparative perspective? The remainder of this introduction addresses these questions, drawing upon the essays in this volume for its illustrations.

WHAT IS A SCIENTIFIC ARCHIVE AND WHO USES IT?

Although the prototypical archive may now be the sort consulted by Ranke at the Vatican—an institutionally based collection of documents not meant for the public eye but sometimes invoked to secure legal claims or adduce precedents, stored in a dedicated space and maintained (however negligently) by a designated staff—, recent work on the history of archives, even of the sort most familiar to historians, reveals how anomalous and anachronistic this image is. Ancient Mesopotamian archives stored primarily receipts of economic transactions, apparently only for a limited period of time; ancient Greek and Roman archives tended in contrast to be repositories of official legal (and some private) documents.[8] Royal and ecclesiastical archives in early medieval Europe were often carted around with their owners, even into battle, so that they might be consulted whenever the need arose. Until the thirteenth century, when the Fourth Lateran Council required that minutes of church councils be preserved and legal procedures for attesting contracts shifted from sworn witnesses to notarial certification, documentation was sparse and haphazard.[9] Only in the early modern period did archives become indispensable instruments of state administration, pinning monarchs like Philip II of Spain and Louis XIV of France to their desks and creating (at least in the bureaucratic imagination) a vast, sedentary machinery of record-keeping and retrieval, with permanent offices as opposed to portable chests in princely households at the Uffizi, Escorial, and Versailles.[10] The physical form of what is archived and therefore storage and retrieval strategies have been as various as the contents: clay tablets, papyrus rolls, parchment codices, leather-bound cartularies and registers but also loose stacks of paper, filing folders and index cards, punch cards, microfilm, hard drives. Contemporary digital information science is only the most recent chapter in a long history of sophisticated archival techniques that make third nature in the sciences possible.[11]

Given the historical heterogeneity of archives, it makes little sense to seek a definition bound by form, location, content, or proprietor. The boundaries between archives and collections (including museums, libraries, and data banks) have also been historically fluid. Scientific archives may consist of digital data stored on servers, paper articles in bound journals stored in libraries, or fossils stored in the drawers of museum cupboards. Functions and usage vary as well, but scientific archives generally share two properties: they are opportunistic and open-ended. These two features stem from the same root cause, the unpredictable development of research agendas. No one knows in advance what questions future historians or climatologists will pose and what traces from the present (and whatever of the past has already been preserved) will be needed to answer them. This is the pathos of time capsules, stuffed with cherished items willed by the present to an amused and bemused future that can make neither head nor tail of them. Long-lived research programs (e.g., planetary astronomy from Ptolemy to Copernicus) may accumulate equally long-lived archives (e.g., observations of planetary positions), but no research tradition lasts forever, or it ceases to be research (Hsia, chapter 1).

Historical consciousness of the mutability of science has its own history, which emerges only from a cross-historical, cross-disciplinary perspective. Archives assembled prior to the early modern period tended to be highly selective: only opinions of philosophers worth remembering were recorded by doxographers (Taub, chapter 4); only observations by certified authorities that jibed with the best mathematical models were handed down by astronomers (Hsia, chapter 1). The sheer cost in time, money, and labor involved in copying manuscripts promoted parsimony (and large gaps) in centuries-long chains of transmission. The printing press eased these constraints without removing them: the earliest scientific journals and other published collections of observations seemed copious in comparison to manuscript predecessors but scanty by the standards of nineteenth-century scientific compendia, which aimed for exhaustive documentation (Daston, chapter 6; Sepkoski, chapter 2). Archive use had always been decontextualizing, ferreting out whatever nugget—quotation, observation, measurement—serves the present occasion and neglecting everything around it. But by the mid-nineteenth century, awareness of the revisability of science, both its questions and answers, had become so acute that opportunism was celebrated: it was precisely the flexibility of the archives, which could be enlisted in the service of rival historical narratives or medical nosologies or astronomical theories, that proved their lasting value (Marchand, chapter 5; Mendelsohn, chapter 3)—and stoked deliberately unselective ambitions to save everything for posterity in ever more compact media, from index card to microfiche (Lemov, chapter 10). The dream of the infinite archive—a nightmare for early modern savants paralyzed by "too much to know"—was born.

Opportunism means that scientific archives must be reconfigured to serve new lines of inquiry, over and over again. Geophysicists who suspect seismic periodicities may ransack medieval chronicles for dated reports of earthquakes, just as tree rings may become part of the meteorological archive of rainfall records. Roman inscriptions written on mundane topics in ungrammatical Latin transcribed by local antiquarians were of little interest to classical philologists—until they shifted their attention from the Ciceronian and Horatian literary texts that had made the classics classical to questions about everyday life in the ancient Mediterranean world, for which inscriptions were an invaluable source (Daston, chapter 6). New hypotheses create new archives.

They also retool old ones. Fossil collections have been a scientific archive since at least the eighteenth century, but mid-twentieth-century paleontologists exploring the possibility of mass extinctions put nineteenth-century fossil compendia to new statistical uses (Sepkoski, chapter 2). The description of an odd case may languish for decades, if not centuries, in a medical journal, only to light up with neon significance once a few others like it are perceived as a syndrome rather than a singularity (Mendelsohn, chapter 3). In a darker key, the members of the Havasupai Tribe of New Mexico were dismayed to learn that blood data collected by geneticists allegedly for the exclusive purpose of investigating the genetic basis of diabetes had later been used to assert abnormally high rates of schizophrenia and depression in their community (Gere, chapter 8). The open-ended scientific archive carries a retrospective empirical cachet: because its contents were not originally assembled with a later hypothesis in mind, no suspicion of stacking the evidentiary deck taints confirmations based on its holdings.

Open-ended exploitation of scientific archives often shifts the dividing line between collections accidentally and intentionally assembled. DNA and the fossils embedded in the earth's stratigraphy, metaphorically described as nature's archives, count as accidental from the standpoint of human agency. Analogously, archives intentionally constituted for one purpose but later used for an entirely different one are often treated as accidental, on an epistemic par with nature's own archives. This shift can also occur in the purely human archives, as historians happily reassemble the archival "mosaic stones" prospected by their predecessors into entirely different narratives (Marchand, chapter 5). However, these reappropriations ignore at their peril the ways in which the original context leaves durable if invisible fingerprints on the archives, e.g., when metadata is lost in the migration from an archive of physical specimens to one of data points (Sepkoski, chapter 2).[12]

The most obvious of these traces are the technologies of collection and transcription. Although astronomers are always looking for ways to lengthen the baseline of observations for phenomena that occur on a superhuman timescale (e.g., stellar motion), they often balk at the inclusion of observations before the seventeenth-century invention of telescopic sights (Hsia, chapter 1). Less obvious, especially at more than a decade's distance, are the ways in which the archive has been constructed to serve specific users with specific tools. Climatological data collected under the auspices of the US National Climate Program is tailored for certain clients and purged of artifacts by software that in turn depends on specific models, despite keen awareness that this archive must serve future and as yet unforeseen uses as researchers trawl for patterns and trends (Janković, chapter 9). Making data commensurable is a labor of Hercules: methods, instruments, records, and observers must be calibrated across polities, epochs, and genres: how to stitch together the weather observations from a Victorian ship captain's log with those of Soviet lay observers in Siberia—and both of these with the latest satellite feed?[13] These caveats hold in spades for commercially generated archives and the techniques used to mine them: marketing researchers and academic statisticians, for example, do not see eye-to-eye on retrieval algorithms (Jones, chapter 12).

The users of scientific archives (and potential scientific archives, since opportunism makes almost anything fair game) are as miscellaneous as the uses. Today's enthusiasts for digitalizing every waking moment of their lives may fondly imagine their grandchildren replaying the scene in which grandpa meets grandma, but the fate of Buckminster Fuller's self-avowed experiment in self-archiving—now held by Stanford University—suggests a rather different, more scholarly clientele (Lemov, chapter 10). "Clientele" is increasingly the mot juste for such archives, designed from the outset for commercial purposes (Jones, chapter 12; Janković, chapter 9). Politics as well as economics plays a role in determining who may use scientific archives, and who owns them. Historically, public access to archives has been rare, and secrecy governed their administration.[14] The case of early modern Venice—which did not permit even a doge to enter the state archive unaccompanied and stipulated that the archivist be illiterate[15]—is extreme but not exceptional. Nor was science any different. Early modern astronomers such as Tycho Brahe asserted ownership over their observations, and astronomers' heirs were still selling them off for tidy sums throughout the eighteenth century (Hsia, chapter 1). The ethos of public access to state archives and of scientific communality (or at least reciprocity) in data sharing may turn out to have been a historical oddity of the nineteenth and twentieth centuries. Only with the utmost reluctance do molecular biologists contribute their data to the Protein Data Bank, despite sanctions and scolding (Strasser, chapter 7); the right of indigenous peoples to access the data of the Genographic Atlas came only after acrimonious battles waged over other scientific archives cataloging so-called "vanishing peoples" (Gere, chapter 8). And then there is the great unknown of future users: who will they be, and what will they be looking for, and how? Or will they be looking at all? It is not unknown for laboriously, expensively constructed scientific archives, even recent ones, to fall into a Sleeping Beauty slumber, visited only by the occasional historian.[16]

THE PRACTICES OF SCIENTIFIC ARCHIVES

It is practices, not uses or users, that compose the basso continuo of scientific archives. Despite a long, ever-changing series of material media and technologies for keeping records, many of these archival practices have proved remarkably durable and fungible, as a longue durée perspective reveals. The excerpting, annotating, and collating practices that Renaissance humanists applied to books were seamlessly adapted by early modern naturalists;[17] in the twentieth century, collection practices in biology survived the transition from field to laboratory.[18] The practices of scientific archives are in turn part of a much larger history of scribal practices in general.[19] We might roughly divide these practices into four categories: acquisition, retrieval, reconfiguration, and transcription.

Of necessity, all archives, even those that aspire to exhaustive totality, are selective. But scientific archives filter their holdings more radically than most, both prospectively and retrospectively. Retrospectively assembled archives, like the Babylonian astronomical diaries edited by Otto Neugebauer and his collaborators in the mid-twentieth century, involve emendation of ancient observations through philological analysis and mathematical computation in order to standardize entries in a form usable by modern astronomers and historians (Hsia, chapter 1). Prospectively assembled archives, like the nineteenth-century compendia of all fossils held in major collections or of all stars down to the fourteenth magnitude, were even more attentive to the standardization of instruments, description and measurements, and published presentation (Sepkoski, chapter 2; Daston, chapter 6). Still stricter criteria regulate what can and cannot enter into contemporary scientific archives-in-the-making such as the US National Climate Program and the Protein Data Bank (Janković, chapter 9; Strasser, chapter 7). The data allowed into a scientific archive are not "givens"; they are groomed and pruned like topiary.

As anyone who has ever used any kind of archive knows, it is only as good as its finding aids. Indices, concordances, and hand lists are among the older tools of the retrieval trade; stop lists and search algorithms, among the newer. Yet the continuities are sometimes as striking as the contrasts between old parchment and flickering computer screen. One of the pioneering projects in what is now called the digital humanities, a concordance of the works of Thomas Aquinas using IBM punch cards and a modern scriptorium of women engaged to type in the entries, was the brainchild of the Italian Jesuit Roberto Busa, whose 1946 doctoral dissertation tracked instances of the preposition "in" throughout the Thomist corpus using the time-honored manual methods of biblical concordances (Rosenberg, chapter 11). Yet the gargantuan scale of the Web, as well as the vaulting ambition of commercial firms like Google, have swamped even the mechanized versions of the older finding aids, as well as the algorithms of artificial intelligence and machine learning. New strategies that elevate quick-and-dirty efficiency over exhaustive rigor govern today's search algorithms (Jones, chapter 12). Total recall is not always a desideratum: just as archival acquisition is selective, retrieval is deliberately imperfect and incomplete. Third nature cannot be allowed to collapse into second nature. In the archive of medical cases, even the most thorough bibliography permits serendipitous discoveries, and bibliography itself is made possible by institutionalized forgetting of the vast majority of cases in hospital archives (Mendelsohn, chapter 3).

The contents of archives exist to be reconfigured—into patterns, periodicities, taxonomies, and generalizations, but also as narratives, revelations, and arguments. Archives are often likened to artificial memory, more capacious and durable than the human kind, but working on the same principle. Such metaphors make the digitized self project plausible: if the self is the sum total of a person's memories, and digital records are the equivalent of memories (only better), then the digitized self is a simulacrum of the flesh-and-blood self (only better) (Lemov, chapter 10). But the analogy is misleading. Decontextualized fragments are rarely memorable; humans habitually structure their recollections as stories rather than as combinations and permutations. The ancient natural philosophical archive seems to have invited reconfiguration more along the lines of natural memory. Although the doxographies of Aristotle and Galen now serve historians as an archive of views of philosophical schools whose works have not survived, for ancient—and indeed, modern—philosophers, to review the positions of one's predecessors is the springboard onto developing new positions of one's own (Taub, chapter 4). This ancient archival practice has been preserved like a fly in amber for millennia in the teaching traditions of philosophy.

Long-lived scientific archives straddle media epochs and survive the transition only if the discipline succeeds in transcribing the contents from one medium to another. Astronomy is the paradigm case, with observations stretching in a chain from cuneiform tablets to papyrus rolls to parchment codices to paper books to digital databases. Transcription is anything but mechanical: each moment of transcription is an occasion for commensuration of old and new disciplinary standards for reliable data—but also for loss of metadata as well as the detection of old errors or the insinuation of new ones. NASA's *Five Millennium* canons (2006, 2009) of lunar and solar eclipses transmits, corrects, and computes its dates and regions of visibility from 2000 BCE to 3000 CE, a massive effort to transcribe the astronomical archive (Hsia, chapter 1). Without transcription, archives remain scattered and secluded, even within a single medium and epoch. The nineteenth-century compendia of the world's major fossil collections or all known Latin inscriptions from Antiquity created new, more portable print archives accessible to anyone with a good library. But the new, handier archives do not supplant the old ones; layers of transcription simply accumulate.

In addition to the fossils, handwritten catalogs of museum collections, nineteenth-century printed compendia, and twentieth-century electronic databases continue to coexist, just as the photographic plates, the printed engravings, and now the digitalisation of the Great Sky Map images do (Sepkoski, chapter 2; Daston, chapter 6). At any moment, a query could send a researcher burrowing down through the layers of transcription in search of some overlooked detail that has suddenly become crucial.

These shared practices of scientific archives become visible only when the historical angle widens; otherwise, the hype of the present moment outshouts extrapolations from the past and into the future. The point is not that there's nothing new under the sun but rather that every novelty requires that the archive be reconceived and often remade to guarantee its integrity. The archive is not and cannot be unchanging. But its usable past must be spliced and respliced with a mutable present in order to guarantee a usable future. Just because continuity is the essence of the scientific archive, to overlook the investment in making the resources of the past, present, and future commensurable, especially at moments of rupture—whether caused by new discoveries, new theories, new instruments, or new media—is to miss the main point of Third Nature: to annihilate time.

ARCHIVE PHANTASMAGORIA

Why do archives haunt the contemporary imagination? Influential essays by Michel Foucault and Jacques Derrida have propagated the archive metaphor far and wide in the humanities as a figure for the grid of possible utterances or the psychoanalytic quest for origins (and patriarchal command), respectively.[20] Following this lead, proliferating metaphors of the archive accent its ominous, coercive character, both explicit (Derrida's sense of the archive as jussive) and implicit (Foucault's sense of the archive as the tacit rules governing what can and cannot count as truth). New technologies—in the sixteenth and seventeenth centuries the printing press,[21] in the early twentieth century film,[22] and now the World Wide Web[23]—certainly breed both utopian and dystopian archive fantasies. Every new medium conjures its own all-encompassing yet compact documentary projects: long before the Internet, the printing press, photography, cinema, the microfiche, the punch card, and other technologies inspired plans for grand encyclopedias, universal bibliographies, planetary film archives, indexes of everything, and other attempts to "document the world."[24] Visions of a brave new world of data mining are simply the obverse of specters of a brave new world of total surveillance. Yet despite the screaming contrasts between these material media and technologies of then and now, not to mention between the actual archives they support, there is a curious sameness to the oversized fantasies they spawn.[25] The same images of "avalanches," "deluges," and "floods" that overwhelm; of gigantic "granaries," "warehouses," and "data bases" to be preserved and exploited; of bird's-eye overview and minute-by-minute micromanagement made possible by such stores; of "neutral" facts and "raw" data eternally relevant and endlessly recombinable; and, above all, of the transcendence of time, of past, present, and future merged in the archives.

Fantasies of the last sort are particularly vivid among the proponents of scientific archives. For some sciences, such as astronomy and paleontology, the superhuman timescale of their objects of inquiry invites yearnings for what one booster called a "panoramic cosmic cinematography" of celestial events (Hsia, chapter 1); the spindle diagrams of the waning and waxing of biological taxa over eons also feed the desire to encompass huge swathes of time at a glance (Sepkoski, chapter 2). For others, such as philosophy and history but also human genetics, it is the continuity of traditions that is at stake, Aristotle and Herodotus enlisted as witnesses to ideas and cultures that have otherwise been swallowed up by time and decay (Taub, chapter 4; Marchand, chapter 5) or peoples alleged to be disappearing (Gere, chapter 8). For still others, the allegedly flat landscape of data, past and present, enlarges the scope of pattern seeking, whether of disease symptoms, word clusters, climate trends, or suspicious associations, and promises a science of predictions without waiting for explanations (Mendelsohn, chapter 3; Rosenberg, chapter 11; Janković, chapter 9; Jones, chapter 12). Finally, archives seem to hold out the promise of immortality, both personal (Lemov, chapter 10) and disciplinary (Daston, chapter 6). The fact that actual archives are inevitably also sites of disciplinary strife (Strasser, chapter 7) does not seem to weaken the hold of these resurgent fantasies. Hovering over the concrete archives examined in this volume, each firmly anchored in its time and place, is a vision of timelessness, science's third nature aspiring to first nature's own scale.

NOTES

1. Bruno J. Strasser, "Collecting Nature: Practices, Styles, and Narratives," *Osiris*, 2nd ser., 27 (2012): 303–40, on 320. See also the chapters by David Sepkoski (chapter 2) and Andrew Mendelsohn (chapter 3) in this volume.

2. On the fears of irretrievable loss and the museal and archival responses it pro-

vokes, see Fernando Vidal and Nélia Dias, "Introduction: The Endangerment Sensibility," in Vidal and Dias, eds., *Endangerment, Biodiversity, and Culture* (London: Routledge, 2016), 1–38.

3. See for example Jerome McGann, "Information Technology and the Troubled Humanities," 49–66, and Julia Flanders, "The Productive Unease of 21st-century Digital Scholarship," 205–18, both in Melissa Terras, Julianne Nyhan, and Edward Vanhoutte, eds., *The Digitial Humanities: A Reader* (Burlington, VT: Ashgate, 2013), for analyses of this ambivalence in the humanities—mutely expressed by the very format of the Reader, which publishes mostly blog posts in an old-fashioned hardcover book.

4. Ann Blair, *Too Much to Know: Managing Scholarly Information before the Modern Age* (New Haven: Yale University Press, 2010).

5. The term "human sciences" is here used as shorthand for the humanities and social sciences, roughly corresponding to the French *sciences humaines*. Disciplinary archives, both past and present, offer a striking example of practices that cut across the divisions of knowledge.

6. On the rites of the archives for historians, see Carolyn Steedman, *Dust: The Archive and Cultural History* (Manchester: Manchester University Press, 2001), and Arlette Farge, *The Allure of the Archives*, trans. Thomas Scott-Railton (New Haven: Yale University Press, [1989] 2013); on Ranke's influence, see Kaspar Eskildsen, "Leopold Ranke's Archival Turn: Location and Evidence in Modern Historiography," *Modern Intellectual History* 5 (2008): 425–53. See also Suzanne Marchand's chapter 5 in this volume.

7. Geoffrey C. Bowker, *Memory Practices in the Sciences* (Cambridge, MA: MIT Press, 2005), 9.

8. Maria Brosius, "Ancient Archives and Concepts of Record-Keeping: An Introduction," in Maria Brosius, ed., *Ancient Archives and Archival Traditions* (Oxford: Oxford University Press, [2003] 2011), 1–16.

9. Markus Friedrich, *Die Geburt des Archivs: Eine Wissensgeschichte* (Munich: Oldenbourg, 2013), 32–52; Dietmar Schenk, *"Aufheben, was nicht vergessen werden darf." Archive vom alten Europa bis zur digitalen Welt* (Stuttgart: Franz Steiner Verlag, 2013), 58–75.

10. Peter Burke, "Reflections on the Information State," in Arndt Brendecke, Markus Friedrich, and Susanne Friedrich, eds., *Information in der Frühen Neuzeit: Status, Bestände, Strategien* (Berlin: LIT Verlag, 2008), 51–64; Jacob Soll, *The Information Master: Jean-Baptiste Colbert's Secret State Intelligence System* (Ann Arbor: University of Michigan Press, 2009). Historians of these early modern archives repeatedly warn against conflating the fantasies of their boosters such as Gottfried Wilhelm Leibniz or Jean-Baptiste Colbert with the messy, inefficient, rat-infested reality: Markus Friedrich, "Archiv und Verwaltung im frühneuzeitlichen Europa: Das Beispiel der Gesellschaft Jesu," *Zeitschrift für Historische Forschung* 35 (2008): 369–403.

11. For a recent overview of how digital methods are transforming research in a range of sciences, see Kirsty Williamson and Graeme Johanson, eds., *Research, Information, Systems, and Contexts* (Prahran, Australia: Tilde, 2013).

12. Bowker, *Memory Practices*, 183.

13. Paul N. Edwards, *A Vast Machine: Computer Models, Climate Data, and the Politics of Global Warming* (Cambridge, MA: MIT Press, 2010), 12–25.

14. So closely were archives associated with the secrets of state power that when Parisian crowds stormed the Bastille on 14 July 1789, they made a point of not only

freeing the prisoners but also pillaging the archives: Lara Jennifer Moore, *Restoring Order: The École des Chartes and the Organization of Archives and Libraries in France, 1820–1870* (Duluth, MN: Litwin Books, 2001), 1–16.

15. Burke, "Reflections," 58.

16. Rebecca Lemov, "Filing the Total Human: Anthropological Archives at Mid-Twentieth Century," in Charles Camic, Neil Gross, and Michèle Lamont, eds., *Social Knowledge in the Making* (Chicago: University of Chicago Press, 2011), 119–50.

17. Brian W. Ogilvie, *The Science of Describing: Natural History in Renaissance Europe* (Chicago: University of Chicago Press, 2006); Isabelle Charmantier and Staffan Müller-Wille, "Carl Linnaeus's Botanical Paper Slips (1767–1773)," *Intellectual History Review* 24 (2014): 1–24; Lorraine Daston, "The Sciences of the Archives," *Osiris* 27 (2012): 156–87.

18. Strasser, "Collecting Nature: Practices, Styles, and Narratives."

19. There is a large and scattered literature on this subject covering diverse epochs and cultures. In addition to works already cited, studies of particular relevance to archival practices in the sciences include: Helmut Zedelmaier, *Bibliotheca Universalis und Bibliotheca Selecta: Das Problem der Ordnung des gelehrten Wissens in der Frühen Neuzeit* (Cologne: Böhlau, 1992); Cornelia Vismann, *Akten: Medientechnik und Recht* (Frankfurt am Main: Fischer Verlag, 2000); Peter Becker and William Clark, *Little Tools of Knowledge: Historical Essays on Academic and Bureaucratic Practices* (Ann Arbor: University of Michigan Press, 2001); Anke te Heesen and Emma Spary, eds., *Sammeln als Wissen: Das Sammeln und seine wissenschaftsgeschichtliche Bedeutung* (Göttingen: Wallstein Verlag, 2002); Ann Blair and Jennifer Mulligan, eds., *Toward a Cultural History of the Archives,* special issue, *Archival Science* 7 (2007); Delphine Gardey, *Écrire, calculer, classer. Comment une révolution de papier a transformé les sociétés contemporaines (1800–1940)* (Paris: Éditions de la Découverte, 2008); Blair, *Too Much to Know*; and Markus Krajewski, *Paper Machines: About Cards and Catalogs, 1548–1929*, trans. Peter Krapp (Cambridge, MA: MIT Press, [2002] 2011).

20. Michel Foucault, *L'Archéologie du savoir* (Paris: Gallimard, 1969); Jacques Derrida, *Mal d'archive: Une impression freudienne* (Paris: Galilée, 1995). Friedrich, *Geburt*, 21–23, offers a brief, cogent overview of the recent proliferation of archive metaphors in cultural studies.

21. Blair, *Too Much to Know.*

22. Paula Amad, *Counter-Archive: Film, the Everyday, and Albert Kahn's Archives de la Planète* (New York: Columbia University Press, 2010).

23. David Weinberger, *Too Big to Know: Rethinking Knowledge Now That the Facts Aren't the Facts, Experts Are Everywhere, and the Smartest Person in the Room Is the Room* (New York: Basic Books, 2011).

24. See for example Arndt Brendecke, Markus Friedrich, and Susanne Friedrich, eds., *Information in der Frühen Neuzeit: Status, Bestände, Strategien* (Berlin: LIT Verlag, 2008); Gregg Mitman and Kelley Wilder, eds., *Documenting the World* (Chicago: University of Chicago Press, in press); Alex Wright, *Cataloging the World: Paul Otlet and the Birth of the Information Age* (Oxford: Oxford University Press, 2014); Amad, *Counter-Archive*; Krajewski, *Paper Machines*; Ronald Day, *Indexing It All: The Subject in the Age of Documentation, Information, and Data* (Cambridge, MA: MIT Press, 2014).

25. On the long history of data revolutions in the sciences, see Elena Aronova, Christine von Oertzen, and David Sepkoski, eds., *Data Histories, Osiris* 32 (2017), forthcoming.

PART I

Nature's Own Canon: Archives of the Historical Sciences

ONE

Astronomy after the Deluge

Florence Hsia

Of all practitioners of the sciences of deep time, astronomers may well be bottom-trawlers of the highest order. Astronomy's empirical basis consists of signals (radiation and high-energy particles) that travel long and far to reach earth-bound observers. The possible meanings of such distant celestial messages are often obscure to those who record them, their decryption and pattern dependent on the hindsight of some future generation of sky watchers. Whether one has in view the orbital motion of our nearest celestial neighbor or the beginnings of the universe itself, the longue durée of astronomical processes seems to demand an archive to match, one that ideally comprehends all observable celestial phenomena.

Plumbing astronomy's archival depths, however, has posed challenges. A 1998 progress report on the inclusion of the "historical literature of astronomy" in NASA's Astrophysics Data System characterized "old" items as those dating from before 1940.[1] Few astronomers today venture into what one of its leading practitioners calls "applied historical astronomy" to identify and extract material from distant disciplinary, linguistic, and cultural substrates.[2] Consider the fragmentary character of cuneiform testimony to the Babylonian astronomical legacy; frustratingly laconic remarks on eclipses and comets buried deep in classical Greek and Roman sources as well as medieval European chronicles; works in Persian, Arabic, Syriac, and Sanskrit composed across the Islamic world; and the possible manipulation of Chinese celestial observations during the imperial period for their political significance: such factors make contemporary data mining difficult, to say the least. The relief experienced by contemporary applied historical astronomers in handling early modern European records is palpable with the inception of the telescopic period, which brought "high quality observations" timed with enough accuracy and described with sufficient specificity to make them "usable."[3]

There surely was a sea change in early modern European astronomical practices that made the resulting material more familiar to modern eyes. Starting in the sixteenth century, reiterated observation of daily as well as long-term periodic phenomena gradually became the norm. Expectations for precision measurement also rose dramatically, a shift accelerated less by the telescope itself than by other innovations: micrometers that ran parallel lines across a telescope's optical field to allow for fine-grained measurements of angular separation; the application of telescopic sights to angle measuring instruments such as quadrants and sextants; and the pendulum clock for timing celestial events.[4] And yet the total astronomical archive with its exhaustive compilation of all observed celestial phenomena is a relatively recent desideratum, one whose logic is hardly inevitable. This essay studies its emergence from the approach to empirical research that dominated astronomical practice from classical Greece through the modern era: the canon of data curated for their probative value. The family resemblances between different modes of presenting astronomical observations hide a multitude of motivations that are often at odds with the modern archival ideal. Teasing them apart will help us better understand the complexities in the truism that "all astronomers use historical data."[5]

THE CANONICAL ARCHIVE

> You noticed, one doesn't write, luminously, on a darkened field. Only the alphabet of stars reveals itself that way, sketched or interrupted: man pursues black on white.
> —Stéphane Mallarmé, "L'Action restreinte" (1897)[6]

The Roman historian Flavius Josephus set the origins of celestial science in the depths of antediluvian time. Claiming in his *Judean Antiquities* that God had given the patriarchs such exceedingly long lives in part so they could make "discoveries" in astronomy and geometry, Josephus recounted how Seth's sons tried to safeguard their hard-won knowledge of the heavenly bodies from Adam's prediction of world-destroying fire and flood by inscribing their discoveries on two stelae, one brick, the other stone. Should the brick pillar be lost to the waves, the stone one would "remain and offer an opportunity to teach men what had been written on it and to reveal that also one of brick had been set up by them. And it remains until today in the land of Seiris."[7] Whether the stone stele told of knowledge forever lost with its brick counterpart in the receding waters, Josephus did not say, nor whether anyone could read its inscriptions following the confusion of tongues. But the threefold moral burden of postdiluvian astronomy for Josephus was unmistakable: to record, preserve, and pass one's findings on to future generations.

Amongst the earliest memory-traces of astronomy's archival dreams are the wedge-shaped marks that Mesopotamian scribes pressed into clay tablets, a form of inscription thought to parallel the "heavenly writing" the gods constellated upon the night sky.[8] Greek and Roman scholars testified wonderingly to the cumulative weight of such ancient star watching. Diodorus Siculus noted in his massive universal history that the Chaldaeans of Babylon had "observed the stars over a long period of time" (and included their claim—scarcely believable, he remarked—to 473,000 years of such study), while Pliny the Elder relayed reports by some that the Babylonians possessed 720,000, others 490,000, years' worth of celestial observations. Rumors of such Babylonian treasures enjoyed a long life in a wide range of textual traditions.[9] But where was this famed archive of 'heavenly writings'? Aristotle referred to the Egyptians and Babylonians, "who have long made observations over a very great number of years and from whom we have many reports about each of the heavenly bodies" to support his characterizations of planetary motion, and asked his nephew Callisthenes, traveling in Alexander the Great's company, to send such records from Babylon where, according to Porphyry, they had been preserved for 31,000 years. Yet the material seems never to have arrived, a point Simplicius deployed centuries later in his commentary on Aristotle's *On the Heavens* to excuse faulty efforts to "save the phenomena."[10] Another millennium passed and scholars still waited for tidings from Babylon. When the French astronomer Jean-Sylvain Bailly repeated claims for the antiquity of the Chaldaean astronomical tradition in his 1775 history of ancient astronomy "from its origin," he summarized learned opinion in giving them little credence, pointedly remarking, "We have few details on the nature of the Chaldaean observations."[11]

It was not until the late nineteenth century that the long-buried celestial riches Pliny described as "inscribed on baked bricks" began to be painstakingly deciphered from cuneiform tablets.[12] These include daily systematic observations of planetary phenomena, extant though far from continuous for the period from 652 to 61 BCE in a corpus of about 1,200 texts now known as the "astronomical diaries," a rendering of the term "regular watching" written on the edges and ends of the tablets.[13] The astronomical diaries, in turn, seem to be the general empirical foundation for other kinds of cuneiform texts, some drawn from the diaries and largely observational in character; others predictive in anticipating future celestial events on the basis of observed periodic phenomena; and still others comprising computed ephemerides of planetary phenomena in tabular form, together with related procedure texts indicating rules and parameters for their construction. Specialists in this area are keenly aware of the partial nature of their object of study. Most of the known texts survive as physical fragments long made inaccessible to scholars by the British Museum. A murmur of this difficult history is audible in an 1974 article by the preeminent scholar of the diaries for most of the twentieth century, Abraham Sachs, in which he published photographs of "two pieces rejoined for a few hours" thanks to the "cooperation of museum officials" from the British Museum and the Wellcome Historical Medical Museum and Library, institutions but "5 minutes' walk" from each other.[14] The laborious reconstruction of physical tablets and textual corpora comprises a significant aspect of scholarship on Babylonian celestial science, notably Otto Neugebauer's and Sachs' editions of nondivinatory cuneiform material in 1955 and Sachs' projected publication of all the known astronomical diaries and related texts, an endeavor continued by Hermann Hunger.[15] Together with editions of divinatory material, such compilations now constitute the archival landscape for Babylonian celestial science.[16]

Just how isomorphic these modern archival formations are with the record-making and record-keeping practices of scribes at work in the complex political and social circumstances of ancient Mesopotamia is unclear.[17] The late nineteenth- and early twentieth-century excavations that uncovered most of the relevant material were largely unsystematic, as were the processes by which the tablets arrived at their present institutional homes; neither their representativeness nor most details of their excavation sites are known.[18] Absent archaeological context, reconstitution of both tablets and texts has depended heavily on physical evidence, philological analysis of linguistic and generic characteristics, mathematical and technical consistencies, and correlations with modern retrodictions of astronomical phenomena. Yet these sundry approaches to reuniting material sundered by historical circumstance have been bound together by an archival conception far broader than the historically contingent and culturally specific bounds of ancient Mesopotamia. Neugebauer announced this ambitious research agenda with the 1962 publication of a curious ephemeris for planets visible to the naked eye that gave their positions from 601 BCE to 1 CE as seen at 7 P.M. "Babylon civil time"—"chosen because of the evening epoch of the Babylonian lunar calendar"—and in the Julian calendar, "which is standard for historical work."[19] Computed by the mathematician Bryant Tuckerman at the Institute for Advanced Study and the IBM Research Center, the ephemeris drew on modern planetary theories and methods of handling selected periodic perturbations to compute positions with a precision "somewhat—but not excessively—better than that to be expected from the material to be studied, which is based upon naked-eye observations."[20] Identifying records of ancient celestial events was but the first step; the ultimate goal was nothing less than the restoration of "a huge archive of well over a thousand texts *to its full usefulness for the astronomer* as well as for the historian of astronomy and the historian of the Hellenistic age . . . in all its aspects, philologically, historically, and astronomically."[21]

While the first installment of the Tuckerman tables opened the door to dating political, military, and economic changes[22] and establishing connections with other textual corpora and genres of Mesopotamian celestial science,[23] the larger relevance of Babylonian observational material was soon made clear. The tables' second volume gave planetary positions through the period in which astronomical work took place at the "important meridians" of Toledo, Greenwich, Hven, Prague, Constantinople, Alexandria, Damascus, Samarkand, and Ujjain; its seventeenth-century terminus "was chosen in order to include the observations of Brahe and Kepler which form the foundation of [Kepler's] 'Astronomia nova.'"[24] Allied efforts further extended the Tuckerman tables' coverage to nearly three millenia, thanks to the collaborative labor of astronomers, computer scientists, and historians.[25] By setting in motion these twinned projects—critical editions of ancient texts on the one hand, and long-term ephemerides based on modern planetary theories and computing technologies on the other—Sachs and Neugebauer put Babylonian astronomical observations on the same footing as those made anywhere or anytime, making an immense corpus of observational data visible to the contemporary astronomer.

As Neugebauer pointed out, the problem of secular accelerations—long-term variations in observed planetary positions resistant to control by gravitational theory—offered a key example of how historical records could advance astronomical knowledge. In 1695 Edmond Halley published his suspicion that the moon was moving faster in its path than it had in the past, based in part on his evaluation of observations made by the medieval Muslim astronomer al-Battānī. Half a century later Richard Dunthorne, Tobias Mayer, and Joseph-Jérôme Lalande drew on ancient and medieval eclipse observations to empirically determine values for the moon's secular acceleration. These early efforts indicated just how modest the acceleration at issue might be, perhaps 9 or 10 arcseconds of celestial longitude per century squared (even the modern value of about 26 arcseconds is a minute fraction of the moon's apparent diameter of about 1,800 arcseconds as seen from earth). Yet they also suggested just how valuable a set of observations made over a long timescale could be, all the more so when efforts to predict the quantity in terms of gravitational theory proved controversial.[26] While attempts through the nineteenth century and into the twentieth to run secular acceleration to ground revealed significant disagreement over the probative value of old observational material, the Tuckerman tables and its successors held open the possibility of its eventual resolution through "the recovery of great masses of detailed eclipse records, accurate data for phenomena like occultations, etc." that would "eventually give reliable early elements for the testing of empirical constants related to the problem of secular acceleration."[27] The Tuckerman tables themselves have long been superseded by ephemerides computed with techniques able to address the accuracy of planetary positions obtained by radar and laser ranging as well as optical measurements.[28] Yet the integration of historical and contemporary astronomical work projected through their gridded columns has developed rapidly. Today's practitioners of applied historical astronomy expect to draw from a totalizing archive of serialized celestial observations spanning centuries, continents, and cultural contexts, the long-term historical analogue to contemporary synoptic sky surveys that sweep the entire sky every few days to create a "panoramic cosmic cinematography" of celestial events.[29]

The successful configuration of Babylonian astronomical material as "the first and longest-running data collection project in world history" capitalized on a form of archival desire concretized in the modern genre of the eclipse canon.[30] Meant for scholars intent on eclipse records as a means of securely dating historical events, Theodor Oppolzer's *Canon of eclipses* (1887) provided instructions for using its tabulated data to calculate the visibilities of 8,000 solar and 5,200 lunar eclipses between 1207 BCE and 2161 CE with precision. In contrast with earlier such aids, Oppolzer chose the term "canon" to signal the work's status as "a standard for all future investigation": "the *Canon*, to justify its title, ought contain *all possible eclipses* taking place within its time span, even though the greater part of them have no relevance."[31] What counted as "relevance" was, to be sure, a product of Oppolzer's skewed sense of both the past and future of celestial observation. In mapping the approximate paths of solar eclipses, Oppolzer declared that "it would be meaningless to graph the eclipses that take place in the southern hemisphere," dismissing the possibility of observers at Santiago, Sydney, the Cape of Good Hope, and other points south.[32] The implied cultural canonicity in Oppolzer's remark was explicit in the work of his close collaborator Friedrich Karl Ginzel, whose 1899 "special canon" paired a table of revised

eclipse elements and visibilities at Babylon, Memphis, Athens, and Rome—
"the principal towns of the *civilized world*" between 900 BCE and 600 CE—
with a collection of a hundred or so eclipse records drawn from cuneiform
tablets and classical texts, each matched to an Oppolzer eclipse number.[33]
Yet the notion of a comprehensive astronomical archive has underwritten
the ongoing production of eclipse canons engaged directly with Oppolzer's
work. Of these, NASA's *Five Millennium* canons are the most spectacular example, listing a total of 11,898 solar and 12,064 lunar eclipses potentially
visible to earth-bound observers between 2000 BCE and 3000 CE.[34] The
canonical archive that such eclipse canons project is slowly being realized
through the patient work of scholars working to produce the historical
counterpart to the federated datasets of today's "virtual observatory."[35]

THE CANONIZED ARCHIVE

Oppolzer's use of "canon" drew on but also departed from its traditional
meanings. Claudius Ptolemy's *prócheiroi kanónes* (*Handy Tables*) comprised
tabular material largely drawn from the *Almagest* and accompanied by an
explanatory text to facilitate its use in solving computational problems.[36]
One particular item in the *Handy Tables* strikingly illustrates the term's rich
connotations: the so-called "royal canon" listing the names and reign periods of successive Babylonian, Assyrian, Persian, Macedonian, Egyptian, and
Roman rulers from the mid-eighth century BCE onwards. When European
scholars rediscovered the text of Ptolemy's royal canon in the early seventeenth century, they deployed it as a key to securely setting ancient realms
and periods in their proper temporal sequence. In publishing the text in
1619 as part of an ambitious work treating chronological cruxes from the
creation of the world to the destruction of the Second Temple, Johannes
Behm, for instance, pragmatically rendered *kanon basileion* as "succession of
kings."[37] Seth Calvisius, however, referred to it as Ptolemy's "mathematical
Canon," a table that took as its starting point not when Nabonassar actually assumed power, nor the official date of his accession, but noon on
February 26 of the first year of Nabonassar in order to count in (Egyptian)
years of 365 days without intercalation—a "form of year most suitable for
Astronomers" in its simplicity.[38] The royal canon thus provided a baseline
for reckoning time with the mathematical precision and computational
facility necessary for an astronomer's purposes, as did Ptolemy in making
the era of Nabonassar the epoch or starting point of the tables of mean motions he gave in the *Almagest* for calculating planetary positions both past
and future. It was, after all, from the historical king Nabonassar's reign that

"the ancient observations are, on the whole, preserved down to our own time." Throughout the work he depended heavily on the Royal Canon's regnal eras, frequently deploying Era Nabonassar to give equivalent dates for observations cited according to other chronological systems.[39] The etymological root of "canon"—a straight rod used for measuring (that is, a ruler)—thus ran through the term's various astronomical usages, at once designating an inscriptive format (tabular), a textual genre (instructions or rules for using instruments, including tables), and a standard for measuring intervals of time and distance.[40]

The royal canon's metrological function applied to both observations of celestial phenomena as well as their computation, making it a normative instrument that "not only registers what is but . . . also prescribes what has to be" in a sense well attested in other ancient works, whether in the context of aesthetics, ethics, medicine, or law.[41] In his exposition of models for planetary motion in the *Almagest*, Ptolemy treated observational material in a manner consistent with this regulative emphasis. In the case of his lunar model, for instance, Ptolemy presented "three ancient eclipses . . . selected from those observed in Babylon" some eight centuries before his time. Elsewhere Ptolemy cited Hipparchus as having used another set of three Babylonian eclipses drawn from "the series brought over from Babylon, and . . . observed there."[42] The kind of selection process hinted at here is critical in two respects. The first is procedural: only certain kinds of observations met the conditions necessary for the geometrical techniques Ptolemy used to construct his models, or that he stipulated in order to isolate particular elements of planetary motion and exclude known sources of error.[43] The second is epistemological: in contemporary terms, the "specific observations" brought to bear stand in for "the primary data or phenomena that the theory is to account for . . . and this account, a quantified geometrical model, is meant to be applicable to all phenomena *of the same kind at any time*."[44] Throughout the *Almagest* the citation of particular observations was characteristically motivated by their relevance for developing the matter at hand, whether as empirical foundation for a planetary model or as a demonstration of its predictive validity.[45] In contrast with the generalized celestial observations on the basis of which "the ancients" and other nonspecified "observers" were said to have come to a collective consensus on the sphericity of both the heavens and earth in the *Almagest*'s opening chapters, the particular observations Ptolemy adduced in its succeeding books worked to both measure and prescribe the limits of the planets' wanderings.[46]

What Ptolemy made visible in the *Almagest* was not the stock of observations on which he drew, but rather his selective curation of an archive already ancient in his own time.[47] Over a millennium later when Nicolaus

Copernicus set the earth in motion around the sun, the observations he poached from the *Almagest* handily outnumbered those he assembled from medieval and contemporary sources, including his own records.[48] Yet even as he read against the *Almagest*'s canonical grain, Copernicus adopted its sparing and motivated approach to the citation of empirical data. Though Georg Joachim Rheticus claimed in his *First Account* (1540) of what would become *On the Revolutions of the Heavenly Spheres* (1543) that his teacher had "always in view the observations of all the ages together with his own, collected in order as in catalogues," Copernicus published but ten specific observations in establishing his lunar theory, where Ptolemy had used fifteen; seven observations for each of the superior planets (Mars, Jupiter, and Saturn), compared to Ptolemy's five; and ten for Mercury to Ptolemy's sixteen.[49] For Copernicus, too, the public archive had to be canonized.

Nor is the comprehensive canonical archive familiar to modern astronomers evident in the print history of Copernicus' sole contemporary empirical source: the work of Bernard Walther.[50] Covering the period from 1475 to 1504, Walther's observations of planetary and stellar positions, eclipses, and altitudes of the midday sun number nearly 1,400 in total. They constitute "the earliest European observations of use to modern astronomy," "the earliest surviving series of nearly continuous observations," and "the most precise extended series of astronomical observations collected in the Latin West" prior to Tycho Brahe.[51] They were, moreover, read and used throughout the early modern period—three different editions in 1544, 1618, and 1666 point towards their continued availability—and well into the twentieth century.[52] But do the fortunes of the Walther observations track the emergence of a recognizably modern archival sensibility at work? Retracing their trail in printer's ink suggests the need for caution in assessing early modern attitudes towards the presentation of observational material. Though Walther's work represents one of the earliest instances of serialized astronomical observations to be circulated via the technology of print, it did so bound to commemorative and patronage agendas, not a new archival ideal. Walther's first editor, Johann Schöner, arrived in Nuremberg in 1526 as the mathematics instructor for a new gymnasium established by Philip Melanchthon. There he printed many works left in manuscript by Johannes Regiomontanus, the author along with his teacher, Georg Peurbach, of the *Epitome of the Almagest* (1496)—at once a translation, abridgement, and exposition from which Copernicus had learned how to read Ptolemy. The Nuremberg city library had acquired Regiomontanus' library as part of Bernhard Walther's estate, and Schöner drew on Walther's material as well.[53] In 1544 Schöner printed a collection of the "writings of the most illustrious mathematician, Johannes Regiomontanus." The small quarto opened with texts by Regiomontanus on instruments, together with Schöner's additions, and concluded with two previously printed works: another instrument text, this time by Peurbach, as well a short work by Regiomontanus on comets that Schöner had published in 1531.[54] Sandwiched in between were first observations made by Regiomontanus, a few in conjunction with Peurbach, followed by those carried out by Walther, identified in the text as Regiomontanus' "pupil." Together these constituted the "treasury of observations" Schöner spoke of so effusively in his dedication to the Nuremberg city council that had safeguarded Regiomontanus' legacy.[55] As great as Walther's labors were—his constituted the overwhelming majority of the observations printed—their role in Schöner's volume was to celebrate Regiomontanus' achievements.

When the professor of mathematics at Leiden, Willebrord Snell, reprinted Walther's observations decades later, he did so to further a family legacy in attracting the attention of his father's patron, the Landgraf Moritz of Hesse. In 1618 Snell published the *Hessen Observations* made by Wilhelm IV, Moritz's father, and by the astronomers Wilhelm and Moritz supported at their court observatory in Kassel.[56] The tome comprised observations made with Hessen support between 1561 and 1597 along with some by Tycho, whose extensive contacts with the Kassel astronomers were well known.[57] To these Snell appended the observations published by Schöner decades earlier, retaining the mise-en-page and even the foliation of the 1544 edition. The sharp contrast with the look of the Hessen and Tychonian material accentuated the earlier observations' venerable aura, an authenticity effect in keeping with Snell's elevation of Wilhelm and, by extension, his son as patrons of astronomy worthy of a lineage that Snell traced back to ancient times.[58] Determined readers were evidently not put off by such difficulties. Snell included a long excerpt from Johannes Kepler's 1604 work on optics in which Kepler closely examined Schöner's 1544 edition of Walther's observations to investigate the problem of refraction, and Kepler would later use Snell's 1618 reprinting to test the *Rudolphine Tables* (1627) prior to publication.[59] The striking lack of material, conceptual, and chronological integration in Snell's collection of observations, however, left the difficult work of technical collation largely to its readers.

CUMULATING DATA

Yet the increasingly unsettled state of early modern European celestial science greatly accelerated the systematic cumulation of observational data. Makers of new planetary tables, for instance, strove to show that their theories saved the most—or at least the most accurately recorded—phenomena.[60] Philips van Lansbergen's 1632 "treasury of astronomical observations"—subdivided by planet and type, and ranging from Antiquity to the author's own time—took up nearly as many pages as the tables they supported. Godefroy Wendelin rejected all reports prior to 1600 except those of Tycho as unreliable in the *Lunar Eclipses* (1644) he published in advance of his projected "Tabulae Atlanticae," instead compiling synchronous observations, as in the twenty timings he took of a lunar eclipse in 1642 (with the help of two young observers with good eyesight) and the multiple sightings of a 1635 lunar eclipse made by Ismail Boulliau in Paris, Jacob Golius in Leiden, Pierre Gassendi in Digne, Nicolas-Claude Fabri de Peiresc in Aix, Peiresc's correspondents in Cairo and Aleppo, and unnamed observers in Rome.[61] Wendelin concluded nearly every set of observations with a glowing remark—variations on "ACCURATISSIME," "EXACTISSIME," and "QUOD ERAT DEMONSTRANDAM" were favored formulations—on the predictive power of his tables.[62]

More fundamentally, the conceptual category of phenomena that needed explaining and refining had undergone tremendous transformation in the decades since Copernicus' canonizing of observations to establish planetary models within a heliostatic framework. The Jesuit mathematician Cristoforo Borri, for instance, tried to catalogue the spectacular efflorescence of what he called the "new appearances observed in our time" in his *Astronomical Collection* of 1631: the new star of 1572; the comets of 1577 and 1618; the telescopic observations made since 1609 that had shown lunar mountains, sunspots, Venus' phases, Mercury transiting the sun, Jupiter's moons and Saturn's odd shape.[63] The technologies that made such "new" celestial phenomena visible drove, in turn, the development of new visual technologies for recording, circulating, and preserving records of their appearances.[64] At the same time, the "common, and ancient appearances [*apparentiis*], which the Mathematicians call Phenomena" also came under greater scrutiny as astronomers grew increasingly uneasy with the fundamental parameters they supported. The question of whether the rate of equinoctial precession or the obliquity of the ecliptic had varied over time, and if so, how, focused attention on observational records over very long periods.[65] Astronomical practices, too, were changing in ways that implicated the integrity of observational records. Tycho Brahe's pioneering approach to systematically correcting the observed positions of celestial bodies for the distorting efforts of atmospheric refraction, for example, raised the concern that perhaps all observations of celestial objects except those at the observer's zenith required refinement.[66]

Early modern astronomers seeking data with which to address these problems relied heavily on colleagues, weaving webs of learned correspondence in which details of instrumentation, observational protocols, data in various forms, and procedures for their reduction were highly prized matters of both jealous and generous exchange. While personal papers constituted an important reservoir of observational material in the period, balancing delicate networks of manuscript circulation was a complex social process.[67] Unsurprisingly, astronomers turned to a number of print genres in which celestial observations were systematically cumulated. The compilation of cometary histories had flourished long before Tycho Brahe elevated comets to celestial status, especially among Lutheran scholars influenced by Philip Melanchthon's interest in astrology for the insights it could give into divine providence. Sifting all sorts of texts for reports of comets and the earth-shattering events that followed in their wake, writers of cometary "catalogues" and "histories" sought to overwhelm the reader with examples of the constant conjunction between comets and calamities.[68] Chronologers, for their part, had begun to see greater value in certain kinds of celestial events for fixing the dates of terrestrial ones. Reports of past solar and lunar eclipses that could be verified by computation on the basis of modern tables could in turn establish intervals of time with a precision and certainty unattainable by historical sources. As Heinrich Bünting proudly wrote in his letter of dedication to the bishop of Halberstadt, his 1590 *Chronology* was "confirmed by the Prutenic calculation," Erasmus Reinhold's *Prutenic Tables* of 1551.[69] Punctuating his timeline of historical events from creation to the present day with comet reports and eclipse computations, Bünting, like other chronologers, compiled catalogues of chronological signposts.[70] Authors of geographical works in the Ptolemaic tradition, too, collected astronomical observations to fix places in terms of longitude and latitude.[71] Constructed in the first instance to secure desiderata in other fields of knowledge—the distance between two cities, synchronisms between the histories of the Assyrians and the Jews, or the relationship between a comet's passage and the outbreak of war—such collections offered astronomers valuable celestial data for their own purposes.

DECANONIZING THE CANON

These trends in the cumulation of observational material were especially prominent in the works of the Jesuit professor of mathematics Giovanni Battista Riccioli, whose *New Almagest* (1651) featured a frontispiece showing Ptolemy supine with his geocentric model, the heliocentric and geoheliocentric systems hanging in Astraea's balance, and hundred-eyed Argos, a telescope to his sighted knee, gazing at celestial phenomena unknown in ancient times. Riccioli amassed "histories" of comets, "new stars," Saturn's bulging figure and Jupiter's shifting appearance in his treatment of these heavenly novelties, but also assembled observations of mundane celestial events (the sun at equinox and solstice, for instance), organized by type and set in chronological order from Antiquity to the present as a necessary prelude to examining the various conclusions that might be drawn from such data.[72] The longest such series consisted of solar and lunar eclipse reports spanning nearly 2,500 years, which Riccioli aptly termed an "astronomical, geographical, and chronological treasury."[73] Yet Riccioli also called his eclipse series a "history of observations," a striking phrase that foreshadowed a turn towards Baconian natural histories of the skies. When Riccioli himself presented a catalogue of lunar eclipse observations suitable for obtaining longitudes in his *Geography and Hydrography Reformed* (1672), he juxtaposed his determinations with those based on eclipse data drawn from Kepler's *Rudolphine Tables* (1627), Wendelin's *Lunar Eclipses* (1644), Boulliau's *Philolaic Tables* (1645), and many other sources, both manuscript and print, in order to provide a thorough review of the resulting discrepancies. The list of geographical locales that Riccioli ultimately tabulated gave longitudes that he thought "certain, or else more plausible to us."[74] Thicker descriptions of celestial events could soothe the epistemological anxieties of scholars more wary than Bünting or Wendelin about founding conclusions on insufficient empirical grounds.

In his *Instruments of a Restored Astronomy* (1598) celebrating the massive quadrants, sextants, armillary spheres, and other instruments housed at his observatory on the island of Hven, Tycho Brahe had assiduously cultivated his public persona as an astute observer dedicated to reconstructing astronomy's foundations. The observations he made there—more numerous, more precise, and more accurate than anything known in the history of the celestial sciences—he guarded, as he wrote in his dedication to Rudolph II, as a "most rare and precious treasure."[75] Kepler depended heavily on Tycho's records for his own work and hoped to have them printed, but astronomers waited long for this observational vault to be opened. It was not until 1666 that the Jesuit rector at the Society's college in Dillingen, Albert Curtz, published the *Celestial History*—the only edition of Tycho's observations to appear in print prior to the modern *Opera Omnia* edited by J. L. E. Dreyer in the early twentieth century. Curtz's labors were not entirely well received. Frederick III of Denmark acquired Tycho's observational protocols and charged Erasmus Bartholin, professor of mathematics at Copenhagen, with preparing another edition. Bartholin signaled the Curtz volume's "very conspicious errors" in an excruciatingly long errata list, taking thirty-six pages to review just the problems with the observations for 1582. In excoriating the 1666 edition, however, Bartholin revealed just how much he had in common with Curtz on the rationale for printing observations seriatim. Collating Curtz's text against the manuscript protocols, Bartholin complained that dates, timings, numbers, and zodiacal and planetary signs had been incorrectly printed. In some places, key observational conditions had been omitted, such as whether the clock had lost time and the state and quality of the instruments used. Elsewhere the material had been laid out so confusingly that information relevant to the preceding observation might be taken to apply to the following one. Finally, many Tychonian observations were glaringly absent, including all those made prior to 1582 as well as those for 1593.[76]

But whether due to the typesetter's carelessness, editorial inattention, or the incomplete character of the copies Curtz used as the basis for the printed text, the problems Bartholin signaled reflected disappointment with the final result, not disagreement over the edition's rationale. Though Curtz was not sure why the Tychonian volume for 1593 had gone missing—he speculated that it had been lost in transit to Kassel—he supplied in its place the Hessen observations for 1593 that Snell had published in 1618, along with the star catalogue that Wilhelm IV had projected.[77] This was no ad hoc editorial decision.[78] Comprehensiveness was Curtz's goal for the edition, which contained extensive "Additions besides Ticho's Observations."[79] Bookending the Tychonian "treasury" were Babylonian, Greek, "Alexandrian," "Syro-Persian," "Nurembergian," "Prutenic," "Miscellaneous" observations made between those of Copernicus and Tycho, and many made in the post-Tychonian period. The bulky tome comprised a set of observational data spanning some 2,300 years. Bernard Walther's material, too, had a place in Curtz's volume. Stripped of the commemorative trappings they had enjoyed in their earlier appearances in print, they appeared alongside celestial events remarked in "Turkish annals," recorded by Abraham Zacut in Portugal, noted by Johannes Werner at Rome, or seen from Jamaica.[80] Curtz acknowledged but did not address the chronological puzzles raised by so much astronomical data. His purpose, as he put it, was "simply to relate what had been seen and observed in the sky."[81] Setting astronomical observations free from the drilled formations into which they had been typically marshaled for public display, Curtz's *Celestial History* presented an uncanonized archive of empirical material to its readers, who could decide for themselves how any one of the observations strewn over a thousand double-columned folio pages might be deployed anew as "givens" for predictive models or cosmological claims.

A comet blazed through the skies from November 1664 through February 1665. The Ursuline nun Marie de l'Incarnation saw it from her mission in Quebec, as did the Jesuit François le Mercier from his, while witnesses in Europe generated "an infinite number of writings, treatises, dissertations, conferences, ephemerides, systems, etc."[82] The renowned Danzig astronomer Johann Hevelius did his part by rushing his observations into print, offering readers both a "history of the comet" that first appeared in 1664 as well as a "forerunner" for his comprehensive "cometography" already at the press.[83] Yet Hevelius' eagerness soon embroiled him in a bruising conflict with Adrian Auzout. His observation of its position on February 18, 1665, differed greatly from those of Auzout and other observers for the bracketing days, a discrepancy that Hevelius' colleagues in the Royal Society eventually adjudicated to his disadvantage. The resulting report in the *Philosophical Transactions* announced that some "English Astronomers" had "Joyntly conclude[d], that, whatever that Appearance was, which was seen near the *First star* of *Aries*, by *Monsieur Hevelius* (the truth of whose relation concerning the same, they do in no wise question) the said *Comet* did not come near that *Star* in the left *Ear* of *Aries*," as Hevelius had claimed.[84] The "Controversie" between Hevelius and Auzout thus concerned a *"Matter of fact"* about the observed position of a particular celestial body, a fine distinction that neatly disentangled the current dispute from other thorny issues, like theories concerning its physical nature or hypotheses plotting its motion. Hevelius responded in kind. His *Cometography* (1668) comprised a massive treatment of the subject that ranged from a fully articulated physical theory to lengthy accounts of particular comets, including the contested one of 1664–65. Yet he concluded the volume with a "History of all comets, from the creation of the world to now" that ran to well over 100 pages. Its purpose, wrote Hevelius, was to foster the establishment or refutation of the "most dissimilar hypotheses," including his own.[85]

The Hevelius–Auzout episode illustrates two of the principal drivers for the quickening pace at which astronomical observations went to press through the latter half of the seventeenth and into the next: competition and controversy. Both fed priority claims and reputations, whether in the form of specific reports made in the pages of new learned journals like the *Journal des savants* (from 1665), *Philosophical Transactions* (1665), *Giornale de' letterati* (1668), *Acta eruditorum* (1683), and *Mémoires de Trévoux* (1701), or in serialized sequences capturing decades of sleepless vigils. Pierre Gassendi's posthumous *Opera omnia* (1658), for instance, arranged his observations in temporal order "nude" and "as they are," culled from various papers and letters. The indefatigable Hevelius flooded the world with observational material even after the disastrous fire in 1679 that consumed his observatory, instruments, much of his library, and copies of the works he had published on his own printing press.[86] Other observational corpora advertised the collective labors of astronomers bound by personal ties, such as the volumes Eustachio Manfredi edited of work conducted by colleagues at the University of Bologna and the great meridian lines in the churches of San Petronio in Bologna and Santa Maria degli Angeli e dei Martiri in Rome.[87] Astronomically minded members of the Society of Jesus were especially eager to publish observational material on the strength of Jesuit scientific networks.[88] Demand for observational data also arose from conditions faced by early modern astronomers that made in situ sightings invaluable. Edmond Halley journeyed to St. Helena where he could observe the southern stars absent from Tycho Brahe's star catalogue.[89] Meridian transits of celestial objects as seen from equatorial locales appeared far closer to the observer's zenith and thus offered an opportunity to reevaluate the effects of atmospheric refraction on observation, while Mercury and Venus transits were phenomena of limited visibility that had been avidly pursued since Kepler's time by astronomers seeking to more accurately determine apparent planetary diameters and, with Halley, solar parallax. An especially welcoming generic home for such material was the capacious format of travel journals in which "physical and astronomical observations" sometimes overwhelmed the narrative of the journey itself.[90]

Competition, controversy, and opportunism converged spectacularly in the mid-1730s with the Paris Academy of Sciences' expeditions to determine the shape of the earth by measuring the length of an arc along a meridian line. Observational material from Lapland and Peru quickly became a key weapon with which to impugn fieldwork, question instrumentation and methods of mathematical analysis, and challenge conclusions concerning the earth's shape: prolate or flattened.[91] Claims concerning unmediated access to observations and the procedures employed for their reduction soon became a matter of controversial tactics. Impatient to lay his work before a public beyond the confines of the academy and unwilling to wait for the academy's annual *Mémoires*, then running a few years behind in printing academicians' papers, Pierre-Louis Moreau de Maupertuis sent his *Figure of the Earth* (1738) to the press complete with "the observations themselves such as they are found" in his own and his fellow expedition members' "registers."[92] He claimed that the observations as printed had not been corrected or averaged, unlike those of his unnamed opponents—understood by his auditors at the Academy of Sciences as their colleagues, the Cassini family of astronomers—whose published results on the meridian line in France could not be checked against their empirical data. Maupertuis' readers, in contrast, were free to evaluate, correct, and reduce the Lapland data anew, a "rhetoric of precision" that threatened to call the bluff of astronomers who vaunted their observational abilities.[93] Yet the rhetoric of the uncanonized archive Maupertuis articulated also proved compelling, as other participants in the affair made clear in practicing that most sincere form of flattery: imitation. The abbé Réginald Outhier noted that although Maupertuis had published their joint observations years earlier, he had decided to print them again in his *Journal of a Voyage to the North* (1744) "such as I wrote them in my Register" in order to assure the "faithfulness of my Journal."[94] César-François Cassini de Thury, for his part, published the definitive account of *The Meridian of the Royal Observatory of Paris* (1744). There the third generation of Cassini astronomers adopted the mathematical techniques, corrections for the aberration of light, and even the shape of the earth that Maupertuis had used to distinguish the Lapland measurements from those the Cassinis had made in France.[95] Yet just as striking was Cassini de Thury's adoption of Maupertuis' archival rhetoric: "The third & last part is only, properly speaking, the extract of our Registers," giving the observations "such as they were made on site. . . . From which one may judge the precision of our operations, and how much one can rely on their exactitude."[96]

The studied deployment in print of manuscript "registers" of observations in the course of these debates is telling. Through the eighteenth century European astronomers treated their notes and logbooks as personal property even in the context of institutional structures like the Royal Observatory in Greenwich, with its often tenuous relation to the Royal Society, as well as the observatory attached to the Paris Academy of Sciences. Disputes over ownership of manuscript material deeply stained the observational legacy of John Flamsteed and one of his successors as Royal Astronomer, James Bradley. Flamsteed saw control wrested from him when fellows of the Royal Society were appointed visitors to the observatory by royal warrant in 1710, a turn of events driven in good part by Isaac Newton's impatience for Flamsteed's closely held data as well as distrust in their reduction.[97] Readers learned from the preface to the *Celestial History* published in 1712 that Flamsteed's manuscript observations had been corrected, emended, selected, and published by Edmond Halley in a form Flamsteed found so unpalatable that he acquired at least 300 of the 400 copies of the Halley edition in order to destroy the offending pages. Flamsteed's widow took possession of his papers and oversaw a new edition in keeping with its original design, though without the section in the autograph preface in which Flamsteed detailed his struggle with Halley and Newton for control of his data.[98]

When James Bradley died in 1762, his executors retained his observational papers, which were passed down through the family. Despite the Crown's claim to the material in virtue of Bradley's post as Royal Astronomer, they were eventually gifted to the University of Oxford, where Bradley had been Savilian professor of astronomy. To sidestep the challenges posed by claims of private property, the Board of Longitude purchased the observations of both Nathaniel Bliss, Savilian professor of geometry and Bradley's successor as Royal Astronomer, and of Bliss' assistant, Charles Green.[99] All were eventually edited for publication by holders of the Savilian chairs at Oxford and printed at the Clarendon Press. It was not until 1774—nearly a century after the Royal Observatory's foundation—that the Royal Society began publishing "at the public expense" the "astronomical observations made at the Royal Observatory at Greenwich," starting with the work of Nevil Maskelyne, who assumed the post of Royal Astronomer after Bliss.[100] Absent reliable institutional mechanisms for taking on the responsibility and expense of sending observational data to the press, print publication of such material through the end of the eighteenth century was largely unsystematic.

CELESTIAL HISTORIES

Across the channel the Paris Academy of Sciences astronomer and Lapland veteran Pierre-Charles Le Monnier was engaged in an ambitious project to publish a "celestial history of France," one that would include "all the Astronomical Observations made in France since 1666" and the founding of the academy.[101] He invoked but two precedents: Albert Curtz's *Celestial History*, with its treasure trove of material from Babylonian times through the seventeenth century, and the 1725 edition of Flamsteed's *Celestial History*, which Le Monnier preferred to that edited by Edmond Halley in 1712 for its chronological breadth. "Much more ample than the first," the 1725 edition included observations made by William Gascoigne and William Crabtree, active in the north of England in the first half of the seventeenth century, as well as Flamsteed's own work prior to his move to Greenwich.[102] Though taking up but a fraction of the three-volume publication, these additions fulfilled both Flamsteed's desire to imitate Curtz's broad editorial vision as well the second edition's titular claim to a "*British* celestial history."[103]

Yet Le Monnier meant to go still further. Halley had selected the "principal and most important" observations, that is, those "utility is recognized," in editing Flamsteed's material for the press. Le Monnier called instead for the publication of celestial histories that contained "the observations such as they were made, such as one finds them in the registers of the Authors who've left them to us."[104] Like Maupertuis, Le Monnier employed the rhetoric of the uncanonized archive for polemical purposes.[105] He himself had drawn strategically on unpublished observations for his own investigations of refraction, stellar aberration, and other uncontrolled variations in apparent stellar positions, all issues critical to Maupertuis' presentation of their joint Lapland observations.[106] When he formally presented the "Project for a celestial history" to his academy colleagues in 1738, Le Monnier warmly praised his fellow astronomers who had promised or given him access to their materials and faintly blamed those who had not. The latter included Jacques Cassini, who had yet to "communicate" the long series of astronomical observations in the "registers" of his father, the illustrious Giovanni Domenico Cassini. Le Monnier was sure that "the Academy would incorporate them into the Celestial History."[107] When he published its first installment appeared a few years later, however, little had changed. The tome covered only two decades' worth of observations—from 1666 to 1686—and reflected the work of but two academy astronomers, Jean Picard and Philippe de La Hire.[108] Nor could Le Monnier obtain material from Joseph-Nicolas Delisle, who had been a student astronomer at the academy and professor of mathematics at the Collège Royale in Paris before heading to the Imperial Academy of Sciences in St. Petersburg; Delisle insisted on Jacques Cassini's cooperation in the project.[109]

In part, the resistance Le Monnier encountered reflected real ambivalence about the deployment of observational resources. Jacques Cassini, for instance, followed his father in comparing "ancient" and "modern" observations to construct planetary theories for his 1740 *Astronomical Tables*, but not when it came to Venus and Mercury: the "ancient" observations of these two planets were made at their greatest elongation, whereas Cassini thought modern observations made at inferior conjunction minimized the problems in establishing their true positions.[110] In a 1749 *Philosophical Transactions* contribution investigating the moon's motion in light of his lunar tables, Richard Dunthorne dismissed the vast majority of the eclipse observations he found in Curtz's *Celestial History* and other sources, as very few could be "at all depended upon." Most were either "too inaccurate to determine any thing from them in this Affair" or "so loosely described, that . . . they are wholly incapable of shewing us how much [the moon's] Acceleration has been."[111] Nor were contemporary observations immune to such critical evaluation. Already estranged before their long-delayed return to France, Pierre Bouguer and Charles-Marie de La Condamine argued bitterly over who had the right to present the operations they had jointly carried out to measure an arc of meridian in Peru. Though the academy's publications committee tried to mediate the conflict, the dispute quickly spilled over from the academy's sessions and *Mémoires* into the print marketplace.[112] Where Maupertuis had accused his opponents of concealing their observational data, Bouguer pilloried La Condamine's competing account of the Peru expedition for giving no guidance on evaluating the relative "exactitude" of a "multitude" of observations, leaving the hapless reader to find "a happy medium among quantities which differ far too considerably." For Bouguer, La Condamine had failed in the astronomer's duty to critically assess the quality of the empirical data he created, distinguish those worth retaining from those that should be discarded, and to show why that choice was an informed one, "neither arbitrary nor the consequence of some Agreement made between the Observers."[113]

Le Monnier himself acknowledged the paradox at the heart of a "universal Celestial History." Given the great number of observations brought into such a project—"repeated day after day with more precision and with more perfect instruments"—it was inevitable that most of them would be "successively destroyed, the ones by the others." This was surely why the ancients hadn't "transmitted [observations] to us with all the detail necessary," instead giving just the results; Le Monnier's fellow academician, Nicolas-Louis de Lacaille, thought that most modern astronomers did the same.[114] Yet Le Monnier's answer to the prospect of a self-destroying archive was to publish "the observations such as they were made" without regard for their present utility. This, he thought, was "the only means of forestalling an infinity of disputes like those that have divided Astronomers for so many centuries."[115] Though Le Monnier only issued one volume of his *Celestial History*, he went on to publish what observations he could, principally his own.[116] More importantly, its vision was of a piece with proposals conceptualized throughout the eighteenth century: Joseph-Nicolas Delisle's youthful ambition to compose "the entire celestial history from the establishment of the Academy up to the present" and much more; Alexandre-Guy Pingré's "annales célestes," laboriously scoured from printed volumes and manuscript papers; Joseph-Jérôme Lalande's "Histoire celeste française," planned in several volumes; and, most strikingly, the ongoing "celestial history" of the Paris observatory into its Revolutionary incarnation as advanced by Jean-Dominique Cassini, the fourth generation of Cassinis to observe there.[117] Producing celestial histories had become an end in itself for the working astronomer. When Jean-Baptiste Joseph Delambre reassessed Ptolemy's parsimonious treatment of empirical material in his magisterial *History of Ancient Astronomy* (1817), he charged the Alexandrian with cherry-picking observational data to support a favored model.[118] Even more fundamental, though, was Ptolemy's failure to "communicate" whatever observations he did possess. "An astronomer who conducts himself today in this manner would be sure to inspire no confidence," wrote Delambre in a condemnation that measured the moral gap between the classical ideal of a canonized archive and its modern reconceptualization.[119]

CONCLUSION

Since the eighteenth century, the notion that an astronomer should have access to "any observation pure and simple, such as it was made" to facilitate its reduction with the "elements" of one's choice has been increasingly realized in successive genres of astronomical publication that, in turn, have swelled today's data deluge.[120] Current projects to scan sky survey photographic plates and digitize related catalogues, for example, seek to convert a visual archive initially conceived in the late nineteenth century into today's material supports and conceptual terms.[121] Textual archives going back to the early nineteenth century—observatory reports, astronomy journals, university and professional society publications, and astronomical catalogues—are being captured by electronic bibliographical databases such as the Centre de données astronomiques de Strasbourg's Set of Identifications, Measurements, and Bibliography for Astronomical Data (SIMBAD) and NASA's Astrophysics Data System (ADS), the latest in a long line of material remediations (punch cards, magnetic tape, microform).[122] Josephus' ancient triad of imperatives has taken on new life as contemporary astronomers struggle to keep afloat in a "data tsunami" estimated in a 2011 article at 1 petabyte of publicly available material in electronic form and increasing by 0.5 PB each year, a rate expected to rise sharply as new telescopes and arrays come online.[123] The primary strategy for weathering the current inundation has been to dematerialize astronomy's earth-bound organizational landscape, whether through cloud computing (such as the SkyNet project operated by the International Centre for Radio Astronomy Research in western Australia) or the federation of astronomical datasets whatever their national or institutional origins into a global "virtual observatory." According to the committee charged with overseeing the Hubble Space Telescope project's transition from operations to data management, "this grail is the ultimate archive."[124] And yet a key challenge for contemporary astronomers working at the leading edge of "data-intensive science"—the so-called "fourth paradigm" of scientific investigation—is one with which their predecessors have long wrestled: how to capture, curate, and analyse empirical givens.[125] If "astronomers have always relied on observational data for their science," it is equally true that they have taken fundamentally different paths in navigating the data-rich seas of a postdiluvian world.[126]

NOTES

1. Michael J. Kurtz and Guenther Eichhorn, "The Historical Literature of Astronomy, via ADS," *Library and Information Services in Astronomy III (LISA III): Proceedings of a Conference Held in Puerto de la Cruz, Tenerife, Spain, April 21–24, 1998*, ed. Uta Grothkopf et al. (San Francisco: Astronomical Society of the Pacific, 1998), 293; see also the discussion of access rates for items in ADS by publication year in Guenther Eichhorn et al., "Current and Future Holdings of the Historical Literature in the ADS," in *Library and Information Services in Astronomy IV (LISA IV): Emerging and Preserving: Providing Astronomical Information in the Digital Age*, ed. Brenda G. Corbin et al. (Washington, DC: US Naval Observatory, 2003), 150.

2. John M. Steele, "Applied Historical Astronomy: An Historical Perspective," *Journal for the History of Astronomy* 35, no. 3 (2004): 337.

3. Fred Espenak and Jean Meeus, NASA Technical Publication TP-2009-214172, *Five Millennium Canon of Lunar Eclipses: −1999 to +3000 (2000 BCE to 3000 CE)* (Greenbelt, MD: NASA, Goddard Space Flight Center, 2009), 18 (quote) and 7, relying on Leslie V. Morrison and F. Richard Stephenson, "Historical Values of the Earth's Clock Error ΔT and the Calculation of Eclipses," *Journal for the History of Astronomy* 35, no. 3 (2004); Stephenson and Morrison, "Long-term Changes in the Rotation of the Earth: 700 B.C. to A.D. 1980," *Philosophical Transactions of the Royal Society of London. Series A, Mathematical and Physical Sciences* 313, no. 1524 (1984): 51 (quote). See Stephenson, *Historical Eclipses and Earth's Rotation* (Cambridge: Cambridge University Press, 1997); John M. Steele, *Observations and Predictions of Eclipse Times by Early Astronomers* (Dordrecht: Kluwer Academic Publishers, 2000); and the works of Robert R. Newton, cited below, n26.

4. See J. W. Olmsted, "The 'Application' of Telescopes to Astronomical Instruments, 1667–1669: A Study in Historical Method," *Isis* 40, no. 3 (1949); Robert M. McKeon, "Les débuts de l'astronomie de précision" *Physis* 13 (1971) and *Physis* 14 (1972); J. H. Leopold, "Christiaan Huygens and His Instrument Makers," in *Studies on Christiaan Huygens*, ed. H. J. M. Bos et al. (Lisse: Swets & Zeitlinger, 1980); Michael S. Mahoney, "Christiaan Huygens, the Measurement of Time and Longitude at Sea," in *Studies on Christiaan Huygens*, ed. H. J. M. Bos et al.; and Guy Picolet, ed., *Jean Picard et les débuts de l'astronomie de précision au XVIIe siècle* (Paris: Editions du CNRS, 1987).

5. Steele, "Applied Historical Astronomy," 337.

6. Stéphane Mallarmé, *Divagations* (Paris: Bibliothéque-Charpentier, 1897), 256.

7. Josephus, *Judean Antiquities* 1.106, 69–71, trans. by Louis H. Feldman, in *Flavius Josephus: Translation and Commentary* (Leiden: Brill, 2000), vol. 3.

8. Francesca Rochberg, *The Heavenly Writing: Divination, Horoscopy, and Astronomy in Mesopotamian Culture* (Cambridge: Cambridge University Press, 2004), 1–2.

9. Diodorus, *Bibliotheca historica*, 2.30.2, 2.31.9, trans. by Charles Henry Oldfather in *Diodorus of Sicily* (Cambridge, MA: Harvard University Press, 1933), vol. 1; Pliny, *Historia naturalis*, 7.193, trans. by H. Rackham in *Natural History* (Cambridge, MA: Harvard University Press, 1942), vol. 1. See Anthony Grafton, "From Apotheosis to Analysis: Some Late Renaissance Histories of Classical Astronomy," in *History and the Disciplines: The Reclassification of Knowledge in Early Modern Europe*, ed. Donald R. Kelley (Rochester, NY: University of Rochester Press, 1997).

10. Aristotle, *De caelo* 291a7–291a9; Simplicius, *In Aristotelis de caelo* 506.12–17; both in Alan C. Bowen, *Simplicius on the Planets and Their Motions: In Defense of a Heresy* (Leiden: Brill, 2013), 119, 169. Elsewhere Simplicius mentioned Babylonian observations for 1,440,000 years and Egyptian observations for 630,000 years; see Bowen, *Simplicius*, 225–26.

11. Jean-Sylvain Bailly, *Histoire de l'astronomie ancienne, depuis son origine jusqu'à l'établissement de l'école d'Alexandrie* (Paris: Frères Debure, 1775), 141 (quote), 12, 144–45, 367–76, 388–89.

12. Pliny, *Historia naturalis*, 7.193. For the development of cuneiform studies of Babylonian celestial science, see N. M. Swerdlow, "Introduction," in *Ancient Astronomy and Celestial Divination*, ed. N. M. Swerdlow (Cambridge, MA: MIT Press, 1999); a recent survey of the state of the field is Hermann Hunger and David Pingree, *Astral Sciences in Mesopotamia* (Boston: Brill, 1999).

13. Abraham J. Sachs bestowed the term "diary" on this predominantly observational corpus in "A Classification of the Babylonian Astronomical Tablets of the Seleucid Period," *Journal of Cuneiform Studies* 2, no. 4 (1948). The diaries also include some predicted celestial phenomena; observations of comets and meteors, weather, and fluctuations in the height of the Euphrates; and reports of commodity prices and historical events. See Sachs, "Babylonian Observational Astronomy," *Philosophical Transactions of the Royal Society of London. Series A, Mathematical and Physical Sciences* 276, no. 1257 (1974), and Hunger and Pingree, *Astral Sciences* (1999), 139–59.

14. Sachs, "Babylonian Observational Astronomy," 49.

15. Otto Neugebauer, *Astronomical Cuneiform Texts: Babylonian Ephemerides of the Seleucid Period for the Motion of the Sun, the Moon, and the Planets* (London: Published for the Institute for Advanced Study by Lund Humphries, 1955); Theophilus G. Pinches and J. N. Strassmaier, *Late Babylonian Astronomical and Related Texts*, ed. A. J. Sachs and J. Schaumberger (Providence: Brown University Press, 1955); and Sachs and Hunger, *Astronomical Diaries and Related Texts from Babylonia* (Vienna: Verlag der Österreichischen Akademie der Wissenschaften, 1988–).

16. Notably the two-tablet compendium MUL.APIN, a text recovered from forty known copies; the series of some 7,000 celestial omens called the *Enuma Anu Enlil* after its incipit; reports concerning celestial omens addressed to the king; and horoscopes. See Erica Reiner, "Babylonian Celestial Divination," in *Ancient Astronomy and Celestial Divination*, ed. N. M. Swerdlow (Cambridge, MA: MIT Press, 1999), and Francesca Rochberg, "Babylonian Horoscopy: The Texts and Their Relations," in *Ancient Astronomy and Celestial Divination*, ed. N. M. Swerdlow (Cambridge, MA: MIT Press, 1999).

17. See David Brown, "What Shaped Our Corpuses of Astral and Mathematical Cuneiform Texts?," in *Looking at It from Asia: The Processes That Shaped the Sources of History of Science*, ed. Florence Bretelle-Establet (Dordrecht: Springer, 2010); Rochberg, *The*

Heavenly Writing, chap. 6; Paul-Alain Beaulieu, "The Astronomers of the Esagil Temple in the Fourth Century BC," in *If a Man Builds a Joyful House*, ed. Erle Leichty and Ann K. Guinan (Leiden: Brill, 2006); and Mathieu Ossendrijver, "Science in Action: Networks in Babylonian Astronomy," in *Babylon: Wissenskultur in Orient und Okzident*, ed. Eva Christiane Cancik-Kirschbaum, Margarete van Ess, and Joachim Marzahn (Berlin: Walter de Gruyter, 2011).

18. Otto Neugebauer, *A History of Ancient Mathematical Astronomy* (Berlin: Springer-Verlag, 1975), 1: 350–53; cf. Mathieu Ossendrijver, *Babylonian Mathematical Astronomy* (Dordrecht: Springer, 2012), 1–10.

19. Bryant Tuckerman, introduction, *Planetary, Lunar, and Solar Positions, 601 B.C. to A.D. 1 at Five-day and Ten-day Intervals* (Philadelphia: American Philosophical Society, 1962), 4 (Babylon civil time), 3 (Julian calendar); Otto Neugebauer, preface, in Tuckerman, *Planetary, Lunar, and Solar Positions, A.D. 2 to A.D. 1649 at Five-day and Ten-day Intervals* (Philadelphia: American Philosophical Society, 1964), v (evening epoch).

20. Tuckerman, introduction, *Planetary, Lunar, and Solar Positions, 601 B.C. to A.D. 1*, 8 (quote), 1; Neugebauer, preface, in Tuckerman, *Planetary, Lunar, and Solar Positions, 601 B.C. to A.D. 1*, v.

21. Neugebauer, preface, in Tuckerman, *Planetary, Lunar, and Solar Positions, 601 B.C. to A.D. 1*, v (emphasis added); see Sachs, "Babylonian Observational Astronomy," 48.

22. See, e.g., Mark J. Geller, "Babylonian Astronomical Diaries and Corrections of Diodorus" *Bulletin of the School of Oriental and African Studies* 53, no. 1 (1990); R. J. van der Spek, "New Evidence from the Babylonian Astronomical Diaries Concerning Seleucid and Arsacid History," *Archiv für Orientforschung* 44/45 (1997/1998); Dov Gera and Wayne Horowitz, "Antiochus IV in Life and Death: Evidence from the Babylonian Astronomical Diaries," *Journal of the American Oriental Society* 117, no. 2 (1997); and Alice Louise Slotsky, *The Bourse of Babylon: Market Quotations in the Astronomical Diaries of Babylonia* (Bethesda, MD: CDL Press, 1997).

23. For the Mesopotamian context alone, see, e.g., Lis Brack-Bernsen, *Zur Entstehung der babylonischen Mondtheorie: Beobachtung und theoretische Berechnung von Mondphasen* (Stuttgart: F. Steiner, 1997); N. M. Swerdlow, *The Babylonian Theory of the Planets* (Princeton: Princeton University Press, 1998); Steele, *Observations and Predictions*; and Rochberg, *The Heavenly Writing*.

24. Neugebauer, preface, in Tuckerman, *Planetary, Lunar, and Solar Positions, A.D. 2 to A.D. 1649*, v.

25. See especially Herman H. Goldstine, *New and Full Moons 1001 B.C. to A.D. 1651* (Philadelphia: American Philosophical Society, 1973), viii; Owen Gingerich and Barbara L. Welther, *Planetary, Lunar, and Solar Positions, New and Full Moons, A.D. 1650–1805* (Philadelphia: American Philosophical Society, 1983), xxvii; and Hermann Hunger and Rudolf Dvořák, *Ephemeriden von Sonne, Mond und hellen Planeten von −1000 bis −601* (Vienna: Verlag der Österreichischen Akademie der Wissenschaften, 1981). On Goldstine's early collaboration with Tuckerman and Neugebauer, see Tuckerman, *Planetary, Lunar, and Solar Positions, 601 B.C. to A.D. 1*, v, 1, and Tuckerman, *Planetary, Lunar, and Solar Positions, A.D. 2 to A.D. 1649*, 1.

26. See Robert R. Newton, *Ancient Astronomical Observations and the Accelerations of the Earth and Moon* (Baltimore: Johns Hopkins University Press, 1970); Newton, *The Moon's Acceleration and Its Physical Origins* (Baltimore: Johns Hopkins University Press,

1979–84); David Kushner, "The Controversy Surrounding the Secular Acceleration of the Moon's Mean Motion," *Archive for History of Exact Sciences* 39, no. 4 (1989); John Phillips Britton, *Models and Precision: The Quality of Ptolemy's Observations and Parameters* (New York: Garland, 1992), appendix 1; Curtis Wilson, *The Hill-Brown Theory of the Moon's Motion: Its Coming-to-be and Short-lived Ascendancy (1877–1984)* (New York: Springer, 2010), 9–30, 237–318; and John M. Steele, *Ancient Astronomical Observations and the Study of the Moon's Motion (1691–1757)* (New York: Springer, 2012).

27. Matthew Stanley, "Predicting the Past: Ancient Eclipses and Airy, Newcomb, and Huxley on the Authority of Science," *Isis* 103, no. 2 (2012); Neugebauer, preface, in Tuckerman, *Planetary, Lunar, and Solar Positions, 601 B.C. to A.D. 1*, v (quotation), and Tuckerman, introduction, in Tuckerman, *Planetary, Lunar, and Solar Positions, 601 B.C. to A.D. 1*, 8.

28. F. Richard Stephenson and M. A. Houlden, "The Accuracy of Tuckerman's Solar and Planetary Tables," *Journal for the History of Astronomy* 12 (1981): 133; Houlden and Stephenson, *A Supplement to the Tuckerman Tables* (Philadelphia: American Philosophical Society, 1986), i; X. X. Newhall, E. M. Standish, and J. G. Williams, "DE 102: A Numerically Integrated Ephemeris of the Moon and Planets Spanning Forty-four Centuries," *Astronomy and Astrophysics* 125, no. 1 (1983); and Andrew J. Butrica, "Redefining Celestial Mechanics in the Space Age: Astrodynamics, Deep-space Navigation, and the Pursuit of Accuracy," in *Exploring the Solar System: The History and Science of Planetary Exploration*, ed. Roger D. Launius (New York: Palgrave Macmillan, 2012). It is not clear whether alternative planetary theories available at the time would have improved the Tuckerman ephemeris's accuracy; see the critical review of Meeus' long-period solar eclipse ephemeris (1966) by Donald H. Sadler, then the superintendent of the British Nautical Almanac Office charged with producing official annual ephemerides in conjunction with its American counterpart: Sadler, "Prediction of Eclipses," *Nature* 211, no. 5054 (1966).

29. S. George Djorgovski et al., "Sky Surveys," in *Planets, Stars, and Stellar Systems*, ed. Terry D. Oswalt and Howard E. Bond (Dordrecht: Springer, 2013), 2: 228. See Robert R. Newton, *Medieval Chronicles and the Rotation of the Earth* (Baltimore: Johns Hopkins University Press, 1972); Newton, *Ancient Planetary Observations and the Validity of Ephemeris Time* (Baltimore: Johns Hopkins University Press, 1976); Newton, *The Moon's Acceleration*; F. Richard Stephenson and David H. Clark, *Applications of Early Astronomical Records* (New York: Oxford University Press, 1978); F. Richard Stephenson and David A. Green, *Historical Supernovae and Their Remnants* (Oxford: Clarendon Press, 2002); and the works by Morrison, Newton, Steele, and Stephenson cited above.

30. Eleanor Robson, "Astronomical Diaries and Related Texts from Babylonia," *Journal of the Royal Asiatic Society* 17 (2007): 61 (quote).

31. Theodor Oppolzer, *Canon der Finsternisse* (Vienna: aus der Kaiserlich-Königlichen Hof- und Staatsdruckerei in Commission bei K. Gerold, 1887). I have used Oppolzer, *Canon of Eclipses: Canon der Finsternisse*, trans. Owen Gingerich (New York: Dover Publications, 1962), ix (emphasis added). Cf. the expansive but still limited efforts by Alexandre-Guy Pingré in "Discours préliminaire" and "Chronologie des eclipses," in Maur-François Dantine et al., *L'art de vérifier les dates de faits historiques, des chartes, des chroniques, et autres anciens monumens depuis la naissance de notre seigneur. . . . Ouvrage nécessaire à ceux qui veulent avoir un parfaite conoissance de l'histoire. Par des religieux benedictins de la congrégation de S. Maur*, 2nd ed. (Paris: Chez G. Desprez, 1770), 39–89, and Pingré,

"Chronologie des éclipses de soleil & de lune qui ont été visibles sur terre, depuis le pôle boréal jusque vers l'equateur, durant les dix siècles qui ont précédé l'ère chrétienne, par M. Pingré," *Histoire de l'Académie royale des inscriptions et belles-lettres [HAIBL 1776–1779]* 42 (1786): 78–150.

32. Oppolzer, *Canon of Eclipses*, xxxvi.

33. Friedrich Karl Ginzel obituary, *Observatory* 49 (1926): 348 (emphasis added); Ginzel, *Spezieller Kanon der Sonnen- und Mondfinsternisse für das Ländergebiet der klassischen Altertumswissenschaften und den Zeitraum von 900 vor Chr. bis 600 nach Chr.* (Berlin: Mayer & Müller, 1899).

34. Espenak and Meeus, *Five Millennium Canon of Lunar Eclipses*, and Espenak and Meeus, NASA Technical Publication TP-2006-214141, *Five Millennium Canon of Solar Eclipses: –1999 to +3000 (2000 BCE to 3000 CE)* (Washington, DC: National Aeronautics and Space Flight Administration, 2006). In addition to the numerous modern canons cited by Espenak and Meeus, see Robert Sewell, Śaṅkara Bālakṛshṇa Dīkshita, and Robert Gustav Schram, *The Indian Calendar: with Tables for the Conversion of Hindu and Muhammadan into A.D. Dates, and Vice Versa: with Tables of Eclipses Visible in India by Dr. Robert Schram of Vienna* (London: Swan Sonnenschein & Co., Ltd., 1896); Paul V. Neugebauer and Richard Hiller, "Spezieller Kanon der Sonnenfinsternisse für Vorderasien und Ägypten für die Zeit von 900 v. Chr. bis 4200 v. Chr.," *Astronomische Abhandlungen* 8, no. 4 (1931); Neugebauer and Hiller, "Spezieller Kanon der Mondfinsternisse für Vorderasien und Ägypten von 3450 bis 1 v. Chr.," *Astronomische Abhandlungen* 9, no. 2 (1934); Homer H. Dubs, "A Canon of Lunar Eclipses for Anyang and China, –1400 to –1000," *Harvard Journal of Asiatic Studies* 10, no. 2 (1947); Robert R. Newton, *A Canon of Lunar Eclipses for the Years –1500 to –1000* (Laurel, MD: Johns Hopkins University, Applied Physics Laboratory, 1977); and M. V. Lukashova and L. I. Rumyantseva, "Canon of Solar Eclipses from 1000 to 2050 for Russia," *Solar System Research* 32, no. 2 (1998).

35. E.g., Pierre Hoang, *Catalogue des éclipses de soleil et de lune relatées dans les documents chinois et collationnées avec le canon de Th. Ritter v. Oppolzer* (Shanghai: Mission Catholique, 1925); Homer H. Dubs, et al., *The History of the Former Han Dynasty* (Baltimore: Waverly Press Inc., 1938–55); Leslie V. Morrison, M. R. Lukac, and F. Richard Stephenson, "Catalogue of Observations of Occultations of Stars by the Moon for the Years 1623 to 1942 and Solar Eclipses for the Years 1621 to 1806," *Royal Greenwich Observatory Bulletins* 186 (1981): 5–7, 36–54; Stephenson, *Historical Eclipses and Earth's Rotation* (1997); Said S. Said, F. Richard Stephenson, and Wafiq Rada, "Records of Solar Eclipses in Arabic Chronicles," *Bulletin of the School of Oriental and African Studies* 52, no. 1 (1989); Stephenson and Said, "Records of Lunar Eclipses in Medieval Arabic Chronicles," *Bulletin of the School of Oriental and African Studies* 60, no. 1 (1997); Zhentao Xu, Yaoting Jing, and David W. Pankenier, *East Asian Archaeoastronomy: Historical Records of Astronomical Observations of China, Japan and Korea* (Amsterdam: Gordon & Breach, 2000); Steele, *Observations and Predictions*; Peter J. Huber and Salvo de Meis, *Babylonian Eclipse Observations from 750 BC to 1 BC* (Milan: Mimesis; IsIAO, 2004); and Pankenier, "On the Reliability of Han Dynasty Solar Eclipse Records," *Journal of Astronomical History and Heritage* 15, no. 3 (2012).

36. The work enjoyed a long and extensive tradition in Greek, Arabic, Syriac, Latin, and other languages. See Anne Tihon and Raymond Mercier, eds., *Ptolemaiou procheiroi kanones* (Louvain-la-Neuve: Université catholique de Louvain, Institut orientaliste, 2011).

37. Johannes Behm, *Chronologica manuductio, & deductio annorum* (Frankfurt am

Main: curantibus Rulandiis, 1619), 259–60; Seth Calvisius, *Opus chronologicum* (Frankfurt an der Oder: Impensis Johannis Thymii, 1620), 74–76. John Bainbridge published another manuscript version of the text as the "canon of kingdoms" in his *Procli sphaera: Ptolemaei de hypothesibus planetarum liber singularis . . . cui accesit ejusdem Ptolemaei canon regnorum* (London: Excudebat Guilielmus Iones, 1620), and Joseph Justus Scaliger gave the text as found in the Byzantine chronicle of George Syncellus as the "mathematical canon of kings" in his "Isagogicorum chronologiae canonum," *Thesaurus temporum* (Amsterdam: Apud Joannem Janssonium, 1658), 291. See Leo Depuydt, "'More Valuable Than All Gold': Ptolemy's Royal Canon and Babylonian Chronology," *Journal of Cuneiform Studies* 47 (1995), and Anthony Grafton, *Joseph Scaliger: A Study in the History of Classical Scholarship* (Oxford: Clarendon Press, 1983–93), 2: 720–28.

38. Calvisius, *Opus chronologicum*, 74, 75.

39. Ptolemy, *Almagest* iii.7; *Ptolemy's Almagest*, trans. G. J. Toomer (London: Duckworth, 1984), 166. On Ptolemy's approach to chronology, see Toomer, introduction, *Ptolemy's Almagest*, 9–14; and Olaf Pedersen, *A Survey of the Almagest*, ed. Alexander Jones (New York: Springer, 2011), 124–28 and appendix A (a list of the ninety-four dated observations in the *Almagest*).

40. See, e.g., Georg Peurbach, "Canones pro compositione et usu gnomonis geometrici," in Johann Schöner, ed., *Scripta clarissimi mathematici M. Ioannis Regiomontani* (Nuremberg: apud Ioannem Montanum & Vlricum Neuber, 1544), 61r–78v, and Erasmus Reinhold, *Prutenicae tabulae coelestium motuum* (Tübingen: Per Ulricum Morhardum, 1551), 67r ("Ordo Canonum").

41. Jan Assmann, *Cultural Memory* (Cambridge: Cambridge University Press, 2011), 87–96; 95 (quote). The interpretation here departs from Assmann's analysis of the modern notion of a (textual) canon and specifically his characterization of astronomical and chronological tables as having only descriptive but not normative power; see Assmann, *Cultural Memory*, 97–110.

42. Ptolemy, *Almagest* iv.6, iv.11, pp. 191, 211 (quotes); for similar language, see Ptolemy, *Almagest* iv.6, iv.9, vi.5, vi.9, pp. 198, 207, 283, 309.

43. For Ptolemy's criteria in choosing lunar eclipse observations, see Ptolemy, *Almagest* iv.1, iv.9, pp. 173–74, 206–7.

44. Bernard R. Goldstein and Alan C. Bowen, "The Role of Observations in Ptolemy's Lunar Theories," in *Ancient Astronomy and Celestial Divination*, ed. N. M. Swerdlow (Cambridge, MA: MIT Press, 1999), 344 (quote, emphasis added), 345–47.

45. See Britton's thorough study, *Models and Precision*, and Bernard R. Goldstein, "What's New in Ptolemy's *Almagest*?" *Nuncius* 22 (2007).

46. Ptolemy, *Almagest* i.3–4, pp. 38–41.

47. Recent efforts to reconstruct Ptolemy's sources of empirical data include Bernard R. Goldstein and Alan C. Bowen, "The Introduction of Dated Observations and Precise Measurement in Greek Astronomy," *Archive for History of Exact Sciences* 43, no. 2 (1991); John M. Steele, "A Re-analysis of the Eclipse Observations in Ptolemy's *Almagest*," *Centaurus* 42 (2000); and Alexander Jones, "Ptolemy's Ancient Planetary Observations," *Annals of Science* 63, no. 3 (2006).

48. Copernicus cited twenty-seven of his own observations in *De revolutionibus orbium coelestium* (Nuremberg: apud Ioh. Petreius, 1543). In addition to observations he took from the *Almagest*, he also used three observations by al-Battānī in Johannes

Regiomontanus and Georg Peurbach, *Epytoma Joa[n]nis de Mo[n]te Regio in almagestu[m] Ptolomei* ([Venice]: [impressionis . . . Johannis ha[m]man], [1496]), and three made by his contemporary Bernard Walther. See N. M. Swerdlow and Otto Neugebauer, *Mathematical Astronomy in Copernicus's De revolutionibus* (New York: Springer-Verlag, 1984), 1: 64–65, 131, 148, 415–16.

49. Georg Joachim Rheticus, *Ad clarissimum virum D. Ioannem Schonerum, de libris revolutionum eruditissimi viri, & mathematici excellentissimi reverendi D. Doctoris Nicolai Copernici Torunnaei, Canonici Varmiensis, per quendam iuvenem, mathematicae studiosum narratio prima* ([Gdańsk]: [Excusum . . . per Franciscum Rhodum], [1540]); I have used the text and modified the translation in Rheticus, *Georgii Joachimi Rhetici Narratio prima*, ed. and trans. Henri Hugonnard-Roche and Jean-Pierre Verdet (Wrocław: Ossolineum, 1982), 68, 123. For Copernicus' and Ptolemy's use of empirical data for constructing planetary models, see Swerdlow and Neugebauer, *Mathematical Astronomy*, 1: 200, 309, 415–16, 374, and Pedersen, *A Survey*, 169–70, 272–73, 309–10, 298–99. Both astronomers used eleven observations each for their respective Venus theories. I set aside the case of the sun for this comparison as Copernicus' treatment departed from Ptolemy's in providing solutions (warranted, he believed, by the observational record from Antiquity to his own time) to certain long-period nonuniform motions perceptible in solar phenomena. Copernicus, of course, imputed these motions to the earth. See Swerdlow and Neugebauer, *Mathematical Astronomy*, 1: 150–66.

50. It is not clear how Copernicus obtained the Walther observations in manuscript that he used for his Mercury model in *On the Revolutions* (1543). Copernicus attributed two of them to Johann Schöner, the addressee of Rheticus' *Narratio prima* (1540); Schöner only published the Walther observations in 1544. Swerdlow suggests that Rheticus was the likely intermediary. See Swerdlow and Neugebauer, *Mathematical Astronomy* (1984), 1: 415–16, 429–41, and Richard L. Kremer, "The Use of Bernard Walther's Astronomical Observations: Theory and Observation in Early Modern Astronomy," *Journal for the History of Astronomy* 12 (1981).

51. N. M. Swerdlow, "Astronomy in the Renaissance," in *Astronomy before the Telescope*, ed. Christopher Walker (London: British Museum Press, 1996, 195); Swerdlow and Neugebauer, *Mathematical Astronomy*, 1: 53; Richard L. Kremer, "Bernard Walther's Astronomical Observations," *Journal for the History of Astronomy* 11 (1980): 174 (quote), 176.

52. Kremer, "The Use of Bernard Walther's Astronomical Observations"; Robert R. Newton, "An Analysis of the Solar Observations of Regiomontanus and Walther," *Quarterly Journal of the Royal Astronomical Society* 23 (1982); and Kremer, "Walther's Solar Observations: A Reply to R. R. Newton," *Quarterly Journal of the Royal Astronomical Society* 24 (1983).

53. Richard L. Kremer, "Text to Trophy: Shifting Representations of Regiomontanus's Library," in *Lost Libraries: The Destruction of Great Book Collections since Antiquity*, ed. James Raven (Houndmills, Basingstoke, Hampshire: Palgrave Macmillan, 2004); Ernst Zinner, *Regiomontanus, His Life and Work* [1938], trans. Ezra Brown (Amsterdam: North-Holland, 1990), 157–72.

54. Georg Peurbach, *Quadratu[m] geometricu[m] praeclarissimi mathematici Georgii Burbachii* ([Nuremberg]: [Impressum . . . per Ioannem Stuchs], [1516]), and Johannes Regiomontanus, *De cometae magnitudine, longitudineq[ue] ac de loco eius vero, problemata XVI* (Nuremberg: apud Fridericum Peypus, 1531).

55. Schöner, ed., *Scripta . . . Regiomontani*, 27r ("eius discipuli"), 27v; a iiiv ("Thesaru[m] observationu[m]"). Regiomontanus' solar observations for 1462–75 appear at 27r–v, followed by Walther's solar observations for 1475–1504 (27v–34r); other observations made by Regiomontanus (the early ones with Peurbach) for 1457–74 appear at 36r–43v, followed by Walther's for 1475–1504 (44r–60v).

56. See Liesbeth C. de Wreede, "Willebrord Snellius (1580–1626): A Humanist Reshaping the Mathematical Sciences" (Ph.D. thesis, Utrecht University, 2007), 145–55; and Bruce T. Moran, "Wilhelm IV of Hesse-Kassel: Informal Communication and the Aristocratic Context of Discovery," in *Scientific Discovery: Case Studies*, ed. Thomas Nickles (Dordrecht: D. Reidel Pub. Co, 1980).

57. Tycho had published their correspondence and dedicated the volume to Moritz. For Tycho's *Epistolarum astronomicarum libri* (1596) and the relations between Kassel and Uraniborg, see Adam Mosley, *Bearing the Heavens: Tycho Brahe and the Astronomical Community of the Late Sixteenth Century* (Cambridge: Cambridge University Press, 2007), 38–115, 125–48. The sheets printed at Uraniborg in 1596 were later reissued with new title pages by printers in Nuremberg (1601) and Frankfurt (1610).

58. Beginning with the Babylonians and Egyptians, Hipparchus, and Ptolemy; Julius Caesar "with his Sosigenes"; then jumping from Alfonso X of Castile to Wilhelm—the only modern figure Snell mentioned. See Willebrord Snell, "Praefatio," in *Coeli & siderum in eo errantium observationes Hassiacae* (Leiden: Apud Iustum Colsterum, 1618).

59. Snell, *Coeli & siderum in eo errantium observationes Hassiacae*, 51v–54v; cf. Kepler, *Ad Vitellionem paralipomena* (Frankfurt: Apud Claudium Marnium & Hæredes Ioannis Aubrii, 1604), 150–56, citing Schöner, ed., *Scripta . . . Regiomontani*, 52v ff. See Kremer, "The Use of Bernard Walther's Astronomical Observations," 128–29.

60. See, e.g., Thomas Streete, *Astronomia Carolina: A New Theory of the Coelestial Motions: Composed According to the Best Observations* (London: Printed for Lodowick Lloyd, 1661), 94–119, and Vincent Wing, *Astronomia Britannica. . . . Cui accesit observationum astronomicarum synopsis compendiaria, ex quâ Astronomiae Britannicae certitudo affatim elucescit* (London: Typis Johannis Macock, 1669), [265]–366. Streete had published a 1667 rejoinder to Wing's 1665 criticism of his *Astronomia Carolina*.

61. Philips van Lansbergen, *Tabulae motuum coelestium perpetuae; ex omnium temporum observationibus constructae, temporumque omnium observationibus consentientes. Item novae et genuinae motuum coelestium theoricae. & astronomicarum observationum thesaurus* (Middelburg: Apud Zachariam Romanum, 1632), 37–186 (for the categories of astronomical observations, see the index, sig. [Qqq5v]); Godefroy Wendelin, *Eclipses lunares ab anno M. D. LXXIII. ad M. DC. XLIII. observatae: quibus Tabulae Atlanticae superstruuntur earumque idea proponitur* (Antwerp: Apud HieronymumVerdussium, 1644), 107–8 (1642 eclipse observations), 102–3 (1635 eclipse observations), and 41–111 (the entire observational series). On Peiresc's efforts to coordinate astronomical observation in this period, including with Wendelin, see Peter N. Miller, "Mapping Peiresc's Mediterranean: Geography and Astronomy, 1610–36," in *Communicating Observations in Early Modern letters (1500–1675): Epistolography and Epistemology in the Age of the Scientific Revolution*, ed. Dirk van Miert (London: Warburg Institute, 2013).

62. Wendelin, *Eclipses lunares*, 69, 70, 73, 75, 76, 86, 88, 92, 106, 109.

63. Cristoforo Borri, *Collecta astronomica: ex doctrina P. Christophori Borri, mediolanensis, ex Societate Iesu* (Lisbon: Apud Matthiam Rodrigues, 1631), 73–160.

64. E.g., Francesco Fontana, *Novae coelestium terrestriumq[ue] rerum observationes, et fortasse hactenus non vulgatae* (Naples: apud Gaffarum, 1646). For an entry point into a large literature, see Robert H. Van Gent and Albert Van Helden, "Lunar, Solar, and Planetary Representations to 1650," in *The History of Cartography*, vol. 3, ed. David Woodward (Chicago: University of Chicago Press, 2007).

65. Borri, *Collecta astronomica*, 8 (quote). See Swerdlow, "Tycho, Longomontanus, and Kepler on Ptolemy's Solar Observations and Theory, Precession of the Equinoxes, and Obliquity of the Ecliptic," in *Ptolemy in Perspective: Use and Criticism of His Work from Antiquity to the Nineteenth Century*, ed. Alexander Jones (Dordrecht: Springer, 2010), and Curtis Wilson, "Predictive Astronomy in the Century after Kepler," in *Planetary Astronomy from the Renaissance to the Rise of Astrophysics*, part A: *Tycho Brahe to Newton*, ed. René Taton and Curtis Wilson (Cambridge: Cambridge University Press, 1989).

66. Albert Van Helden, *Measuring the Universe: Cosmic Dimensions from Aristarchus to Halley* (Chicago: University of Chicago Press, 1985).

67. See especially Adam Mosley, "Reading the Heavens: Observation and Interpretation of Astronomical Phenomena in Learned Letters circa 1600," in *Communicating Observations in Early Modern Letters (1500–1675): Epistolography and Epistemology in the Age of the Scientific Revolution*, ed. Dirk van Miert (London: Warburg Institute, 2013).

68. As shown by Adam Mosley, "Past Portents Predict: Cometary Historiae and Catalogues in the Sixteenth and Seventeenth Centuries," in *Celestial Novelties on the Eve of the Scientific Revolution: 1540–1630*, ed. Dario Tessicini and Patrick Boner (Florence: L. S. Olschki, 2013).

69. Anthony Grafton, "Some Uses of Eclipses in Early Modern Chronology," *Journal of the History of Ideas* 64, no. 2 (2003): 214 (quote), 214–18. For an entry point into the development of technical chronology in early modern Europe, see Grafton, *Joseph Scaliger* (1983–93), vol. 2.

70. Heinrich Bünting, *Chronologia hoc est, omnium temporum et annorum series* (Zerbst: Impressum typis Bonaventurae Fabri, 1590), 499v–501r, chap. 3; see e.g. Denis Petau, *Opus de doctrina temporum* (Paris: Sumptibus Sebastiani Cramoisy, 1627), 1: 770–871; and Johann Heinrich Alsted, *Thesaurus chronologiae*, 2nd ed. (Herborn: [Corvinus Erben], 1628), 53–67, 484–94, and 481–82 (on great conjunctions).

71. See the lists of longitudes given for "notable cities in which celestial observations were made" following each map in Gerhard Mercator's edition of Ptolemy's *Tabulae geographicae* ([Cologne]: [typis Godefridi Kempensis], 1578); Georges Fournier, *Hydrographie contenant la theorie et la practique de toutes les parties de la navigation* (Paris: Chez Michel Soly, 1643), chap. 12.

72. Giovanni Battista Riccioli, *Almagestum novum astronomiam veterem novamque complectens* (Bologna: Ex Typographia Hæredis Victorij Benatij, 1651), 2: 3–23 (comets), 133–39, 166–70 (new stars); 1: 361 (quotes), 361–85 (solar and lunar eclipses, 772 BC–AD 1647); and Riccioli, *Astronomiae reformatae tomi duo* (Bologna: Ex Typographia Hæredis Victorij Benatij, 1665), 1: 7–14, 16–29, 37–38 (solar observations), 95–107 (lunar eclipses, 721 BC–AD 1661), 143–47 (solar eclipses, 585 BC–AD 1661), 364–65 (Saturn), 369–70 (Jupiter).

73. Riccioli, *Almagestum novum*, 1: 361.

74. Riccioli, *Geographiae et hydrographiae reformatae* (Venice: Typis Ioannis La Noù, 1672), 352–57 ("Catalogus Eclipsium Lunae . . . observatarum"), 358–88, 388–409

("Catalogus locorum cum Latitudine, ac Longitudine certa, aut probabiliore nobis. . . ."); see also 267–68, 286–310 (on latitudes).

75. Tycho Brahe, *Astronomiæ instauratæ mechanica* (Wandesburg: [Philippi de Ohr], 1598), F2v ("rarissimi & pretiorissimi thesauri").

76. Erasmus Bartholin, *Specimen recognitionis nuper editarum observationum astronomicarum . . . Tychonis Brahe* (Copenhagen: prostat apud Danielem Paulli: literis Henrici Gödiani, 1668), 7, 9–12.

77. Lucius Barretus [Albert Curtz], ed., *Historia coelestis ex libris commentariis manuscriptis observationum vicennalium viri generosi Tichonis Brahe* ([Augsburg]: [Apud Simonem Utzschneiderum], [1666]), 547, 549–51 (Hessen observations), 553–624 (Hessen star catalogue).

78. Curtz systematically juxtaposed Hessen and Tychonian material in chronological order, and did the same with extended sequences of observations made by Michael Mästlin, Kepler's teacher and professor of mathematics at Tübingen, and by Mästlin's successor, Wilhelm Schickard. "Observationes Hassicae" appear in the sections for the years 1590–97; "Observationes Wirtenbergiae" appear in the sections for 1582–91, 1593–98. Mästlin's 1592 observations seem to have been accidentally omitted from the chronological sequence, but were included in a "conspectus" of the Württemburg observations for 1582–93; see Curtz, ed., *Historia coelestis*, lxxxviii, xcii.

79. John Wallis to Henry Oldenburg, 19 and 21 January 1668/9, *The Correspondence of John Wallis*, ed. Philip Beeley and Christoph J. Scriba (Oxford: Oxford University Press, 2003), 3: 151. The Savilian professor of geometry at Oxford, Wallis wrote to Henry Oldenburg that he agreed with Bartholin's critical evaluation but hoped Bartholin would retain Curtz's "Additions."

80. Curtz, ed., *Historia coelestis*, xlii–lxvii (Peurbach, Regiomontanus, Walther, drawn from Snell's 1618 edition); xlii (Turkish annals), xlvi (Zacut), lx (Werner), lxvii (Jamaica).

81. Curtz, ed., *Historia coelestis*, *3v.

82. Gary W. Kronk, *Cometography: A Catalog of Comets* (Cambridge: Cambridge University Press, 1999), 1: 354–55; Alexandre-Guy Pingré, *Cométographie, ou, traité historique et théorique des comètes* (Paris: De l'Imprimerie royale, 1783), 1: 105 (quote).

83. Johannes Hevelius, *Prodromus cometicus, historia, cometae anno 1664 exorti* (Gdańsk: Auctoris typis, et sumptibus, Imprimebat Simon Reiniger, 1665).

84. "Of the Judgement of Some of the English Astronomers," *Philosophical Transactions* 9 (February 12, 1665/6): 150–51. On the Hevelius–Auzout controversy, see Steven Shapin, *A Social History of Truth: Civility and Science in Seventeenth-Century England* (Chicago: University of Chicago Press, 1994), 266–91.

85. Hevelius, *Cometographia, totam naturam cometarum . . . . Accesit, omnium cometarum, à mundo condito hucusquè ab historicis, philosophis, & astronomis annotatorum, historia* (Gdańsk: Auctoris typis, & sumptibus, imprimebat Simon Reiniger, 1668), 791 (quote); 791–913 (history).

86. Gassendi's "Commentari re rebus caelestibus," cited in Miller, "Mapping Peiresc's Mediterranean," 153. See Pierre Gassendi, *Opera omnia in sex tomos divisa* (Lyon: Sumptibus Laurentii Anisson, & Ioannis Baptistae Devenet, 1658), 4: 75 (quote), a2v–a3v, 75–536, and Gassendi, *Epistolica exercitatio. . . . Cum appendice aliquot observationum coelestium* (Paris: Apud Sebastianum Cramoisy, 1630); Hevelius, *Machinae coelestis pars posterior; rerum uranicarum observationes* (Gdańsk: In aedibus auctoris, eiusq̨ typis, &

sumptibus: imprimebat Simon Reiniger, 1679), and Hevelius, *Annus climactericus, sive rerum uranicarum observationum annus quadragesimus nonus* (Gdańsk: Sumptibus auctoris, typis Dav.-Frid. Rhetii, 1685).

87. Vittorio Francesco Stancari, *Schedae mathematicae post ejus obitum collectae ejusdem Observationes astronomicae*, ed. Eustachio Manfredi (Bologna: Typis Jo: Petri Barbiroli, 1713); Eustachio Manfredi, *De gnomone meridiano bononiensi ad divi petronii deque observationibus astronomicis eo instrumento ab ejus constructione* (Bologna: Ex Typographia Laelii a Vulpe, 1736); and Francesco Bianchini, *Astronomiae, ac geographicae observationes selectae Romae*, ed. Eustachio Manfredi (Verona: Typis Dyonisii Ramanzini, 1737). On the importance of church-sited meridian lines as precision astronomical instruments in this period, see John L. Heilbron, *The Sun in the Church: Cathedrals as Solar Observatories* (Cambridge, MA: Harvard University Press, 1999).

88. Etienne Souciet, ed., *Observations mathématiques, astronomiques, géographiques, chronologiques, et physiques, tirées des anciens livres chinois, ou faites nouvellement aux Indes et a la Chine, par les peres de la compagnie de Jesus* (Paris: Chez Rollin, 1729–32); Ignatius Kögler, *Observationes eclipsium, variorumque caelestium congressuum habitae in Sinis* (Lucca: Typis Salvatoris, & Jo. Dominici Marescandoli, 1745); Christian Mayer, *Solis et lunae eclipseos observatio astronomica . . . facta Schwezingae in specula nova electorali . . . comparata pluribus Europae celebrioribus observationibus* (Mannheim: Ex Typographejo Electorali-Aulico, 1766); Augustin Hallerstein, *Observationes astronomicae ab anno 1717 ad annum 1752*, ed. Maximilian Hell (Vienna: Typis Joannis Thomae, 1768); and Hell's long-running serial, the *Ephemerides astronomicae ad meridianum Vindobonensem* (from 1757). See also the volumes edited by Thomas Gouye (n92); Florence C. Hsia, *Sojourners in a Strange Land: Jesuits and their Scientific Missions in Late Imperial China* (Chicago: University of Chicago Press, 2009), chap. 7; and László Kontler, "The Uses of Knowledge and the Symbolic Map of the Enlightened Monarchy of the Habsburgs: Maximilian Hell as Imperial and Royal Astronomer (1755–1792)," in *Negotiating Knowledge in Early Modern Empires: A Decentered View*, ed. László Kontler, et al. (New York: Palgrave Macmillan, 2014).

89. For each southern star, Halley generally published both its observed distance from two stars in Tycho's catalogue (reduced for the epoch 1678), as well as its derived celestial longitude and latitude in his *Catalogus stellarum australium sive supplementum catalogi tychonici* (London: Typis Thomae James, 1679), which also included a separately paginated account of his 1677 Mercury transit observations. See Alan H. Cook, *Edmond Halley: Charting the Heavens and the Seas* (Oxford: Clarendon Press, 1998), 65–72, 77–83, 219–25; and Van Helden, *Measuring the Universe*, chaps. 12–13.

90. E.g., Académie royale des sciences, *Recueil d'observations faites en plusieurs voyages* (Paris: De l'Imprimerie Royale, 1693); Thomas Goüye, ed., *Observations physiques et mathematiques . . . envoyées de Siam* (Paris: Chez la Veuve d'Edme Martin, Jean Boudot, & Estienne Martin, 1688); Goüye, ed., *Observations physiques et mathematiques . . . envoyées des Indes et de la Chine* (Paris: De l'Imprimerie Royale, 1692); François Noel, *Observationes mathematicæ, et physicæ in India et China* (Prague: typis Universit: Carolo-Fernandeæ, in Collegio Soc. Jesu ad S. Clementem, per Joachimum Joannem Kamenicky Factorem, 1710); Louis Feuillée, *Journal des observations physiques, mathematiques et botaniques . . . sur les côtes orientales de l'Amerique meridionale, & dans les Indes occidentales* (Paris: Chez Pierre Giffart; Jean Mariette, 1714–25); Antoine Francois Laval, *Voyage de la Louisiane, fait par ordre du roy en l'année mil sept cent vingt: dans lequel sont traitées diverses matieres de*

physique, astronomie, géographie & marine (Paris: Chez Jean Mariette, 1728); and Joseph Bernard Chabert, *Voyage fait par ordre du roi en 1750 et 1751, dans l'Amérique septentrionale, pour rectifier les cartes des côtes de l'Acadie, de l'Isle royale & de l'Isle de Terre-Neuve; et pour en fixer les principaux points par des observations astronomiques* (Paris: De l'Imprimerie royale, 1753). See Hsia, *Sojourners*, 88–109.

91. See especially John Leonard Greenberg, *The Problem of the Earth's Shape from Newton to Clairaut: The Rise of Mathematical Science in Eighteenth-Century Paris and the Fall of "Normal" Science* (Cambridge: Cambridge University Press, 1995), and Mary Terrall, *The Man Who Flattened the Earth: Maupertuis and the Sciences in the Enlightenment* (Chicago: University of Chicago Press, 2002), chaps. 4–5.

92. Pierre-Louis Moreau de Maupertuis, *La figure de la terre, déterminée par les observations de Messieurs de Maupertuis, Clairaut, Camus, Le Monnier . . . & de M. l'Abbé Outhier . . . accompagnés de M. Celsius . . . faites par ordre du roy au cercle polaire* (Paris: De l'Imprimerie royale, 1738), iii–iv.

93. As Mary Terrall has argued, relatively few readers were equipped to review the data in the way he suggested, and comparison with an extant manuscript notebook indicates that observations were suppressed, whether because they were deemed incidental or too internally discordant. See Terrall, *The Man Who Flattened the Earth*, 117 (quote), 115–18.

94. Réginald Outhier, *Journal d'un voyage au nord, en 1736. & 1737.* (Paris: Chez Piget . . . Durand, 1744), 203.

95. Terrall, *The Man Who Flattened the Earth*, 152–54.

96. César-François Cassini de Thury, *La meridienne de l'observatoire royal de Paris, vérifiée dans toute l'étendue du royaume par de nouvelles observations* (Paris: Chez Hippolyte-Louis Guerin, & Jacques Guerin, 1744), 28 (quote); see also the separately paginated "Troisieme partie," iii, vi.

97. For a study of the dispute, see Adrian Johns, *The Nature of the Book: Print and Knowledge in the Making* (Chicago: University of Chicago Press, 1998), chap. 8. The warrant also gave the Royal Society control over the Royal Astronomer's future work, including access to his observations.

98. Edmond Halley, "Praefatio," in John Flamsteed, *Historiae coelestis libri duo* (London: Typis J. Matthews, 1712), iv ("chartis"), v; both observations related to Flamsteed's star catalogue as well as observations of other celestial phenomena were implicated. See the royal warrant, Henry St. John to Flamsteed (12 December 1710), and Flamsteed to Abraham Sharp (29 March 1716) in *The Correspondence of John Flamsteed*, ed. Eric G. Forbes, Lesley Murdin, and Frances Willmoth (Bristol, UK: Institute of Physics Pub., 1995), 3: 573–75, 784–85; and Flamsteed, *The preface to John Flamsteed's Historia coelestis Britannica: or British catalogue of the heavens (1725)*, trans. Alison Dione Johnson, ed. Allan Chapman (London: Trustees of the National Maritime Museum, 1982), 177–78. For the omitted text, see Flamsteed, *The preface to John Flamsteed's Historia coelestis Britannica*, 160–79.

99. James Bradley, *Astronomical observations, made at the Royal Observatory at Greenwich, from the year MDCCL to the year MDCCLXII*, ed. Thomas Hornby and Abram Robertson (Oxford: At the Clarendon Press, 1798–1805), 1: i–ii; 2: iii–iv. See the dossier on the conflict over Bradley's observations in the Board of Longitude Papers, RGO 14/4, 230r–272v, and the Confirmed Minutes for 18 June 1768, RGO 14/5, 167–68 and 6 January/1 March 1804, RGO 14/7, 2: 47r–v.

100. Nevil Maskelyne, *Astronomical observations made at the Royal Observatory at Greenwich, in the years 1765, 1766, 1767, 1768, and 1769. . . . Published by the President and Council of the Royal Society, at the public expence, in obedience to His Majesty's Command* (London: Printed by W. and J. Richardson, 1774).

101. Pierre-Charles Le Monnier, "Discours préliminaire," *Histoire celeste, ou Recueil de toutes les observations astronomiques faites par ordre du roy* (Paris: Chez Briasson, 1741), i; "Projet d'une histoire celeste," *Histoire céleste,* [aiii]r.

102. Le Monnier, "Projet d'une histoire celeste," *Histoire céleste,* [aiii]r.

103. John Flamsteed, *Historia coelestis Britannica* (London: Typis H. Meere, 1725), 1: 1–29, 3: 95. See Flamsteed to Newton (draft, 24 October 1725), in *The Correspondence of John Flamsteed,* 3: 230–31; for Newton's opposition, see Johns, *The Nature of the Book,* 588n111.

104. Le Monnier, "Discours préliminaire," *Histoire céleste,* iii–iv. Le Monnier's tone was sharp; he thought this approach suitable only for "an Author who would make a particular history of his own observations."

105. Maupertuis had read the text printed as his preface to *La figure de la terre* (1738) at the academy's public session of 16 April 1738; Le Monnier read the text of the "Projet" in the academy session of 10 May 1738. See Maupertuis, "Préface," *La figure de la terre,* iii, and Le Monnier, "Projet d'une histoire celeste," *Histoire céleste,* [aiii]r.

106. Pierre-Charles Le Monnier, "Recherches sur la hauteur du pole de Paris [14 June 1738]," *Histoire de l'Académie royale des sciences: année [1738]: avec les mémoires de mathématique & de physique . . . tirés des registres de cette académie* (1740), 209, 214–20. Le Monnier read this paper at the academy session of 14 June 1738, a prelude to the fuller discussion in his "Discours préliminaire" to the *Histoire celeste.* See Maupertuis, "Discours qui a été lu dans l'assemblée publique de l'Académie Royale des Sciences, le 13 Novembre 1737," *La figure de la terre,* 41–45, 69–71; compare the corrections to observations made later in the work (121–27, 136). Maupertuis' "Discours" was also published in the academy's 1737 *Mémoires,* which appeared in 1740. For James Bradley's work on aberration, the stellar phenomena he explained in terms of the nutation of the earth's axis, and his correspondence on these issues with Maupertuis through October 1737, see John Fisher, "Conjectures and Reputations: The Composition and Reception of James Bradley's Paper on the Aberration of Light with Some Reference to a Third Unpublished Version," *British Journal for the History of Science* 43, no. 1 (2010).

107. Le Monnier, "Projet d'une histoire celeste," *Histoire céleste,* [aiv]r (quote); [aiv]r–v.

108. Le Monnier, "Discours préliminaire," *Histoire céleste,* i.

109. Sven Widmalm, "A Commerce of Letters: Astronomical Communication in the Eighteenth Century," *Science Studies* 5, no. 2 (1992): 47.

110. Jacques Cassini, "Préface," *Tables astronomiques du soleil, de la lune, des planetes, des etoiles fixes, et des satellites de Jupiter et de Saturne: avec l'explication & l'usage de ces mêmes tables* (Paris: De l'Imprimerie royale, 1740), viii–ix. Observations of the 1631 Mercury transit and the 1639 Venus transit were old enough for Cassini's purposes; see Cassini, *Elemens d'astronomie* (Paris: De l'Imprimerie royale, 1740).

111. Richard Dunthorne, "A Letter from the Rev. Mr. Richard Dunthorne to the Reverend Mr. Richard Mason F. R. S. and Keeper of the Wood-Wardian Museum at Cambridge, concerning the Acceleration of the Moon," *Philosophical Transactions* 46 (1749): 163, 169. For a thorough assessment of Dunthorne's work, see Steele, *Ancient Astronomical*

Observations, chap. 6; for Tobias Mayer's and Joseph-Jérôme Lalande's use of historical observations, see chaps. 7–8.

112. James E. McClellan, *Specialist Control: The Publications Committee of the Académie royale des sciences (Paris), 1700–1793* (Philadelphia: American Philosophical Society, 2003), 51–53.

113. "Avertissement," Pierre Bouguer, *Justification des memoires de l'Académie royale des sciences de 1744: et du livre de la figure de la terre, déterminée par les observations faites au Pérou, sur plusieurs faits qui concernent les opérations des académiciens* (Paris: Chez Charles-Antoine Jombert, 1752), [i], vi–vii, in response to Charles-Marie de La Condamine, *Mesure des trois premiers degrés du méridien dans l'hémisphere austral: tirée des observations de Mrs. de l'Académie royale des sciences, envoyés par le roi sous l'équateur* (Paris: De l'Imprimerie royale, 1751); but see La Condamine's presentation and remarks on his observations in *Mesure*, 121–226. Cassini de Thury took an alternative approach, marking each of the stellar observations tabulated and averaged in *La meridienne* (1744) as "exact," "passable," and "mediocre"; other evaluative terms included "good," "good enough," "in haste," "too late," "dubious," and "weak." See *La meridienne*, lxxiv–lxxv, lxxxiii–c. See Stephen M. Stigler, *The History of Statistics: The Measurement of Uncertainty before 1900* (Cambridge, MA: Belknap Press of Harvard University Press, 1986); Oscar Sheynin, "The Treatment of Observations in Early Astronomy," *Archives for History of Exact Sciences* 46 (1993); and Jed Z. Buchwald, "Discrepant Measurements and Experimental Knowledge in the Early Modern Era," *Archives for History of Exact Sciences* 60, no. 6 (2006).

114. Nicolas-Louis de Lacaille, "Extrait de quelques observations astronomiques, faites au Collège Mazarin pendant l'année 1743," *Histoire de l'Académie royale des sciences: année [1743]: avec les mémoires de mathématique & de physique . . . tirés des registres de cette académie* (1746), 160.

115. Le Monnier, "Discours préliminaire," *Histoire céleste*, iv.

116. Pierre-Charles Le Monnier, *Observations de la lune, du soleil, et des étoiles fixes: pour servir a la physique céleste et aux usages de la navigation* (Paris: De l'Imprimerie royale, 1751–73).

117. Joseph-Nicolas Delisle to Jean-Paul Bignon (26 June 1720), Observatoire de Paris, B1/1–170, 1v, and Delisle to Raymond de Navarre (7 February 1721), ObsParis, B1/2-9, 18r; Guillaume Bigourdan, "Introduction," Alexandre-Guy Pingré, *Annales célestes du dix-septième siècle*, ed. Guillaume Bigourdan (Paris: Gauthier-Villars, 1901), ix (quote), v–xi; Joseph-Jérôme Lalande, *Histoire céleste française, contenant les observations faites par plusieurs astronomes français* (Paris: De l'Imprimerie de la république, 1801), xii; and Jean-Dominique Cassini, *Mémoires pour servir à l'histoire des sciences et a celle de l'Observatoire royal de Paris, suivis de la vie de J.-D. Cassini, écrite par lui-même, et des éloges de plusieurs académiciens morts pendant la revolution* (Paris: Chez Bleuet, 1810), 148 (quote), 12–13, 148–51, 166, 169, 174–76. See Florence C. Hsia, "Chinese Astronomy for the Early Modern European Reader," *Early Science and Medicine* 13, no. 5 (2008).

118. For a modern reiteration of Delambre's criticisms, see Robert R. Newton, *The Crime of Claudius Ptolemy* (Baltimore: Johns Hopkins University Press, 1977). For earlier assessments, see Swerdlow, "Tycho, Longomontanus, and Kepler"; a rigorous review of the issue is given in Britton, *Models and Parameters*.

119. Jean-Baptiste Joseph Delambre, *Histoire de l'astronomie ancienne* (Paris: Mme Ve Courcier, 1817), 1: xxxiv–v (quote); xxv–xxxv.

120. Cassini, *Mémoires*, 150, of his "celestial history of the royal Observatory of Paris," proposed in 1774.

121. Gretchen Greene, Brian McLean, and Barry Lasker, "Development of the Astronomical Image Archive and Catalog Database for Production of GSC-II," *Future Generation Computer Systems* 16, no. 1 (1999); Derek Jones, "The Scientific Value of the Carte du Ciel," *Astronomy & Geophysics* 41, no. 5 (2000); M. Tsvetkov et al., eds., *Proceedings of the International Workshop on Virtual Observatory: Plate Content Digitization, Archive Mining [and] Image Sequence Processing* ([Sofia], Bulgaria: Heron Press, 2005); S. Laycock et al., "Digital Access to a Sky Century at Harvard: Initial Photometry and Astrometry," *Astronomical Journal* 140, no. 4 (2010); and Charlotte Bigg, "Photography and Labour History of Astrometry: The Carte du Ciel," in *The Role of Visual Representations in Astronomy: History and Research Practice: Contributions to a Colloquium Held at Göttingen in 1999*, ed. Klaus Hentschel and Axel D. Wittmann (Thun: Verlag Harri Deutsch, 2000).

122. B. G. Corbin and D. J. Coletti, "Digitization of Historical Astronomical Literature," *Vistas in Astronomy* 39, no. 2 (1995), and Eichhorn et al., "Current and Future Holdings of the Historical Literature in the ADS."

123. G. Bruce Berriman and Steven L. Groom, "How Will Astronomy Archives Survive the Data Tsunami?" *Queue* 9, no. 10 (2011).

124. Hubble Second Decade Committee, *The Hubble Data Archive: Towards the Ultimate Union Archive of Astronomy* (Baltimore: Space Telescope Science Institute, 2000), 20. See A. S. Szalay and R. J. Brunner, "Astronomical Archives of the Future: A Virtual Observatory," *Future Generation Computer Systems* 16, no. 1 (1999); Raymond L. Plante et al., "Building Archives in the Virtual Observatory Era," *Software and Cyberinfrastructure for Astronomy: 27–30 June 2010, San Diego, California, United States*, ed. Nicole M. Radziwill and Alan Bridger (Bellingham, WA: SPIE, 2010).

125. Gordon Bell, Tony Hey, and Alex Szalay, "Beyond the Data Deluge," *Science* 323, no. 5919 (2009); Gordon Bell, foreword to *The Fourth Paradigm: Data-Intensive Scientific Discovery*, ed. Tony Hey, Stewart Tansley, and Kristin Tolle (Redmond, WA: Microsoft Research, 2009), xiii.

126. Kirk Borne, "Virtual Observatories, Data Mining, and Astroinformatics," in *Planets, Stars, and Stellar Systems*, ed. Terry D. Oswalt and Howard E. Bond (Dordrecht: Springer, 2013), 2: 406.

TWO

The Earth as Archive: Contingency, Narrative, and the History of Life

David Sepkoski

In the early 1970s, the late evolutionary paleontologist Stephen Jay Gould joined a small group of colleagues in conducting a campaign to upgrade the status of paleontology—for much of the twentieth century considered a descriptive "handmaiden" to evolutionary biology—by introducing a quantitative, theoretical agenda capable of producing independent contributions to evolutionary theory. Gould and his compatriots called this approach "paleobiology," and their central conceit was that the fossil record— the archive of the history of past life—was a reliable source of knowledge about the patterns and processes of evolution.[1] One of the signature early moments in this emerging movement was the publication of Gould and Niles Eldredge's model of "punctuated equilibria," a hypothesis that hinged on the notion that the notorious incompleteness of the fossil record was often overstated: "Many breaks in the fossil record are real," they wrote, and "they express the way in which evolution occurs, not the fragments of an imperfect record."[2] Gould considered this a vital intervention in a discipline that had long held a deeply internalized pessimism about the quality of its data—a pessimism that can be traced all the way back to Darwin, who wrote that "we have no right to expect to find in our geological formations, an infinite number of those fine transitional forms, which on my theory assuredly have connected all the past and present species of the same group into one long and branching chain of life."[3] On the contrary, Eldredge and Gould asserted, accepting the fossil record at face value "would release us from a self-imposed status of inferiority among the evolutionary sciences," and escape the "collective gut-reaction [that] leads us to view almost any anomaly as an artifact imposed by our collective millstone—an imperfect fossil record."[4]

Gould's insistence on a literal reading of the fossil record conflicted at times, however, with one of the other major premises of his new paleobiology, which was that paleontology could be "nomothetic" or law-producing. Gould wanted to free paleontologists from being mere "stamp collectors" who spent their lives amassing and describing individual fossils, and to encourage his colleagues to pursue laws of evolutionary development that could be broadly generalized for the history of life. This agenda also closely depended on viewing the fossil record as a reliable source of data, but it drew on the use of mathematical generalizations that privileged the statistical regularity of patterns of data above individual historical events. This underlined a conundrum facing Gould and his colleagues: should the fossil record be treated as an archive of distinct historical events, whose patterns could be "read literally" to uncover the empirical pattern of life's history, or is it best read as a source of statistical generalizations in which the individual event is far less significant than the ensemble properties of its constituent "molecules"?[5]

Ultimately, Gould backed away from the vision of a purely nomothetic paleontology because he came to realize that the essence of paleontology—or of any historical science—is contingency, and that paleontological explanations must ultimately confront the role of individual, contingent events in shaping history's patterns. This view, which he developed through the late 1970s and early 1980s, found its most famous elaboration in his 1989 *Wonderful Life*, where Gould explained that "a historical explanation does not rest on direct deductions from laws of nature, but on an unpredictable sequence of antecedent states, where any major change in any step of the sequence would have altered the final result. This final result is therefore dependent, or contingent, upon everything that came before—the unerasable and determining signature of history."[6] Because of this "determining signature of history," historical explanations, Gould maintained, also take the form of narratives: "The resolution of history must be rooted in the reconstruction of past events themselves—in their own terms—based on narrative evidence of their own unique phenomena."[7] The narrativity of historical explanations derives from the contingency of the events of which they are composed. The history of life—like human history—did not *have* to unfold the way it did. Unique events that occurred at particular times have shaped this history in irreducible ways. Another way of putting it is that while history is not adeterministic—in retrospect, cause and effect are often very clear—at any given moment the future is underdetermined—at least so far as our knowledge is concerned—by existing conditions. Contingency is not the same thing as randomness, but an unexpected or chance event might produce unpredictable consequences which set a new sequence of events in motion.

Because of this contingency—unpredictability and path-dependency—historical explanations depend crucially on archives, which are generally understood to be collections of unique historical "documents" that record contingent historical events. Archives have a special epistemic status in disciplines—such as paleontology—that cannot depend on axioms or laws of nature to produce explanations: in other words, the history of life cannot be deductively reconstructed simply from a knowledge of present biota and the rules of natural selection. Historical explanations are, Gould maintained, a special kind of scientific explanation. "The narrative sequences of history are, in no sense, incomprehensible or inaccessible to science," but they "differ from the standard experimental model" for two reasons: First, because they document "events that can occur but one time in all their detailed glory," and secondly, because they describe "events that cannot—in principle, and not merely as a limitation of attainable human knowledge—be predicted from antecedent conditions, but only explained after their probabilistic occurrence."[8] The only access a paleontologist has to these unique events is through the archive of the history of life: the fossil record.

The fossil record is paleontology's ur-archive, and has been for more than two centuries: it contains records of distinct events—phenomena that are, in their temporal sequence and contingent relationships to other phenomena, unique. This is why Gould insisted that evolutionary explanations must reflect "the archive of life's actual history as displayed in the fossil record."[9] But that archive known as the "fossil record" has not been a stable object throughout the history of paleontology as a professional discipline. In the first instance, the "original" archive is the earth itself: the fossil record exists in a "natural" state as an archive organized in the rock strata that preserves the temporal and ecological relationships between once-living organisms. The discovery of stratigraphy in the early nineteenth century opened the realization that not only does the earth have a history, but that it has preserved, in a systematic order, the "relics" of that history—much as the specimen drawers and fossil compendia organize the fossil archive maintained by humans.[10]

But at another level, the fossil record is also the archive of data that has been accumulated and curated by paleontologists for over two centuries, and used as a source for interpretations and explanations of the history of life. In this sense the fossil record has had not one but a series of archival reconfigurations, each of which is a kind of "second nature" in which natural objects have been removed from their original contexts and reembedded in new systems of relations that mirror, but do not reproduce exactly, their "natural" relationships.[11] The first such reconfiguration of the natural archive was to organize fossils in specimen collections, but subsequent iterations of the fossil record have included illustrated fossil atlases, taxonomic compendia and tables, and ultimately digital databases. Each successive archival reconfiguration (and, I will argue, they were sequential; we can call them $archive_1$, $archive_2$, etc.) bears some relationship to the "original" archive of the earth ($archive_0$), but reflects most directly the organizing principles and epistemic considerations of the version of the archive most directly preceding it. Thus, as this paper will explore, the initial human-built fossil archives (physical collections and illustrated atlases) attempted to fairly directly transcribe the principles of arrangement of fossils in the earth's strata into physical and visual genres of organization. Later archival conceptualizations—taxonomic compendia and tables of data, and eventually databases—were abstracted not directly from the earth itself, but from organized specimen collections and atlases. In this sense, archives are often (and perhaps always) based on other archives; or, to put it another way, if each archival presentation is a second nature, then successive versions have had an increasingly distant relationship to the original "first" nature.

But in each case, a central function of paleontological archives has been to preserve a record of the contingent history of life on the earth. By the era of digital fossil databases this sense of contingency had become tenuous, a development that precipitated the epistemic dilemma Gould and his colleagues found themselves confronting during the 1970s. Ultimately, as Gould realized (and as the conclusion of this chapter will show), to lose this sense of contingency is to lose the essential historicity of life. Historical science without the contingency preserved in the archive is impossible, and attempts to formulate a purely nomothetic paleontology were impossible precisely because no history can be told without an appreciation for the role of unique events. Archives in historical sciences like paleontology preserve the historicity of the phenomena being studied themselves, and as such are the chief epistemic domain for knowledge claims in those disciplines.

THE EARTH AS ARCHIVE

How the archive came to have a central role in paleontology has everything to do with the context in which paleontology developed its historiographic sensibility in the early nineteenth century. Naturalists have been collecting fossils since classical Antiquity, but before the late eighteenth century, these collections were not archives. While sixteenth- and seventeenth-century pictorial atlases sometimes grouped fossils into categories, there was rarely a sense of the temporal, geographical, or taxonomic relationships between the organisms of which they were remnants (nor indeed any consensus that they *were* organisms). Fossils were considered marvels of nature, and in many ways these atlases were printed versions of contemporary cabinets of curiosities. However, these works did establish several important conventions, including mechanical reproduction of realistic images and dissemination of knowledge about local collections to broader scientific networks, that would continue into the era of the archival sensibility in paleontology.[12]

A major epistemic break took place, however, in the early nineteenth century, with the discovery of stratigraphy. At the heart of this shift was the simultaneous realization—by Georges Cuvier and Alexandre Brongniart in France, and by William Smith in England—that distinctive kinds of fossils are associated with particular layers (strata) of the earth's crust in a nonarbitrary way. The earth, in other words, came to be seen as having a deep history, which could be "read" in the succession of fossils embedded in the strata of the earth's crust. In this sense, the earth is its own archive, and fossils are the documents preserved in the drawers of its cabinet (the strata). It is not a coincidence that this development took place when ideas about the proper way to study and document human history were being hotly debated. As Martin Rudwick has compellingly demonstrated, "the sciences of the earth became historical by borrowing ideas, concepts, and methods from human historiography."[13] In particular, Rudwick shows that eighteenth-century "geohistorians," especially in France and Germany, developed the sense that the earth has a history that can be interpreted by naturalists from the "relics" and "documents" left behind in the rocks, just as human history is pieced together from texts, coins, monuments, and other artifacts. What was particularly significant about geology as it emerged in the late eighteenth century, he argues, is that its practitioners began to conceive of earth's history as a contingent sequence of events—a narrative—and that this narrative could be deciphered in much the same way that erudite historians and antiquarians reconstructed human history from the detritus left behind by civilizations.

Even before the discovery of stratigraphy—from the late seventeenth century onwards—it was a common trope in geology to compare fossils to human historical artifacts. For example, in his "A Discourse of Earthquakes" (published posthumously in 1705 but likely presented to the Royal Society in the late 1660s), Robert Hooke reflected that "there is no Coin can so well inform an Antiquary that there has been such or such a place subject to such a Prince, as these [fossils] will certify a Natural Antiquary, that such and such places have been under water, that there have been such and such kind of Animals, that there have been such and such preceding Alterations and Changes in the superficial Parts of the Earth."[14] While Hooke, like others of his time, was not certain that these objects were organic remains, he nonetheless speculated that "Providence does seem to have design'd these permanent shapes, as Monuments and Records to instruct succeeding Ages of what past in preceding." Hooke's analogy was part of a much larger context, and drew directly from a contemporary epistemic tradition that was developing in the antiquarian study of human history. The distinction between "antiquarian" and "historical" genres dates back to classical Rome, but during much of the Middle Ages and Renaissance the favored mode of historical explanation in Europe was the learned political or religious history.[15] History was meant to provide political or moral lessons, and historical accounts were drawn mostly from earlier authoritative narratives by Greek or Roman authors. While Varro and other Roman authors had developed a comprehensive study of *antiquitates* that sought to recover the details of past civilizations through the remnants of their material cultures, this approach fell out of favor until it was recovered during the later Renaissance. By the seventeenth century, though, "antiquarianism" as practiced by empirically minded authors like Nicolas-Claude Fabri de Peiresc was seen as an antidote to biased, sectarian religious histories, and shared epistemic space with the empirical natural philosophy and Pyhrronian skepticism found in the works of Galileo, Pierre Gassendi, and others. Central here was the notion that nonliterary sources—coins, inscriptions, artifacts, sculptures, charters, etc.—were more reliable than authoritative secondhand accounts. This antiquarian tradition also departed from classical historiography in that it did not assume a necessary philosophical framework for human history. The aim of antiquarians was to produce history that was "factual" and avoided bias or a priori assumptions about how things "must have been"—to leave more space for contingency.[16]

For these reasons, antiquarian history had a close relationship with the natural history of the late seventeenth and early eighteenth centuries. A number of early members of the Royal Society of London had strong interests in both natural and human-made antiquities and tended to discuss them in similar terms. Martin Lister, for example, was a naturalist and physician who wrote important treatises on living and fossil bivalves, *Historiae conchyliorum* (1685), and *Conchyliorum bivalvium* (1696), as well as several papers in the *Philosophical Transactions* about Roman antiquities in Britain.[17] Likewise, both Edward Lhwyd and John Woodward—two of the more important British naturalists of the late seventeenth and early eighteenth centuries—applied a crude kind of "stratigraphy" both to studies of fossils and to archaeological sites in Britain. The notion that the location of an object in layers of sediment is a clue to the relative age of its surroundings was a profound realization for geologists in the next century, and appears to have developed at least as early in the study of human antiquities as it did in geology. From this point forward it became increasingly common to compare fossils to coins, monuments, or inscriptions, and even to refer to their deposits as "archives." For example, the Swedish chemist Torbern Olof Bergman's 1766 treatise *A Physical Description of the Earth* proclaimed that fossils "are actually medallions of a sort, which were laid down on the originating earth surface, whose layers are archives older than all [human] annals, and which appropriately investigated give much light on the natural history of this our dwelling place."[18] Similarly, the French naturalist Jean-Louis Giraud-Soulavie compared the geologist to "the writer or gene-alogist who rummages around in libraries and archives," while the great German savant Johann Friedrich Blumenbach described fossils as "the most infallible certificates [*Urkunden*] in the archive of nature."[19] And François-Xavier Burtin, who frequently referred to fossils as "coins," "documents," and "monuments," also described the history of the earth as "written in a majestic language . . . carved in permanent characters in the great code of nature."[20]

These examples could be multiplied many times over, and demonstrate that antiquarian and archival analogies had deep cultural currency for late eighteenth- and early nineteenth-century naturalists.[21] The language of archive and antiquarianism in early geology is interesting for more than just the evocative metaphors it provides; this "archive talk," I argue, signals the recognition that contingency is an essential feature of historical reconstruction. This was a period, as Rudwick has shown, when grand "geotheoretical" speculations, which tended to depict the earth's history as the deterministic unfolding of universal laws on the Newtonian model, were increasingly challenged by attempts rather to see the history of the earth as a sequence of specific events linked by contingent circumstances that formed a particular narrative.[22] It was also, as Arnaldo Momigliano has argued, a time when so-called philosophical histories were under similar attack by antiquarians, and when a new historiographic approach that applied the emerging lessons and sources of antiquarianism to more traditional historical subjects—such as the decline of ancient Rome—was being pioneered by historians like Edward Gibbon and Johann Joachim Winckelmann. These developments were central to the emergence of geol-ogy as a distinct scientific discipline: the historicization of the earth was accompanied by a dramatic increase in geological interest and activity among European savants, and by the establishment of professional societies and journals devoted to geology. Writing in 1789, the French geologist François-Dominique de Montlosier proclaimed that from the moment Buffon and others revealed the deep history of the earth, "erudition has won the archives of nature: savants have come from everywhere to the provinces to interrogate its monuments and to search its memoirs, and geology has finally become a major science, in which mineralogy, assaying, and chemistry have the honor to be subordinate."[23] The archive was coming to be seen, in other words, as the central resource for the study of history—whether natural or human.

ARCHIVES AS SECOND NATURES

The notion that the earth is an archive—and that the organization of geological specimens and data should in some way be modeled on this natural archival arrangement—was made much more tangible with the emergence of stratigraphy in the early nineteenth century. There are two major reasons for this. First, the recognition that the history of life appears to be directional reinforced the sense that geology has an essentially genealogical component. Second, the physical arrangement of the strata, layered on top of one another in precise chronological sequence, closely resembled the pages of a book or the drawers of a specimen cabinet. Thus both in a conceptual sense—as a repository of genealogical information—and in a physical sense—as an organized cabinet of information—the geological record appeared to be a natural archive.

During the first decade of the nineteenth century, Georges Cuvier collaborated with Alexandre Brongniart in a comprehensive survey of the geology of the Paris region, with the aim of determining the order and arrangement of the characteristic strata of the Paris Basin. The product of this work, published in 1811, was a report that reconstructed an "ideal section" of the strata, in which a cross-section of the formation is depicted as a series of successive layers of rock, containing distinct fossil flora and fauna in each stratum (fig. 2.1).

This work was continued by Brongniart, who in 1822 published an updated schematic of the Paris formations, and in 1829 a much larger grand "Theoretical tableau" depicting the succession of strata for all of Europe. These schematic, idealized stratigraphical images became a standard iconography for representing the narrative of geological and paleontological history. They could be "read," from bottom to top, as depicting the his-

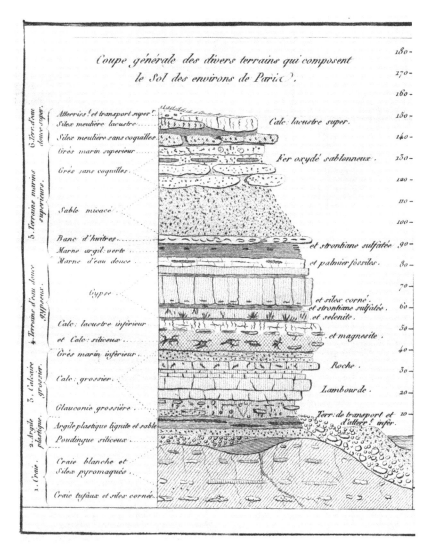

Fig. 2.1 Cuvier and Brongniart's idealized strata of the Paris basin, in Cuvier, *Ossemens fossiles* (Paris, 1812).

torical succession of the strata, and the universality of stratigraphy ensured that, as Cuvier put it, "nature everywhere maintains the same language."[24]

In the widely influential "Preliminary Discourse" to his collected *Researches on Fossil Bones* of 1812, Cuvier explained that "as a new species of antiquarian, I have had to learn to decipher and restore these monuments, and to recognize and reassemble in their original order the scattered and mutilated fragments of which they are composed."[25] In addition to invoking the now-familiar analogy between the geologist and the antiquarian, this passage is also notable because it highlights the task of the geologist or paleontologist as one of reconstruction. The earth has preserved fossils in a chronological sequence, but the fossils themselves are often fragmentary and disordered, and nature's original archive (archive$_0$) is scattered and incomplete. The fossils themselves must be interpreted and classified based on principles of comparative anatomy and taxonomy (for which Cuvier himself was becoming justifiably famous), but paleontologists must also reassemble and reconstruct the original archive—either in physical collections or illustrated depictions—in order to produce knowledge about the past.

For this reason, the principles of stratigraphy became the tools for the first archival transcription of the fossil record. While Cuvier and Brongniart's ideal stratigraphic sections were influential, these representations were schematic rather than archival since they did not attempt to record all of the actual fossils found in particular locations. At the same time in Britain, however, the surveyor William Smith was attempting to develop a system by which characteristic fossils of individual strata could be used to produce general stratigraphical maps for England and, eventually, for all of Europe. These maps were two-dimensional spatial projections, and unlike Cuvier's ideal sections showed what strata were exposed at the earth's surface in an individual location (fig. 2.2).

Smith produced these maps by collecting fossils from particular localities and strata, precisely noting their geographical and stratigraphical placement, and identifying analogous strata in other locations by finding similar fossils. In order to do this, he began amassing a large fossil collection around 1804, which by 1815 (when his first stratigraphical maps were published) contained over 2,600 specimens. This collection, in itself, was not unprecedented for its time, but what was especially noteworthy was the way Smith organized it: he displayed his fossils (in his London home) "arranged on sloping shelves one above the other with each shelf corresponding to a particular stratum."[26] In other words, Smith's collection was an archive of fossils arranged not by taxonomic group (as would become common practice in museums by the mid-nineteenth century), but rather according to stratigraphy, as in one of Cuvier's ideal sections: it was a transcription from the original archive (archive$_0$) to a physical collection (archive$_1$—a second nature) that preserved the organizational principles of nature. This transcription was heuristic rather than literal, since the collection did not attempt to contain *all* the fossils found in a given stratum. But as a second nature it served as the basis for the production of knowledge about the earth—in this case, Smith's pioneering stratigraphical maps.

Smith's collection (or any physical collection of fossils organized sys-

Fig. 2.2 William Smith's stratigraphic map, in William Smith, *A Memoir to the Map and Delineation of the Strata of England and Wales, With Part of Scotland* (London: John Cary, 1815).

tematically) was a first-order archival transcription. It preserved fossils as physical entities, but removed them from their original context and placed them in a new system of relations. In Smith's case, individual fossils represented particular strata; in more common taxonomic collections one or more fossils stand for an entire species or genus. But physical archival collections quickly became the basis for a further transcription, this time to the printed page. In 1816, facing crushing debts from his self-financed geological researches, Smith was forced to sell his collection to the British Museum, where it later became part of the collection at the Natural History Museum of London.[27] Before he parted with his specimens, however, he prepared a short book—with help from his nephew, John Phillips—titled *Strata Identified by Organized Fossils* (1816), wherein each of Britain's typical strata (e.g., "Lower or Hard Chalk") was described, along with a listing and detailed illustration of its representative fossils.[28] A year later, he published a longer work intended as a catalog to his collection, where he adopted a tabular format for presentation and omitted illustrations (he commented that the public display of his collection "seemed to render figures of them unnecessary").[29] These works can be seen as further archival transcriptions (archive$_2$) based not directly on the natural archive but rather on organized physical collections (archive$_1$). The extent to which Smith himself saw his works this way—as well as to which he remained influenced by the historiographic conventions of his day—can be seen in his introductions to these works. As he put it in the introduction to the 1816 *Strata Identified by Organized Fossils*, "The organized Fossils (which might be called the antiquities of Nature,) and their localities also, may be understood by all, even the most illiterate: for they are so fixed in the earth as not to be mistaken or misplaced; and may be as readily referred to in any part of the course of the Stratum which contains them, as in the cabinets of the curious."[30] Smith's clear message was that there is a correspondence between the earth, the physical collections ("cabinets of the curious") and the published works, which he reinforced in 1817, claiming that quarries exhibiting geological strata could be consulted "with as much certainty of finding the characteristic Fossils of the respective rocks, as if they were on the shelves of their cabinets."[31] Furthermore, he underlined the analogy between human and natural history, writing that "organized Fossils are to the naturalist as coins to the antiquary; they are the antiquities of the earth; and very distinctly show its gradual regular formation, with the various changes of inhabitants in the watery element."[32]

Smith's works were not the first fossil catalogs—such volumes had been published fairly continuously since the early eighteenth century—but they feature two important departures from earlier efforts. First, they explicitly employ "universal" stratigraphic principles that are derived from the natural archive of the earth as "metadata" for organizing information abstracted from earlier archival configurations. The printed works are archives of information, not of things, but those metadata maintain an epistemic continuity between the objects contained in archive$_0$ (the earth itself), archive$_1$ (organized fossils), and archive$_2$ (illustrated atlases). Second, the stratigraphic (temporal) organization underwrites the sense that fossils are part of a historical and genealogical sequence. That sequence could, quite directly, be transcribed from the rocks themselves to the printed page, both in pictorial form and in text. Shortly after the publication of Smith's catalogs this approach became common in British and European paleontology. For example, Smith's nephew John Phillips produced several regional surveys of fossils, such as his *Illustrations of the Geology of Yorkshire* (1829), which was an early attempt to demonstrate the succession of strata and their characteristic fossils in one region of England. In this early work Phillips' main ambition was to assemble a reliable stratigraphical sequence for Yorkshire fossils based on a comprehensive survey. As he put it in the introduction, "No pains have been spared to copy the natural sections of this coast as perfectly as possible," and he reproduced the order of the strata by depicting them in ideal sequence, much as Cuvier and Brongniart had done for the Paris region (fig. 2.3a).[33] However, the work also contained a "Synoptic Table of the Fossils," in which the characteristic species found in the strata were arranged in tabular form (fig. 2.3b). The primary organization for the fossils was taxonomic, but Phillips also included additional metadata for each taxon: these included references to their descriptions in the literature, to the strata with which the fossils were associated, to other localities where they could be found (both within Britain and abroad), and to accompanying illustrations in the text (fig. 2.3c).

The synoptic format (archive$_3$) essentially transcribed the relevant stratigraphic information directly from the formations to the text, while accompanying illustrations presented the reader with a virtual "specimen drawer" in which generalized type specimens for individual fossil species could be consulted as if the reader was viewing the actual physical collections.

By the 1830s Phillips' approach was the standard practice in dozens of fossil compendia published in Britain, France, Germany, and elsewhere. Early examples of these compendia usually focused on a particular region and/or taxonomic group in order to reconstruct the local stratigraphic record.[34] However, at the same time, more ambitious projects were under way to compile the entire fossil record for larger regions and, eventually, for the entire globe. Brongniart's *Prodrome d'un histoire des végétaux fossiles* (1828), John Woodward's *A Synoptical Table of British Organic Remains* (1830), Alcide d'Orbigny's *Paléontologie française* (1840), G. G. Giebel's *Fauna der Vorwelt*

a

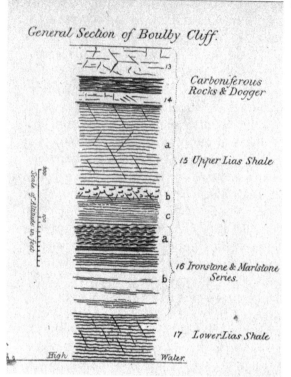

b

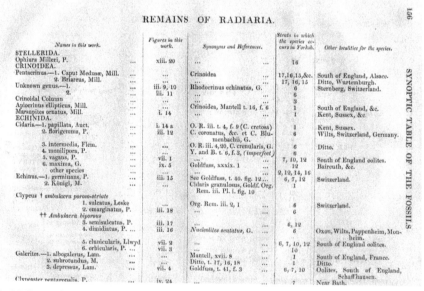

Fig. 2.3 Three different represenations of the fossil archive from John Phillips, *Illustrations of the Geology of Yorkshire* (York, 1829). (a) Depicts an idealized stratigraphic column; (b) presents the archive as a taxonomic list; and (c) illustrates type specimens in a "virtual specimen drawer."

PLATE 1.

C

CHALK.

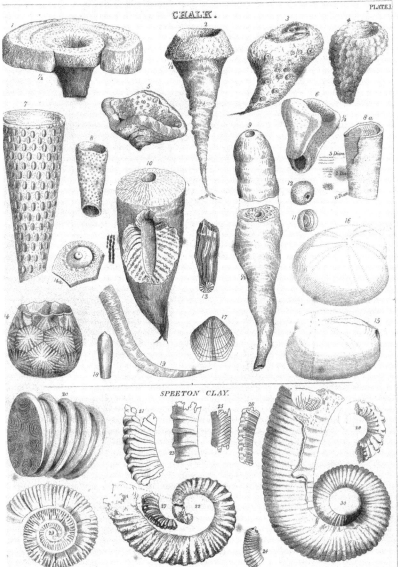

Drawn by J. Phillips.

Dawson & Brown. sc. York.

(1847), and Deshayes' *Description animaux sans vertèbres* (1860) are influential examples of such compendia, in some cases comprising thousands of records spread across multiple volumes. The most ambitious and substantial of these were produced in Germany during the middle decades of the nineteenth century; Bronn's *Lethaea Geognostica* (1835–38) and later *Index Palaeontologicus* (1848–49) were systematic attempts to summarize the entire known fossil record for plants and animals in comprehensive multivolume chronological synopses.[35]

The most obvious purpose of these massive compendia—which have been continuously produced by paleontologists ever since—was to provide a meta-archive of the fossil record, bringing together information from individual museum collections, regional surveys, and monographic literature into a centralized resource that could be consulted by any paleontologist with access to a library.[36] As with previous iterations of the fossil archive, the most ambitious of these compendia were built on previous archives: for his *Index Palaeontologicus* Bronn, for example, extracted data from more than 500 regional compendia and taxonomic publications (fig. 2.4). In these comprehensive compendia we also see an explicit effort to reconstruct the natural archive of the fossil record as a textual archive of data. The *Index Palaeontologicus* was divided into two parts. The first—"Nomenclator palaeontologicus"—is essentially a catalog of genus and species names, organized taxonomically and formatted much like Phillips' or Smith's earlier works (archive$_3$). But the second part, "Enumerator palaeontologicus," reorganizes the taxonomic information into some 500 pages of tables that present the stratigraphic context of fossil species in highly abstracted form.

This instantiation of the fossil archive (which I will call archive$_4$) is essentially an archive of numerical data in which an individual entry bears only a distant relationship to physical objects found in the earth. Nonetheless, a very clear genealogy can be traced back from this through previous archival configurations. Furthermore, Bronn's numerical approach maintained the earlier commitment to historicizing the objects of paleontological study: by distilling only the basic taxonomic and stratigraphic information, Bronn and others hoped to produce quantitative analyses of the history of life that took the form of innovative pictorial narratives.

In his theoretical treatment of the history of life, *Untersuchungen über die Entwicklungs-Gesetze der organischen Welt* ("Investigations into the Laws of Development of the Organic World"), published just a year before Darwin's *Origin*, Bronn described the earth "as a great book, her layers are the leaves of the same, and fossils, the letters of the alphabet with which it is written." But he also noted that "those pages are incomplete, broken, jumbled up and faded before us; we need to organize them and to search to

Bonennungen	Weltgegend. (E S P M U)	KohlenP. (a b c d e f g)	SalzP. (h i k l)	OolithP. (m n o p)	Krei-deP. (q r [)	MolasseP. (s t u v w x)	Neu (y z)

(Weltgegend.: Europa. Asien. Afrika. Amerika. Australia. — E S P M U.
KohlenP.: U.-Silur. O.-Silur. Devon-F. Bergkalk. Kohlen-F. Todtlieg. Zechstein. — a b c d e f g.
SalzP.: St.Cassian Buntsand. Muschelk. Keuper. — h i k l.
OolithP.: Lias. Unter-Jura. Ober-Jura. Wealden. — m n o p.
Krei-deP.: Neocomien Grünsand. Kreide. — q r [.
MolasseP.: Numm.-G. Untre Mittle (Molasse). Obere Dituvial. — s t u v w x.
Neu: Alluvial. Lebend. — y z.)*

III. ARTHRODEA EB.

A. APODA.
(nuda)

Genera multa viventia speciebus   · · · · · · · · · · · · · · ·   60

B. CHAETOPODA BLV.

1. TERRICOLAE CUV.

Genera multa viventia speciebus   · · · · · · · · · · · · ·   50

Tubifex LK. 1. · · · · · · · · · · ·   .2

? antiquus PLIEN. · · · · · · · · · 1 · · · ·

2. TUBICOLAE CUV.

Bonennungen	marks
Arenicola LK. 0. .	∞
Clymene SAV. 0. .	∞
Terebella CUV. 1.	n^5 ∞
lapilloides MÜ. . . .	∞
Pectinaria LK. 0.	∞
Amphitrite LK. 0.	∞
Sabella CUV. 0. . .	∞
Ditrypa BERKELEY 4	
plana FORB. · · · · · ·	t
gadus LYELL · · · · E^2 .M^2.	t u.w. z
polita WOOD · · · ·	u. .
subulata BRKL. · · ·	u.w. z ∞
Spirorbis LK. 33. .	
Lewisi SOW. · · · ·	b
tenuis MURCH. · · ·	b
ammonius EDW. · ·	c
† gracilis SANDB. · · ·	c
omphalodes EDW. ·	c d
minutus PORTL. ·	d
Valvata EDW. · · ·	k
complanatus MÜ. · ·	m
planorbiformis EDW.	n^5 r f
rotula EDW. · · · ·	r f
conulus · · · · · · · ·	f
anfractus EDW. · ·	f
lituitis DEFR. · · · ·	s
subcarinatus EDW. ·	t
conoideus LK. · · ·	

Fig. 2.4 Data tables from Bronn's *Index Paleontologicus*. H. G. Bronn, *Index Palaeontologicus, oder, Übersicht der bis jetzt kekannten Fossilen Organismen* (Stuttgart: E. Schweizerbart, 1848), 546–47.

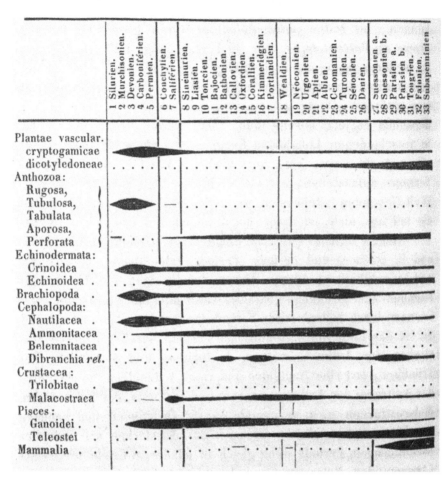

Fig. 2.5 "Spindle" diagrams from Bronn's *Untersuchungen über die Entwicklungsgesetze der organischen Welt während der Bildungs-Zeit unserer Erdoberfläche* (Stuttgart: E. Schweizerbart, 1858), 312.

supplement what is missing."[37] Bronn's preferred method of reconstruction was statistical analysis of the data tables he compiled in works such as *Index Paleontologicus*; in the *Untersuchung* he presented several hundred pages of such analysis, where he tabulated numerical ratios comparing relative diversity of particular kinds of organisms over time, extinct versus extant taxa in particular formations, and a variety of other relationships.[38] While the numbers themselves tell a kind of "story," the end product of this analysis was a series of images depicting patterns of diversity through time—a type of graph known today as a "spindle diagram"—which in a clear, simple way presented a summary narrative of the history of life (fig. 2.5).

It is worth noting that this kind of analysis is still performed by pale-ontologists today, and that these spindle diagrams are an iconic format for representing historical patterns of evolution and diversification in the fossil record. One of the useful features of this kind of data analysis is that a large enough sample size can provide at least a low-resolution narrative summary, even if some data are missing from the record.

Other paleontologists contemporary to Bronn also explored this quantitative/visual narrative approach. Phillips, whose earlier regional stratigraphic works have already been discussed, published a slim volume titled *Life on the Earth* in 1860, where he discussed broad historical questions about the origin and succession of life. Phillips drew on John Morris' *Catalogue of English Fossils*—a typical regional compendium of its day—for data on the relative diversity of species and higher taxa found in English strata through time. Like Bronn, he counted these and computed ratios, then depicted the results in several different visual modes. These included spindle diagrams, but also a striking line graph representing changes in relative diversity of all taxa through time, from the earliest Paleozoic to the most recent fossils (the divisions Paleozoic, Mesozoic, and Cenozoic were Phillips' own invention) (fig. 2.6).[39]

Phillips' graph is likely the first attempt to visually depict the entire history of life in a single summarizing image. One final point to make about these visual narratives, though, is that while they were groundbreaking for their time in natural history, they reflected a somewhat earlier precedent in the study of human history. For example, Joseph Priestly's 1769 *A New Chart of History* depicted the historical succession of empires in a timeline where time is represented on the horizontal axis and area of occupation on the vertical (fig. 2.7).

The resulting image is essentially a spindle diagram of human civilization, and emphasizes the contingency of historical events while at the same time providing a broad overview of historical development in schematic format.

CONTINGENCY LOST AND FOUND

We have now followed the fossil archive through a series of reconfigurations—successive "second naturings"—from $archive_0$ (the earth), through $archive_1$ (organized fossil collections), $archive_2$ (pictorial atlases of fossils), $archive_3$ (text-only catalogs), and $archive_4$ (an archive of numerical data). This last iteration of the archive shows that by the late nineteenth century the fossil record had become the basis for quantitative analysis that pro-

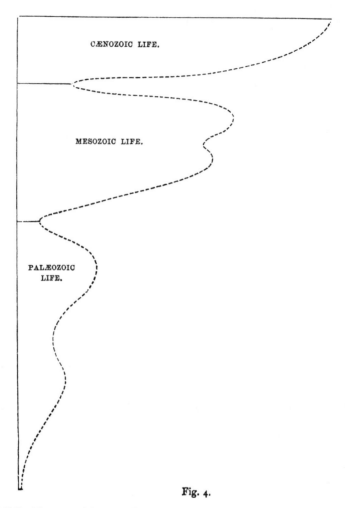

CÆNOZOIC LIFE.

MESOZOIC LIFE.

PALÆOZOIC
LIFE.

Fig. 4.

Fig. 2.6 Phillips' illustration of changes in diversity over time. Note that the original image is oriented vertically (as it is here), beginning at the bottom (earliest) and proceeding to the top (recent). This orientation mirrors public domain presentations of the stratigraphic column. John Phillips, *Life on the Earth: Its Origin and Succession* (Cambridge: Macmillan and Co., 1860).

duced narrative arguments about the history of life. One of the features of the analytic/visual approach of Bronn and Phillips is that individual events became subsumed into a larger narrative, and individual data lost some of their particularity and uniqueness. While data points in an archive of numerical information are not strictly speaking interchangeable—it still mattered, for example, in which geological period a particular group appeared or disappeared from the record—the individual histories of each

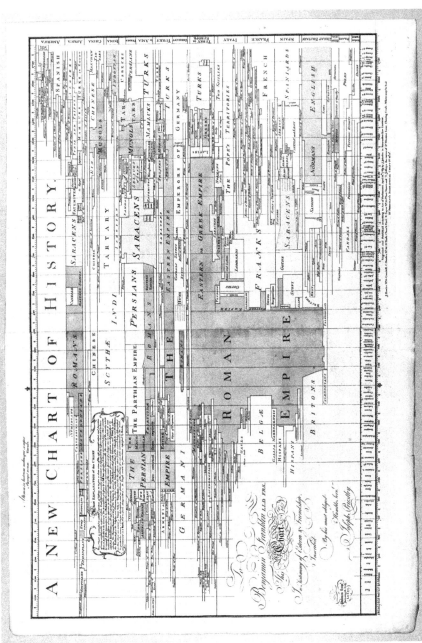

Fig. 2.7 Joseph Priestly's "spindle" diagram of human history. Joseph Priestly, *A New Chart of History* (1769). The Library Company of Philadelphia.

group were less important than the collective story they told (for instance, as a pattern of diversification over time). Another way to put it is to say that each item in the collection had the same significance as any other—data was made homogeneous and commensurable. This, as Lev Manovich has argued, is characteristic of a "database," although the term may strike some as anachronistic when applied to the nineteenth century.[40]

The database or "data collection" format was relatively uncommon in paleontology during the later nineteenth and early twentieth centuries, but with the advent of more sophisticated statistical techniques and the arrival of digital computers it eventually became the standard mode of presentation for paleontological information. Indeed, in the earliest attempts to migrate the archival collection of fossil data to a digital format in the late 1960s and 1970s, it was the database—and not the organized collections of fossils that preceded it—that increasingly came to be understood as "the fossil record." One such effort was a 1967 collaboration sponsored by the British Palaentological Association and the Geological Society of London that collected representative data at the taxonomic level of order and suborder for the entire marine and terrestrial fossil record, titled simply *The Fossil Record*.[41] Although the data were not initially entered into a digital database, paleontologist Michael Benton has nonetheless described the work as "the first published comprehensive data base designed specifically for studies of the nature of the history of life."[42] Benton himself was the author of a follow-up effort in the 1990s that published a much more comprehensive data collection both digitally and as a print volume, which he titled *The Fossil Record 2*.[43]

The advent of the electronic database as the most recent instantiation of the fossil archive (archive$_5$) had a number of important consequences, but it did not replace earlier archival presentations of the fossil record. Paleontologists still, to this day, regularly work with organized collections of specimens, illustrated atlases of type specimens (usually containing photographs), taxonomic catalogs, and printed compendia of numerical data. Of course, they also still regularly consult the original archive, the earth itself, by performing fieldwork and stratigraphical surveys. It is not the case that successive transcriptions of the fossil archive necessarily render previous versions obsolete: for example, in collections practices in many natural history museums, early handwritten catalogs are kept alongside modern electronic databases. At one such museum, the Field Museum of Natural History in Chicago, the vertebrate paleontology collections manager explained to me that it was often necessary to return to these earlier ledgers—which despite having broken spines and crumbling pages from years of use were still frequently consulted—because important metadata (such as field observations) were often lost in the transition to subsequent versions, whether analog or digital. Another way to put it is that, much as Lorraine Daston and Peter Galison have argued in the case of "epistemic virtues," instantiations of the fossil record "do not replace one another like a succession of kings. Rather, they accumulate into a repertoire of possible forms of knowing."[44] But the "form of knowing" that had become dominant in the later twentieth century, at least in evolutionary studies of the fossil record, was statistical analysis of data contained in large digital databases.[45] This approach is now considerably more sophisticated (in addition to being faster and easier) than the one taken by Bronn and Phillips, but from an epistemic perspective is not terribly different from the analysis of data collections in the nineteenth century.[46]

One particular consequence of the transcription of the archive of the fossil record to digital databases has been the potential loss of historicity, contingency, and narrative that was present in previous archival forms. As Manovich has argued, "Database and narrative are natural enemies," since "database represents the world as a list of items, and it refuses to order this list," while "a narrative creates a cause-and-effect trajectory of seemingly unordered items (events)."[47] That "cause-and-effect trajectory" of unordered items is the essence of historical contingency: historical explanations take the form of narratives because they reconstruct an irreducibly contingent series of events that did not have to unfold the way they did. The central importance of the archive for paleontologists, I have argued, is that the fossil archive—in each of its various second natures—preserves the context that illuminates contingent historical relationships. It is important to stress, however, that while databases may seem unlikely sources of narrative, it is not the case that historical contingency is entirely lost in the transition from an archive of things to one of data. As Manovich continues, while "there is nothing in the logic of the medium [of the database] that would foster its generation," nonetheless "a database can support narrative."[48] This is seen most obviously in the products of data analysis—the spindle diagrams or line graphs of diversification produced by paleontologists from Bronn and Phillips to the present day, for example, are essentially narrative summaries of data—and also in the retention of metadata (stratigraphical or geographical) that preserve the temporal and spatial context of the objects from which the data are abstracted. But it is a major problem if the metadata are lost, or if data are allowed to become entirely interchangeable. In the first case, paleontologists have been very careful to migrate metadata from one archival format to the next, but there are also inevitable consequences in moving from archives of things to information. Each successive iteration of the archive is essentially an abstraction from the previous one, and there is the possibility of loss at each step. Printed taxonomic atlases and compendia often contain pages of descriptive metadata for each taxonomic group that simply cannot be translated to the next version—one might call this the "thinning" of metadata. A collection of information without metadata cannot be an archive, so to preserve the archival characteristics of the fossil record, efforts have to be made when compiling databases to sometimes recover lost metadata.

The second case—the interchangeability of data points—is potentially more problematic. This was the issue encountered when Gould and his colleagues began to explore the possibility of a purely nomothetic paleontology during the 1970s, and its resolution ultimately influenced Gould's recognition of the importance of contingency in reconstructing the history of life. One of Gould's close collaborators was the paleontologist Thomas J. M. Schopf, who was especially drawn to a purely stochastic view of the history of life, which he often referred to as a model that treated "species as particles" based on "gas laws" of paleontology. In particular, Schopf was keen to argue that causal, deterministic explanations for the success and failure of evolutionary lineages should be replaced with generalizing equations that treated individual taxonomic groups as "particles" behaving stochastically in a volume of "gas" (e.g., in time). As Schopf put it in a 1979 paper,

> The proper type of theory to apply to large statistical summaries is some form or another of stochastic theory, such as occurs in chemistry (the gas laws), population biology (demography), and physics (the Heisenberg uncertainty principle). The fate of any given molecule, or individual animal, is of no concern per se. Rather, the ensemble statistical properties of the particles and the types of predictions which those properties allow are what is of interest.[49]

Schopf's major concern was to eliminate what he understood as "determinism" from evolutionary accounts, which he argued had overly influenced paleontologists to seek adaptive explanations for all patterns of diversification and extinction in the history of life. To some extent, this view did not depart from Gould's own evolving position, particularly Schopf's emphasis that "there was nothing inevitable either in evolution or in history that a priori determined either the present state of affairs, or any specific past configuration."[50] However, in private correspondence Schopf revealed a more radical stance: as he put it in a letter to another colleague, "in my view, all of paleontology, i.e., all of those fossils, is (are) simply a metaphore [sic] for what is really the statistical mechanics of a series of interacting hollow curves."[51]

This was, for Gould, taking things too far. After reading a draft of Schopf's paper, Gould wrote to Schopf that

> you continually conflate (though they are not unrelated of course) the notions of predictability and stochasticity. Stochastic models can, of course, lead to a high degree of predictability, at least for general patterns of events. Of course I agree that the most fascinating aspect of life on earth is that it would probably play itself out in a totally different way if we started again from the same initial conditions—but this metaphysic (which I share with you) is not the essence of maintaining a stochastic perspective in paleontological theory.[52]

In other words, while Gould agreed with Schopf that future historical outcomes could not be predicted (at least on the basis of available data) by present conditions at any moment in time, he rejected the notion that the data contained in the fossil archive was merely a "metaphor." He also argued that Schopf was simply replacing one kind of determinism—the inevitability of selection in predicting evolutionary outcomes—with another—an appeal to a purely nomothetic, internalist explanation for historical change.

One result of Gould's disagreement with Schopf was that Gould ultimately developed his distinctive stance on the role of contingency in the fossil record, including his now-famous thought experiment, presented most fully in *Wonderful Life*, where he imagined that if one could "replay the tape of life" from the same initial starting conditions, a very different outcome might obtain. Gould became increasingly convinced of this because, in part, of what the fossil archive revealed. Years earlier, in 1973, Gould had asked his graduate student Jack Sepkoski to compile a new database of marine fossil orders, families, and genera. This stemmed from a concern that available fossil databases (like *The Fossil Record*) contained incomplete, low-resolution, or redundant data, which he regarded as an obstacle to meaningful analysis of patterns in the history of life. Sepkoski's task eventually expanded to become the first truly digital database of the fossil record, *A Compendium of Marine Fossil Families*, which was published in 1982.[53] One of the most important early results of Sepkoski's analysis of this database was the realization that (a) patterns of diversity in the history of life were not stable—they tended to fit a series of overlapping logistic growth curves, with steep rises and sharp drops in standing diversity at particular times, and (b) some of the sharp drops in diversity indicated that major mass extinctions were a regular feature of the history of life (fig. 2.8).[54]

These mass extinctions were features that could be explained neither purely nomothetically—i.e., as intrinsic properties of population pressures—nor using the logic of Darwinian selection alone. They were too

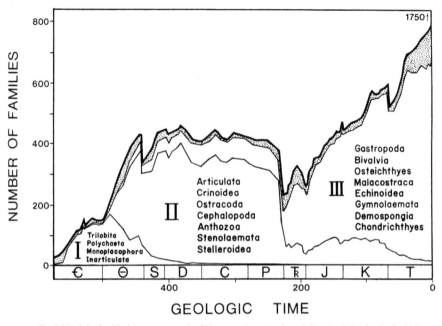

Fig. 2.8 Jack Sepkoski's three-curve graph of Phanerozoic marine faunal diversity, in J. John Sepkoski, Jr., "A Kinetic Model of Phanerozoic Taxonomic Diversity: III. Post-Paleozoic Families and Mass Extinctions," *Paleobiology* 10 (1984).

big, affected too many very different groups, and happened too suddenly to be explained by anything other than some kind of distinctive triggering event. Sepkoski's analysis coincided with the discovery, by Luis and Walter Alvarez, of an anomalous layer of the element iridium at the stratigraphic boundary between the Cretaceous and Tertiary periods—exactly when one of the major mass extinctions (the one that killed off the dinosaurs) appeared in the record. The Alvarezes' explanation was that this iridium layer was produced by an enormous bolide impact (a comet or asteroid), a hypothesis that was later confirmed by the discovery of an impact crater off the coast of the Yucatán Peninsula.[55]

What the work of both Sepkoski and the Alvarezes provided, then, was evidence from two different instantiations of the fossil archive (the record of data and the stratigraphic record) that singular events had drastically altered the course of evolution several times in the history of life. These events could not have been predicted from prior conditions, nor could their consequences be understood in any way other than by examining the actual record of subsequent diversification of life. In his writings on the theme of contingency in the history of life, mass extinctions became

Gould's central evidence for "the unerasable and determining signature of history."[56] The essence of such historical explanations, Gould argued, is that they are composed of events "that cannot, both by laws of probability and time's arrow of irreversibility, occur together again"—in other words, they are irreducibly contingent.[57] That contingency, Gould pointed out, is an essential feature of the fossil archive, and would be undetectable without that archive. Indeed, it was only after Sepkoski had compiled and analyzed his database that paleontologists generally began to accept the role mass extinctions have played in the history of life. As Gould put it,

> The history of life is not a continuum of development, but a record punctuated by brief, sometimes geologically instantaneous, episodes of mass extinction and subsequent diversification. The geological time scale maps this history, for fossils provide our chief criterion in fixing the temporal order of rocks. The divisions of the time scale are set at these major punctuations because extinctions and rapid diversifications leave such clear signatures in the fossil record. Hence, the time scale is not a devil's ploy for torturing students, but a chronicle of key moments in life's history.[58]

Despite the conversion of the fossil record to an abstract archive of data, the contingency of individual events was a signal that could not be erased or ignored. The central message of the archive—whether composed of physical specimens or numerical data—is that history is composed of unique, unrepeatable events whose explanations—even in the form of complex statistical analyses—take the form of narratives.

The ultimate goal of historical paleontological analysis may well be, as the great twentieth-century German paleontologist Otto Schindewolf argued, "to present the grand scheme of evolution truly and, in keeping with the goal of any history, 'to describe the way it was' (Leopold von Ranke)."[59] However, few paleontologists in any century would have adopted such an idealistic Rankean perspective. Archives are not history; history is produced by interpreting and reconstructing information that is contained in its archives. For paleontologists, as a recent textbook put it, "the fossil record is our one true chronicle of the history of life. As such, it preserves a rich and varied archive containing abundant information on the modes, habits, and evolution of prehistoric life."[60] The importance of the fossil record is that it is an archive of particular historical events, from which sequential and contingent narratives can be constructed and compared with other possible narrative accounts. This is no different from the way we historians research and construct our own narrative explanations, with one major exception: the fossil record is both a natural and a human-made archive. The interpretive strategies of paleontology evolved out of the recognition of the fundamental archival value of the earth itself, and from the very beginning paleontology has been—in its methodology, its modes of visual representation, its organization of data, and its "ways of knowing"—a paradigmatic "science of the archive."

NOTES

1. David Sepkoski, *Rereading the Fossil Record: The Growth of Paleobiology as an Evolutionary Discipline* (Chicago: University of Chicago Press, 2012).

2. Niles Eldredge and Stephen Jay Gould, "Punctuated Equilibria: An Alternative to Phyletic Gradualism," in *Models in Paleobiology.* (San Francisco: Freeman, Cooper & Co., 1972), 96.

3. Charles Darwin, *On the Origin of Species* (London: J. Murray, 1859), 301.

4. Eldredge and Gould, "Punctuated Equilibria," 96.

5. On the efforts of Gould and others to pursue a "nomothetic paleobiology," see Sepkoski, *Rereading the Fossil Record*, ch. 7.

6. Stephen Jay Gould, *Wonderful Life: The Burgess Shale and the Nature of History* (New York: W. W. Norton, 1989), 283.

7. Gould, *Wonderful Life*, 278.

8. Stephen Jay Gould, *The Structure of Evolutionary Theory* (Cambridge, MA: Belknap Press, 2002), 195–96.

9. Gould, *Structure*, 755.

10. As a recent discussion of the fossil record puts it, "The term 'fossil record' is used in two ways: either the totality of fossils preserved in all rocks or the sum of human knowledge of those fossils. In either case, the term carries the connotation also of the geologic context of the fossils—their distribution in time and space and their relationship to the enclosing rock." Susan M. Kidwell and J. John Sepkoski, Jr., "The Nature of the Fossil Record," in *Evolution: Investigating the Evidence*, ed. Judy Scotchmoor and Dale A. Springer (Pittsburgh: Paleontological Society Special Publication, 1999), 61.

11. I am using the term "second nature" here in the sense discussed in Bruno Strasser, "Collections," in *Eine Naturgeschichte für das 21. Jahrhundert: Hommage à, zu Ehren von, in honor of Hans-Jörg Rheinberger,* ed. Safia Azzouni et al. (Berlin: Max-Planck-Institut für Wissenschaftsgeschichte, 2011). It is from second natures, and not "nature" itself, Strasser argues, that natural scientists produce knowledge.

12. David Sepkoski, "Towards 'a Natural History of Data': Evolving Practices and Epistemologies of Data in Paleontology, 1800–2000," *Journal of the History of Biology* 46, no. 3 (2013).

13. Martin J. S. Rudwick, *Bursting the Limits of Time: The Reconstruction of Geohistory in the Age of Revolution* (Chicago: University of Chicago Press, 2005), 181.

14. Robert Hooke, *The Posthumous Works of Robert Hooke, Containing His Cutlerian*

Lectures, and Other Discourses, Read at the Meetings of the Illustrious Royal Society (London: printed by Sam, 1705), 321.

15. See, for example, Arnaldo Momigliano, "Ancient History and the Antiquarian," *Journal of the Warburg and Courtauld Institutes* 13, no. 3/4 (1950), and Momigliano, *The Classical Foundations of Modern Historiography* (Berkeley: University of California Press, 1990), ch. 3.

16. Momigliano, *Classical Foundations*, 57.

17. Michael Hunter, *Science and the Shape of Orthodoxy* (Suffolk, UK: Boydell Press, 1995), 183.

18. Torbern Olof Bergman, *Physical Description of the Earth* (*Physisk beskrifning ofvert jordklotet*) (Uppsala: 1766), quoted in James F. Aber, "Torbern Olaf [*sic*] Bergman," http://academic.emporia.edu/aberjame/histgeol/bergman/bergman.htm, accessed April 2016.

19. Jean-Louis Giraud Soulavie, *Histoire naturelle de la France méridionale, ou recherches sur la minéralogie du vivarais* ([Nismes]: Belle, 1780–84), 33; Johann Friedrich Blumenbach, "Beyträge zur Naturgeschichte der Vorwelt," *Magazin für den neuste aus der Physik und Naturgeschichte* 4 (1790): 1–2.

20. François-Xavier Burtin, "Réponse a la question physique, proposé par la société de teyler, sur les révolutions générales, qu'a subies la surface de la terre, et sur l'ancienneté de notre globe," in *Algemeene Omkeeringen* (Haarlem, 1789), 216.

21. Indeed, Charles Lyell—perhaps the most influential geologist of the nineteenth century—made frequent references to the earth as an archive in his monumental *Principles of Geology* (1830–33). For example, in the 1853 second edition of *Principles* he wrote that "it may undoubtedly be said, that strata have always been forming somewhere, and therefore at every moment of past time nature has added a page to her archives." Charles Lyell, *Principles of Geology*, 2nd ed. (London: John Murray, 1853), 191–92. This reinforced his comment in the first edition that "as the historian receives assistance from the antiquary . . . so the geologist should avail himself of the aid of many naturalists." Charles Lyell, *Principles of Geology*, vol. 1 (London: John Murray, 1830), 3.

22. Rudwick, *Bursting the Limits of Time*, 219.

23. François-Dominique de Reynaud Montlosier, *Essai sur la théorie des volcans d'auvergne* (Paris: Imprimerie de Landriot et Rousset, 1789 [1802]), 6.

24. Georges Cuvier, "Preliminary Discourse" to *Recherches sur les ossemens fossiles de quadrupedes* (1812), in Martin J. S. Rudwick, *Georges Cuvier, Fossil Bones, and Geological Catastrophes* (Chicago: University of Chicago Press, 1997), 185.

25. Cuvier, "Preliminary Discourse," 183.

26. Jack Morrell, *John Phillips and the Business of Victorian Science* (Aldershot, UK: Ashgate, 2005), 17.

27. Simon Winchester, *The Map that Changed the World: William Smith and the Birth of Modern Geology* (New York: HarperCollins, 2001), 247–48.

28. William Smith, *Strata Identified by Organized Fossils* (London: W. Arding, 1816), 1.

29. William Smith, *Stratigraphical System of Organized Fossils with Reference to the Specimens of the Original Geological Collection in the British Museum* (London, 1817), iii.

30. Smith, *Strata Identified by Organized Fossils*, introduction.

31. Smith, *Stratigraphical System of Organized Fossils*, v.

32. Smith, *Stratigraphical System of Organized Fossils*, ix–x.

33. John Phillips, *Illustrations of the Geology of Yorkshire; or a Description of the Strata and Organic Remains of the Yorkshire* (York: Thomas Wilson and Sons, 1829), xiv.

34. See, for example, Gideon Algernon Mantell, *The Fossils of the South Downs; or, Illustrations of the Geology of Sussex* (London: Lupton Relfe, 1822); Gérard P. Deshayes, *Description des coquilles fossiles des environs de Paris* (Paris: L'auteur, chez Bechet jeune, 1824); and H. G. Bronn, *Italiens Tertiär-Gebilde und deren organische Einschlüsse* (Heidelberg: K. Groos, 1831).

35. H. G. Bronn, *Lethaea Geognostica, oder Abbildungen und Beschreibungen der für die Gebirgs-Formationen bezeichnendsten Versteinerungen* (Stuttgart: E. Schweizerbart, 1835); H. G. Bronn, *Index Palaeontologicus, oder, Übersicht der bis jetzt bekannten Fossilen Organismen* (Stuttgart: E. Schweizerbart, 1848).

36. On the similar archival function of compendia in other disciplines, see Daston, chapter 6, this volume.

37. H. G. Bronn, *Untersuchungen über die Entwicklungsgesetze der organischen Welt während der Bildungs-zeit unserer Erdoberfläche* (Stuttgart: E. Schweizerbart, 1858), 7.

38. On Bronn's innovative quantitative approach to the fossil record, see Sepkoski, "Towards 'a Natural History of Data.'"

39. John Phillips, *Life on the Earth: Its Origin and Succession* (London: Macmillan and Co., 1860).

40. Lev Manovich, "Database as Symbolic Form," in Victoria Vesna, ed., *Database Aesthetics: Art in the Age of Information Overflow* (Minneapolis: University of Minnesota Press, 2007), 39. Whether the term "database" can be applied anachronistically in this way is not a question I will address here, but I discuss it in more detail in a related article. It is sufficient for the present purposes to note that digital databases have at least an analogical relationship with predigital data collections, which contain many of the features of an electronic database (i.e., they are "structured collection[s] of data") although perhaps not an identical one (paper data collections lack automated sorting, search, and retrieval functions). See David Sepkoski, "The Database before the Computer?," *Osiris* 32 (forthcoming).

41. W. B. Harland et al., eds., *The Fossil Record* (London: Geological Society, 1967).

42. Michael Benton, "The History of Life: Large Databases in Palaeontology," in D. A. T. Harper, ed., *Numerical Palaeobiology: Computer-based Modelling and Analysis of Fossils and Their Distributions* (New York: John Wiley & Sons, 1999).

43. Michael Benton, *The Fossil Record 2* (London: Chapman & Hall, 1993). See also http://www.fossilrecord.net/fossilrecord/.

44. Lorraine Daston and Peter Galison, *Objectivity* (New York: Zone Books, 2007), 113.

45. Sepkoski, *Rereading the Fossil Record*, ch. 8.

46. Sepkoski, "Towards 'A Natural History of Data.'"

47. Manovich, "Database as Symbolic Form," 44.

48. Manovich, "Database as Symbolic Form," 47.

49. Thomas J. M. Schopf, "Evolving Paleontological Views on Deterministic and Stochastic Approaches," *Paleobiology* 5, no. 3 (1979): 343.

50. Schopf, "Evolving Paleontological Views," 343.

51. Thomas J. M. Schopf to David M. Raup, February 8, 1979, Thomas J. M. Schopf Papers, Smithsonian Institution Archives, Box 3, Folder 30.

52. Stephen Jay Gould to Thomas J. M. Schopf, June 25, 1979, Schopf Papers, Box 5, Folder 59.

53. J. John Sepkoski, Jr., *A Compendium of Fossil Marine Families* (Milwaukee: Milwaukee Public Museum, 1982).

54. On the history of studies of diversification and mass extinction in the fossil record, see Sepkoski, *Rereading the Fossil Record*, chs. 8 and 9.

55. Two "insider" accounts of this story are Walter Alvarez, *T. Rex and the Crater of Doom* (Princeton: Princeton University Press, 1997), and David M. Raup, *The Nemesis Affair: A Story of the Death of Dinosaurs and the Ways of Science* (New York: W. W. Norton & Co., 1986).

56. Gould, *Wonderful Life*, 283.

57. Gould, *Wonderful Life*, 278.

58. Gould, *Wonderful Life*, 54.

59. Otto H. Schindewolf, *Basic Questions in Paleontology: Geologic Time, Organic Evolution, and Biological Systematics*, trans. Judith Schaefer (Chicago: University of Chicago Press, 1993), 4.

60. Bruce S. Lieberman and Roger Kaesler, *Prehistoric Life: Evolution and the Fossil Record* (Oxford: Blackwell Publishing, 2010), 51.

THREE

Empiricism in the Library: Medicine's Case Histories

J. Andrew Mendelsohn

I intended to start with a case. It would be the first or "classic" case of one of the diseases or conditions that was defined in the great age of modern internal medicine. Leukemia, coronary occlusion and myocardial infarction ("heart attack"), Hodgkin's disease—there were several good options. I settled on Stokes-Adams disease, characterized by extreme slowness of the pulse accompanied by attacks of fainting or epileptic-like seizure, which was eventually attributed to a cardiac electrical disturbance, or "heart block," but can have a variety of other causes. But I found myself unable to determine which case to start with. After some consternation and a couple of false starts, I realized that this problem, this indeterminacy was of course the point. There was no "first case," or even a single classic description, and so no line of accumulation. Clustering of cases—none the same or equivalently framed, from several authors, each with a different understanding of his material—into a named condition, Stokes-Adams disease, and clear specification of its triad of features, first occurred in a textbook on heart disease by a relatively obscure Paris clinician, Henri Huchard. That was 1889. It had taken sixty years and the work of many physicians for this to happen, sixty years since Dublin surgeon Robert Adams had published the story of a sixty-eight-year-old revenue officer of remarkably slow pulse and prone to apoplectic fits. Or one could say it had taken 200 years for this to happen, I realized, reading more widely and finding reference in the nineteenth and twentieth centuries to cases from the eighteenth and seventeenth.

This was the second surprise. I had expected a much closer relationship—temporal, spatial, cognitive—between observing and knowing. Yet this distance, too, I came to see, was the point. Between observing and knowing stood a library. It was not only a library of books and articles—of investigations and classifications, results and conclusions. It was a vast library of what we might call data, of description in the form of cases. And, more than a record of observations, it was a record of readings of those observations, readings and rereadings ad infinitum as well as unread cases waiting, in fairytale-like sleep, to be awakened. Medicine, maker of some of our most powerful ideals and practices of firsthand scientific experience—of "autopsy" (from *autopsia*, "I see") and uncompromising "clinical" observation (from *klínē*, "bed" as in bedside)—was through that same activity a vast engine of *secondhand* experience of uncertain meaning and relevance, a written second nature of cases. The modern medicine that began with a clarion call of autopsia and empiricism, and a renaissance of observation and *observationes* (case histories) after 1500, inaugurated equally and through the very same move, to put an anachronistic point on it, 500 years of data mining.[1] Of course modern medicine has relied on observation and experiment, clinic and laboratory. Yet fundamentally medical research has been library research, and its library research has been *empirical* research. That is the argument I wish to advance in this chapter, which will explore what exactly that activity was, how it worked, what sorts of things it did and produced.

The library, I will show, was not only, as we usually say, a library of knowledge. It was equally a library of the unknown. This is no mere figure of speech. No one knew quite what was in the library of cases, of disease, of interventions and their effects. This was not the information overload and management problem that busied scholars for centuries (including many physicians). The unknown of the library of cases was opportunity rather than obstacle.[2] It was the vast investigable version of nature that could otherwise be met only bed by bed. Physicians and surgeons dug into it to test a conjecture, not knowing which way the test would go. Or they read to answer a question to which there was no previous answer because the question had not been asked or because answers conflicted. Or they read what others had seen to find out what they had seen and what to note of it.

But first, why is my claim more about published cases than, say, doctors' casebooks and patient files? Medicine abounded in these, too, after all. Hospital records, measured in hundreds of linear meters; autopsy books filling bookcases with descriptions of what was seen in bodies after disease and death; or, no less staggering, the shelves of casebooks a single physician could fill in a lifetime of practice: these, I initially thought, were medicine's working "archive" of disease, ever ready to be used retrospectively in new ways and with unforeseen aims and results. Yet most recorded medical observation, I came to see, passes into disuse or indeed vanishes. Not without a trace: it vanishes into various forms of exemplification and generalization.

Temperature charts provide a good example. Taking and noting hospital patients' temperatures several times a day is emblematic of modern medicine as a vast data-generating, record-keeping enterprise. Before around 1850, it had not been obvious that diseases could be differentiated by fever curve. To see this and to establish type curves for a range of common diseases required two million temperature measurements using the cumbersomely large, slowly registering thermometers of the mid nineteenth century on 200,000 patients and plotting the measurements on printed blank charts over a period of many years. The Leipzig hospital where this work was carried out under Carl Wunderlich was thus like Tycho Brahe's island observatory, Uraniborg, generating a massive flow of data on paper through highly organized, exacting, collective work.[3] Unlike astronomical records, which were carefully preserved and printed, however, the thermometry data is gone. This is not because the records were carefully kept, long used, and thrown away a century later when medical science had moved on. It is because they were not kept long at the time. What was left are the type curves. These were printed ad infinitum. For perhaps a century, they led the way in which physicians learned to grasp diseases. Each chapter of William Osler's *Principles and Practices of Medicine*, in its many editions from 1892 to the 1940s, features the curve for the disease. What survived in these curves was not even plotted averages of data, just the shapes. Medicine, in short, relies on vast data recording yet works through idealizations from it. It was precisely the curve's idealization, the *loss* of data, that made it applicable in practice, that is, to the task of differentiating kinds of illness and thereby deciding treatment. Forgetting, in this structured manner, allowed physicians to move knowledgeably from one patient to the next. And since medicine was about that task, no one missed—unimaginable in astronomy or paleontology or genetics or indeed lifelogging—the lost data.[4] Medicine has been as much an art of forgetting as an art of memory.

This is because it has been an art, and still is. By art, I do not mean not science, but having a practical goal. This art, moreover, deals in particulars. It always starts from and comes back to the patient, who is classifiable yet unique. Total recall would be paralysis. Amnesia, not recommended either. Between them lies that peculiar form of information, at once particular and already generalized, usually called the *case*.[5] Cases therefore are what medicine has preserved best, made accessible, and revisited—through print. Exactly how, though, and with what consequences and changes over time remains murky and hardly studied. Scholarship concentrates on the narrative form of cases and how and why they were produced.[6] My subject is their actual use rather than production and purpose; my object of study, case literatures rather than case forms.[7]

For what was published added up. Published cases represented a small fraction of those written as "observations" or "casebook" entries. These, in turn, represented a still smaller fraction of the patient information recorded in hospitals and private practice, which, finally, represented a fraction of all illnesses treated. Yet this published fraction of a fraction of a fraction, this tip of the iceberg of medical practice, grew to continental size over the past 500 years and especially over the past two centuries. The sheer size of medicine, which dwarfs all sciences, ensured that a very large number of case reports has been made public and permanent in print—and now in electronic form. Physicians use this sprawling library every day. So do historians and natural and social scientists. The tools developed for using it and coping with its size are paradigmatic. Yet we have little idea how it works, much less how medical knowledge thereby takes shape and changes. This library, more than the hospital and military medical archives and files that have arrested historians' attention, is the lasting, working medical record.[8] This chapter is an attempt to begin to grasp what it is and how it is worked with and worked upon and with what consequences for thinking about studying the written world as a way of knowing the physical one. To that end, I examine (1) case literatures of "new" diseases and pathologies of modern medicine since around 1800, such as leukemia, coronary occlusion, Hodgkin's disease, Graves' disease or *Basedow'sche Krankheit*, stomach perforation, and Stokes-Adams disease, comparing (2) the encyclopedic compilations in which most case reports found their place in the sixteenth through eighteenth centuries with (3) the periodicals in which case reports tended to appear in the nineteenth and twentieth. This survey includes publications from major centers of modern medicine, such as Edinburgh, Paris, London, Dublin, Vienna, and Berlin, as well as by early modern physicians all over Europe.

CASE HISTORIES, OR SLEEPING IN THE LIBRARY

"Wondrously inconstant pulse" was the title Dr. Marko Gerbec gave to a patient history he submitted to the Academy of the Curious. A member of the Leopoldina, Gerbezius was the most prominent Slovenian physician of his time. It appeared as Observatio LXIII in a volume of the academy's periodical *Miscellanea curiosa* in 1692, alongside three other patient histories by Gerbezius, the rest of the volume consisting of 208 other curious *observationes*, ranging across medicine and natural history.[9] No one paid

Gerbec's case any attention for many years, so far as we know (remember, the library is a big place), until the Padua professor of anatomy Giovanni Battista Morgagni dug it up sometime in the eighteenth century.

Morgagni was prompted to rummage in the library by a case of his own. Extremely slow pulse was, for Morgagni as for Gerbec, no disease or kind of illness. Morgagni noted "the greatest slowness of pulse" in a patient whose history illustrated "epilepsy," the category of illness described in the ninth letter of book I, the book on the head, of his epochal work on pathological anatomy. He might have given this slowness of pulse little or no notice. "Perhaps," he addressed his reader, "you will suspect, whether the rarity of the pulse be, in fact, a very uncommon symptom to remain after an epilepsy." This was another way of saying: Perhaps, reader, you may deem it essentially irrelevant to medical practice and knowledge. Had he not found another such case, the one reported by Gerbec, Morgagni implied, he might have felt the same and given it no notice. Still, its noteworthiness remained that of an aside: "But to return to my subject," epilepsy, Morgagni went on.[10] And that was that. Until another half century later, the same process of iterative reading repeated itself. Dr. William Burnett, a member of the Royal College of Physicians, waited three years before publishing a case of epilepsy with remarkably slow pulse. Clearly he was in no haste to publish this case. The library appears to have made it publishable and perhaps prompted him to take extensive notes at the time. Burnett wrote that he had not met this "train of symptoms" before "in epilepsy." He must have taken detailed notes, perhaps in a casebook, for years later he was able to publish the case in seven pages of detail.[11] His working method in the library seems to have been simply to look up "epilepsy" in Morgagni. There he found Morgagni's remark regarding slow pulse, an aside from his "subject," epilepsy. Morgagni could have left the detail about the pulse in his notes, or failed to note it at all. Ditto Burnett. What seems to have made slow pulse notable, reportable, publishable, is that someone else had already done so. But then how do we explain Gerbec publishing in 1692 a slow pulse case with no antecedent? Gerbec could publish his singular case because he lived in an earlier time, the age of curiosity. Without it, this and other disease bibliographies—diseases, in effect—might never have got started. The growth of the modern library in which it was no longer acceptable to publish single curiosities—"mira"—nonetheless depended on exactly that having happened, on the premodern medical library as cabinet of curiosities.

The medical library of cases survived medical revolutions—a longue durée pattern that gives pause about how revolutionary such moments or periods were. A monograph on spontaneous perforation of the stomach published in Paris in 1803 by Alexandre Gérard, for instance, puts "Paris medicine" and the "birth of the clinic" in this different, library light.[12] I assumed that its seventeen "observations" would be cases from Gérard's own experience in the newly reorganized Paris hospitals and clinical instruction program. Instead, there is only one such, witnessed in no less than the new medical school's clinic presided over by Jean-Nicolas Corvisart at the Charité Hospital. Fifteen of the sixteen other cases reported are from the old *observationes* literature of the sixteenth through eighteenth centuries. What caused the small round stomach opening with its red border?, Gérard asked of the first case recounted in the monograph. This case turns out to have been reported by his father. Neither foreign body, nor sharp material, nor worms were the cause, Gérard reasoned from the autopsy results. Most probable, it seemed to him, was that perforation followed from a small abscess that grew between the stomach membranes—most probable because he found this explanation in a case with similar symptoms reported earlier: in the *observationes* of Charles Lepoix published in 1618.

The point is not only that old cases never die, though they may long sleep—that medical knowing relies on permanent retrospectively usable description. The point is that even a Baconian storehouse of such description, a well-organized, catalogued archive of cases would have been unusable. For Dr. Gérard to have found Dr. Lepoix's case in such a repository would have taken a finding aid of far higher resolution than in archives up to this day, or indeed demanded today's full-text searchability. Luckily, however, the old cases belonged to a literature—of readings, citations, and rewritings. Gérard found his way to this and other cases indirectly—through generations of readers and writers. And this is also how he could work out what to do with them. The point is that the sleepy old library of cases was also a lively library.[13] It consisted not merely of cases, but of case literatures—empirical research on the written bedside unknown, forever in medias res. This was the power, not the problem, of too much to know. They are two sides of a coin. And surprisingly we shall see *both* grow, not decline, as we move into the nineteenth century and medicine enters its age of the hospital, bureaucratic organization, and technology.

It is perhaps easy to accept the idea that inquiry such as Gérard's into correlation between symptoms and lesions was as much library as bedside labor (even if this could significantly alter our understanding of clinical revolution, "Paris medicine," and "hospital medicine" generally). Likewise disease classification, or nosology, is plausibly a matter of vast reading and writing.[14] Surely, however, none of this can hold for the "discovery" of new diseases, the identification of previously unrecognized forms of illness, like Stokes-Adams disease with which we began, which characterized medicine from the late eighteenth to the twentieth century. Surely that activity was necessarily mainly an activity of observation and dependent on the rise of rigorous physical examination and comprehensive postmortem dissection and on the advent of new techniques and technologies. Medicine entered the reign of technology, as S. J. Reiser put it in his classic study, with chapters on the stethoscope, diagnostic microscopy, thermometry, graphic recording devices, x-rays, and so on.[15] In what follows, I shall argue instead that the recognition and characterization of diseases derived equally from library change broadly conceived, new forms of publication and reading. Diseases *were* bibliographies as much as they were type curves or diagnostic test results or specific combinations of symptoms. The meaning and relative status of all of the latter depended on and changed with the changing "literature" as well as affecting it.

Having left slow pulse with fits in a literature related to observations of patients in seventeenth-century Slovenia, eighteenth-century Padua, and nineteenth-century London, let us pick up—no, restart—the story from a case in Dublin. Back to Robert Adams on the sixty-eight-year-old revenue officer: "What most attracted my attention was the inequality of his breathing"—a feature that would not become part of the disease identity—"and remarkable slowness of the pulse." These "attention"-attracting features distinguished this case from countless others and lifted it out of unreportable routine and variation into the 100 pages of "cases of diseases of the heart" that Adams found "remarkable" enough to publish in the serial *Dublin Hospital Reports* in 1827. There it lay buried—such was the capacity of the case library to hide as well as to highlight—until William Stokes found Adams's case almost twenty years later when Stokes was prompted to publish "Observations on some cases of permanently slow pulse" (1846). How had he found Adams's case? He too was a Dublin physician contributing to the Dublin medical journals, and Adams told him about it.[16]

Although these two descriptions would later be enough to give a disease a name, meanwhile and apparently unbeknownst to the Dubliners, there had already appeared a number of other reports later deemed related: Dr. Burnett's "epilepsy" case in the London *Medical-Chirurgical Transactions* (1827), then a set of eight cases of "slowness of the pulse" in the *London Medical Gazette* (1838), which seems to have prompted "Case of Fits with Very Slow Pulse" in the same journal the following year, and "Case of Slow Pulse with Fainting Fits, Which First Came on Two Years after an Injury of the Neck from a Fall and Proved Fatal Five Years and Three Months after the Accident" in the *Lancet* in 1840, followed the next year in that same journal—again a local prompting—by a case of "Remarkable Slowness of the Pulse."[17] Strikingly, none of these authors knew of Adams's report, and few knew of each others' reports, but one of them dug up a "History of a

Case in Which There Took Place a Remarkable Slowness of the Pulse" from a forty-year-old volume of the Edinburgh journal *Medical and Philosophical Commentaries* of 1793,[18] doubtless via some paper instrument of memory (index or abstract or medical library catalogue or footnote in a treatise, we need to know more here) and maybe a social one ("I seem to recall reading . . . ," library equivalent of "There was this one guy . . .").[19] Did this outpouring mean that the combination of fits and remarkably slow pulse was there for all to see or remember once prompted? It seems not, for we saw that Morgagni in the eighteenth century had nothing to cite except the decades-old *observatio* from Slovenia, and Burnett in the early nineteenth had not previously "met with" such a case and had only Morgagni to cite. In 1840, it was unfamiliar to most physicians who joined the discussion of the first *Lancet* case at the Royal Medical and Chirurgical Society. Only in the 1850s did one physician decide to go looking systematically through past volumes of a journal, the *Transactions* of the Pathological Society of London. This was bibliography as empirical research. It yielded, in tell-tale sociable style, "Remarks" (on ten cases) appended to a single case report with the still fragmented title "Fibrinous Deposit Infiltrated, and in Masses within the Substance of the Walls of the Heart. Tendency to the Formation of an Aneurismal Pouch. Peculiarities in the Pulse."[20] In the 1880s, one American reporter of "three cases of remarkably slow pulse" was apparently encouraged to see these as *cases of* when he visited the Surgeon General's Library, looked up everything indexed under "slow pulse," and found so many that he could select ninety-three of them to abstract in the article he published in 1889.[21] Slow pulse was, however, not the same as its combinations with fainting and fits, the "triad" by which the French clinician Huchard defined Stokes-Adams disease that same year.[22]

Sprawling, prolix, ever in need of work to make it cohere in one way or another, the disease as case literature was virtually the opposite of the disease as type curve or as laboratory test result or as single classic description. On the one hand, there were and are "classic" descriptions, exemplary cases (which cannot exist except through forgetting practices), yet also, on the other hand, vast bibliography and case literature (which cannot exist except through remembering practices). Had the identification of a set of clinical criteria alone sufficed for the medical art (the triad in Stokes-Adams), or had etiologic consensus been able to replace it ("heart block" in Stokes-Adams), generations of physicians would hardly have created and perpetuated the laborious, never-ending bibliographic construction of the disease. Take, for another example, the ailment classically described by Graves in England in 1835 and (differently) by Basedow in Germany in 1840 and usually known as Graves' disease or as *die Basedow'sche Krankheit*. Its major clinical discussion, a two-part treatise published in 1909–10, boasted a bibliography of 3,000 titles.[23] Much of this was a vast case literature, as much characterizing illness as working out mechanisms and causes. All this for a relatively rare disease, and one for which there were no published descriptions until the third decade of the nineteenth century.

There is a further point to be made here. Such literatures were not footnotes to the "classic case," not its mere confirmation and application. Had they done no more than confirm and apply, their meticulous and resource-costly production would hardly have been worth the effort. The Stokes-Adams literature shows this. It grew backward in time as well as forward; in every direction or, from some points, not at all. It was as much an archipelago of mutual ignorance, from which the fog later lifted, as a network of citations. No forward-proceeding trajectory of reading and writing, it did not even have an origin.

Looking for that origin had initially kept me from sketching its story. For there is no such thing as a first case. Or so it now seems to me. "First case," even exemplary "classic" case, is a retrospective honor. It is even a contradiction in terms, since *case* implies *case of* or at least the potential to be a *case of*, hence more than one, or the existence of grounds for expecting more than one. "Case of Disease of the Spleen, in which Death Took Place in Consequence of the Presence of Purulent Matter in the Blood," published by David Craigie in the *Edinburgh Medical and Surgical Journal* in 1845, has a claim on being the first description of leukemia and is cited as such. Yet it was "Case 1" of two cases published together, the other by Craigie's better-known colleague John Hughes Bennett under the title "Case of Hypertrophy of the Spleen and Liver in Which Death Took Place from Suppuration of the Blood." So the first case citation in the leukemia literature is two cases. Moreover, the meaningfulness and reportability of the earlier of the two cases depended on the appearance of the later one and vice versa, either because Dr. Craigie could not report his mere "speculation" about "only one case" or because the "significance of the change in the blood was not appreciated, or indeed its occurrence remembered, by Dr. Craigie, until it was found in the second case by Dr. Bennett."[24] In short, two's a case.

Why does this point about case knowing matter for thinking about the archive of knowledge? It matters because it is how, in medicine, the empirically obtained units that are stored are nothing in themselves but depend on the existence of the other units, yet in an open-ended way. Their meaning is not merely relative, but expectant or retrospectively endowed or both. In this respect, the sleeping metaphor misleads: the case that awakens is seldom the one that dozed off. The kiss, the reading that awakens the *belle au bois* also tends to transport it to another part of the woods. Thus cases that showed Morgagni and Burnett an occasional feature of brain disease, epilepsy, showed later authors an occasional feature of heart disease. Cases in 1845 that showed the Edinburgh doctors suppuration in the blood would soon show other doctors different pathological processes that they then aimed to define, an act enabled by reading still older cases they found in the library, which, once upon a time, a hundred years before, had shown doctors changes in the blood less definite than suppuration.[25] Such shifts of pathological meaning—by reading more than by observing—make up much of what changes in medical knowledge. Retrospective use can move cases around among categories and create new ones. Yet cases stay together in literatures, or if and when they are split up, these acts of separation remain recorded in the literature: for future practical use and future writing, the one called "differential diagnosis" and the other "research" or "science" though they consist in many of the same analytic and synthetic practices of reading and comparing and may be continuous in their pursuit. What, then, keeps cases together at all when they are not, or not yet, or not yet not, "cases of" some disease in the process of being defined? The answer seems to lie in the power of "remarkable" single features of uncertain significance, like slow pulse, which did not itself become a disease, despite "case of slow pulse" publications that headed in that direction. Remarkable features not only get literatures started; they do much to sustain them. We will shortly see that this power of the remarkable belongs to wider social histories.

"Ideas re how diseases crystallize" was the heading I originally gave my notes on these phenomena, but it became increasingly unclear to me that diseases—and the ways doctors know and treat patients—ever actually crystallize out of cases. They live as literatures. Much of medicine makes do with remarkable lack of diagnostic and classificatory clarity and standardization. Even for an apparently clearly defined form of illness with its own unmistakeable name, like Hodgkin's disease, thereby tied to a classic description and exemplary set of cases published by a single author (in 1832), it turns out that as late as the 1930s Hodgkin's was only the dominant designation in a literature of some fifty-five names, with their associated case reports and considerable range of opinion on exactly what constituted what, in a clinical literature upwards of 750 titles.[26] None of this is usually noticed. Attention falls instead on classificatory diagnostic manuals. Or on "medical classics" that allegedly gave diseases definitive form. Or on milestones of diagnostic differentiation, such as typhoid/typhus.

Very different histories are at stake. On the one hand, existing historiography features a story of vast irregular (ultimately stagnant) premodern variety before the nineteenth century—notably the fever names ad infinitum—giving way to social and scientific construction of relatively regular and well-defined disease identities, or disagreements over rival versions of these; and a story of a succession of ways of doing this—new approaches, "schools," technologies, and so on. On the other hand, taking the life of the medical literature seriously suggests a story of a practical art and body of natural knowledge subsisting in long continuities of reading and writing practice and changing clusterings and categorizations of cases.

Subsisting but also evolving. Having too much to know was productive for knowing in at least three ways. First, it allowed many to publish locally, in blissful ignorance of each other, exactly as we saw with "slow pulse" reports in the nineteenth century, whose geography was fragmented—and sometimes bridged—between London, Dublin, Paris, Washington, and elsewhere. Had each physician been able to know that "his case" had already been reported, he would have been less likely to publish. The blessing of this mutual ignorance: a much bigger library of cases, hence a far more empirically investigable one. And one that maximized the variety inevitable among infinitely different patients and observers: this is the second way in which too much to know was productive for knowing. The size and labyrinthine geography of medical library sprawl, over decades and centuries, kept the library open, worked against standardization of disease categories (witness the various and fragmented titles and categorizations of case reports around phenomena like slow pulse), and thus maximized variety in case reports even when physicians were unaware of particulars of that variety and these awaited later notice and investigation. A physician who refrained from publishing because he thought he could see "his case" already in the literature deprived later physicians of the opportunity of seeing something in his case that he had not and that could point observing and reading in some possibly productive new direction. Third, the fact that a case literature was often a foggy archipelago, rather than a well-ordered and unified encyclopedia, created an experience of independent corroboration, of the reality of what was reported, when patterns were discovered. At Middlesex Hospital in mid-nineteenth-century London, for example, records made and kept by the Registrar in his routine duties were reread later for "cases of bronzed skin," responding to a call for cases by the *Medical Times & Gazette* to test a symptom/lesion correlation published by Thomas Addison. At least one case published from the Middlesex Hospital was "transcribed from notes," the registrar reported, "taken at a time when I was not aware that any connexion . . . had been observed" between anemic fatalities with skin darkening and disease of the suprarenal capsules.[27] The registrar expressed the objectivity of recording yet equally its archivality: information of uncertain meaning preserved for future interpretation.

Even when medical writers hoped that through compiling similar cases "an important and well-characterized disease [would be] added to the nosological list,"[28] their efforts did something more: bring into being the case literature itself—a written and read world that was continuous with observation at the bedside. The only difference was in the lapse of time (and in the range of modes of writing employed). "Bedside observation" was in fact writing—fresh writing. And the "literature" was in fact bedside observation—older observation, older writing. Both were also reading, not in the figurative sense of "reading" bodily signs of disease, but in the literal sense of the reading that fed into any act of writing, whether that act took place without delay in scrawl at the bedside or appeared finely tuned in print. In sum, each of these three modes of productivity of the library of cases, of the known unknowns, is another facet of written second nature, each another way in which that world on paper was vast and messy and foggy enough to need investigating, like the world not on paper, yet simplified and reduced enough to be investigable. So second nature was this second nature to physicians that it is difficult to find them either crying information overload as many a scholar did or whispering, in the words of today's big data miners, "Volume is Our Friend."[29]

SOCIAL HISTORIES

What made all this possible? To what bigger histories does library empiricism belong? Social, or more precisely, sociable histories, in a word: the history of forms of collective life and communication involving medicine and beyond.

To begin when the modern library of cases began, there is a surprisingly ready answer to this question, a reinterpretation of what we already usually say about the renaissance of observation and empiricism in Europe, of stories of revolutionary valorization of firsthand experience over the authority of received knowledge—in medicine, from books to bedsides. The reinterpretation is this: the new *reliance* on experience in knowing the natural world was, in one of its key forms, equally a *renunciation* of knowing. That renunciative key form was the case history. The *observatio* was as archival (for secondhand knowing, not now but someday) as it was empirical (firsthand knowing, here and now). To write and collect cases and publish them literally by the hundreds—so-called "centuriae"—in no other order than the chronology in which the patients came across one's doorstep was, on the one hand, to prove one's vast experience and assert its value. On the other hand, it was to say: I don't know—what exactly these histories mean, what they add up to. And yet just as importantly: Someday (through this very activity) you will, so I write them to you, whoever you are, out there in the future of medicine. This renunciation was not in the elegiac positivist mode of big science projects like the Carte du Ciel—the map of all stars—centuries later.[30] It was expectant. In founding the library of cases, physicians did more than become newly empirical. They gave medicine a future—not a future in some future innovation, but a future in the (practices of the) present.

How can this reinterpretation of the renaissance of medical observation be a historical explanation? The answer is that this renunciative stance could not be taken alone. To act in this producing yet postponing way was to be, above all, a member of a collective (indeed, by writing, to invoke it): the collective of medical practice—past, present, future. This, I suggest, is the descriptive and explanatory power of the sociological fact that the early writers of case histories shared them avidly and that they tended to be town physicians rather than professors: the collective of practice rather than that of lectures.[31] Why practice? If knowing had to be collective because it exceeded the capacities of individual physicians in the present (and future) and practice was what all physicians shared, then practice was what had to be recorded, indeed studied, in order to know and, as they saw it, to know better. This was not the only medical empiricism of the time. William Harvey, for one, did his experiments alone, and he not only experienced, he knew. Accordingly he published books that were fully worked out new accounts of life processes, not hundreds of particular histories. But that is a different story from the archival one told here, whose historical specificity and therefore explanation thereby become more apparent.

What happened next can be summed up in a word: encyclopedism. Whereas the vogue of publishing *centuriae* launched the permanent open library of reinterpretable case reports, encyclopedism made it navigable at the price of closing down its future for a while. Encyclopedic compilers in the seventeenth and eighteenth centuries found a place for each patient history—or each bit of a patient history, if they cut them to pieces—in the overall *historia* of the sick body. That place was usually anatomical location, of symptoms or lesions or both, arranged chapter by chapter from head to toe, a book structure adopted from the *practica* genre. This library of cases was thus epistemically rather closed. Everything had its place. It put the known past and present more than the open future on the shelf. Thus in Morgagni's encyclopedic *On the Seats and Causes of Disease* of 1761, our Slovenian Dr. Gerbec's wondrously inconstant pulse *observatio* of 1692 ended up a case of a classical disease, epilepsy.[32]

This structure of meaning began to loosen with the proliferation of journals in the later eighteenth century. Unlike the encyclopedic *historia*, a publication such as *Recueil périodique d'observations de médecine, de chirurgie et de pharmacie*, as the first major French medical journal was titled when it began in 1754, was open. It was mainly future. The encyclopedic compilation was a book of commonplaces. A journal was a book—or rather conversation—of uncommonplaces. The journal enabled publication of cases without places, without context. Had not the earlier *centuriae* of patient histories already done so? Yes, but when cases were excerpted from the *centuriae*, they often wound up in the commonplaces of encyclopedic compilations. Twice decontextualized (published in merely chronological order and then excerpted), moved about and reshuffled on a sheet or slip of paper, case histories tended nonetheless to be reintegrated into compiled *historia* that were open to addition but not much novelty and rearrangement.[33]

Periodical publication encouraged seeing and reporting the remarkable in the routine, as opposed to either the common and well known or the rare and bizarre. To record an illness day by day was to be comprehensive, so reliable and machinelike that the activity had to be minutely regulated by hospitals and medical schools, as in the detailed rule book for observing and recording in the Paris clinic.[34] In contrast, to write up a patient's illness was to select. To publish a write up was to be still more selective—and selected by professors or editors—even when one published a hundred pages of heart ailment cases, as Adams did. The contents of the *Dublin Hospital Reports* could not be the routine of the Dublin hospital archives. Some physicians were so worried about boring their peers with uninteresting cases that they published none, only the generalizations, as William Heberden did expressly for that reason in his *Commentaries on the History and Cure of Diseases* (first in Latin in 1802), though these were closely based on his extensive bedside notes and large commonplace books of cases, whose structure was perpetuated in the published book's alphabetically arranged "titles" (the *tituli* of *loci communes*), which were mostly disease names, under which appeared compressed extracted information and no case histories.[35] This was an option for an alphabetical handbook like Heberden's, but not an option for publishing in a periodical to which subscribers and other readers turned not for something they sort of knew, or had forgotten and needed to check, or should have known, but for things neither they nor almost anyone else knew: news. So when the same Heberden had in 1767 written a preface to launch a new medical journal, he called for the publication of cases and made rules for being interesting but not too interesting. Report cases of rare distempers but not extraordinary cases, he admonished in clearly moral tones.[36] These were not the naturalists' virtues and practices of long attention and exact description.[37] They were not specifically observational virtues at all. They were library virtues (and sociable ones, as we are about to see). For Heberden's publication program required knowing not only one's own practice but also those of others. You could hardly know what was boring and what was interesting unless you read everything, critically. Distinct from the "art of observing," as it had come to be called in Heberden's time,[38] the art of being medically interesting—and getting cases published in journals—was mastery of noteworthiness.

This meant that the new literature of medicine, in the age of journals, would be made of the exceptional. Where *observationes* author-compilers had until recently aimed to map the body in all its usual morbid appearances, exceptions now expanded into being the recorded body of internal medicine. This puts a twist in the story that usually runs from early modern rarity fascinations to modern uniformities—a twist that invites a fresh look at routine. Imagine the voyaging naturalist, instead of encountering the new at every turn, attending the same hospital, running the same private practice, seeing the same population of patients every day, year in, year out. And yet still generating novelty. Take a condition like Graves' or *Basedow'sche* disease, one of whose frequent symptoms was bulging eyes and therefore another of whose German names was *Glotzaugen* disease. That is, one would think it was hard to miss. But as Armand Trousseau told the Paris clinical students in 1860 when he introduced them to cases of *goître exophthalmique* or *la Maladie de Basedow* at the Hôpital Hotel-Dieu: "There are diseases that come into notice all at once; they pass unperceived, an observer calls them to the attention of physicians, and soon examples multiply ad infinitum."[39] Pathologies that would soon be very familiar, such as hardening of the coronary arteries, could also start with the same sort of remarkability: when in the course of a dissection Edward Jenner's knife reached those vessels, the blade hit something so hard and gritty that it was nicked and, knowing nothing of such calcification, Jenner looked up at the ceiling, which happened to have been rather dilapidated, thinking some plaster must have fallen unnoticed into the cadaver.[40]

The keyword was "remarkable." It appears in countless case reports and their titles. Medical knowledge was not simply streamlined and standardized by the rise of the hospital and the "birth of the clinic." Previously, the publishing of *observationes* in the early modern period had shifted from emphasis on the rare and unusual—the written *Wunderkammer* of disease—to emphasis on documenting *all* observed disease.[41] Yet after hospital and clinic had become institutions of routine documenting, a great new age of the remarkable began. Remarkability and the new forms of collectivity on which it depended—constant communication in journals and societies— became the engine of novelty, of growth in medicine. Entire new diseases emerged as the unusual out of the same—and became extensive case literatures *despite* the republishable exemplarity of one or two "classic" descriptions. Hodgkin's disease began as a few "remarkable" observations, which, thanks to library medicine, soon grew to hundreds of published cases. It did not matter that during the century after Hodgkin published his set of seven cases four of them were reclassified as other conditions.[42] What histopathology took apart, the library kept—forever will keep—together and productively so, rather than merely erroneously: the literature sprouted from the fact that Hodgkin could plausibly claim so many examples and from the consequent continuing investigation of similarities and differences. Library subsumed laboratory.

And hospital. The hospital has given us images of large-scale data generation as inevitably machinelike and bureaucratic. There are good examples of hospital-based bureaucratic construction of disease, such as Pierre Louis' redefinition of typhoid fever through hospital-register-style tabularization of cases, which was so unsociable that in the 977 pages of his famous book he makes not a single reference to any other case report or any other physicians' work at all.[43] Yet this mode of medical knowing is marginal to my story—to the archive of medicine—because its archives did not last, though they were made out of the formally archival, record-keeping functions of institutions. Of course physicians did and do have to fill in forms with blanks for "disease" designation. Hence the "tyranny of diagnosis" by which Charles Rosenberg has characterized medicine in the modern bureaucratic age.[44] Yet bureaucracy is only half the story of modern medicine whose other, opposing half is the library. That same form-filling physician could step outside the tyranny of naming and classifying, with its administrative and economic matrix, when he wrote up a case not of this or that defined disease but titled "Fibrinous Deposit Infiltrated, and in Masses within the Substance of the Walls of the Heart. Tendency to the Formation of an Aneurismal Pouch. Peculiarities in the Pulse," and stepped into the library of the pathological society to read it aloud—and hear what his colleagues thought. All of which could later be read, and was.

The nineteenth century saw the emergence of journals that were styled as newly "scientific" and distinct from clinical practice. These tended to bear the name *Archive*. Journals that were, on the other hand, closest to practice and bedside observation bore titles like the German *Wochenschrift* (weekly), the French *Bulletin* or indeed *Journal* in the original daily sense, the English *Gazette* or *Circular* or *Times*. Their titles announced not the permanent archive of knowledge, but news. These journals are not even indexed in the *Royal Society Catalogue of Scientific Papers*. The revealing twist is that these newsy weeklies made up the core of medicine's permanent scientific archive, the library of case reports. Behind them were the medical societies, whose publishing outlet the weeklies often were, unless they had their own journal as did, to take an example we have already seen, the Pathological Society of London. Much of its *Transactions* consisted of case reports read to society meetings, organized anatomically by organ system.[45] The chatty communication of news in the great age of associational life and cheap periodical print that stretched from the later eighteenth century into the twentieth is closer to the historical dynamics of the formation of new diseases as case literatures—call it sociable construction—than is the sedimentation of files in an archive. Another word for diagnostic category or disease identity could be: old news. When conversation circled and news repeated, a new disease began to emerge. Medicine's library of cases consisted more of letters to the editor—a popular form of case report—in its fleeting weeklies than original research articles in its eminent journals. Precisely the ephemeral *Gazette*, as opposed to the august *Archive*, made the archive: by enabling physicians to publish cases in extenso and out of context and to leave their meaning and utility up to future readers. This runs against usual ways of thinking that oppose the timely and the timeless, lively communication and deadening accretion. Data in this bibliographic rather than bureaucratic mode of medicine was not daily recording, because that was little shared. Data was chat—case chat—made permanent for all in print. The livelier, denser (more urban), more sociable, and accelerated medicine became, the more novel data (cases) it archived. The reading and writing nineteenth-century medical collective was like a twenty-first-century search engine learning—empirically—from the activity of all its users, but with no Google-like monopolizing center of storage.[46] And what was interesting enough to discuss in person or print was what just might change medicine, ever so much, or very much, or not at all, only time—and the future of the collective—would tell.

The rules of this game remain yet to be fully understood. That it exists at all and that medicine partly consists of it seems to have to do as much with the history of forms of collective life and communication and their associated values as with any history specific to medicine or the sciences. The rise of dynamic "modern medicine" and its new and changing diseases was the rise of the weekly rag and the club. This argument can be compared to the argument that developments around mass periodicals in the nineteenth century made scientific periodical information retrievable.[47] Yet, as we saw in the previous section, medicine's library empiricism depended on the limits as well as the powers of generic ways of searching and organizing published information, from cataloguing to cross-referencing. More fundamental and characteristic, in any case, was another generic development: the review.

The review article is not usually thought of as a major epistemic and ontological instrument. Historians generally follow scientists in valuing "original research" articles much more highly. Yet there are reasons to believe that the practice of reviewing the literature shapes and reshapes knowledge at least as much and possibly more consistently over time. And when the literature reviewed consists partly or importantly of cases, then review, too, can be a form of empirical research. Consider, again, leukemia, one of the paradigmatic new disease constructions of medicine since the birth of the clinic. It is well known that Rudolf Virchow coined this term. Yet he did so not in the case report he published in 1845 under the title "White Blood," but in subsequent articles in which he reprinted and reviewed cases from the literature, including his own and the ones from Edinburgh that we met above and that had been reported as suppuration, an old category that Virchow now rejected for these cases. What occasioned and patterned this iterative process of empirical research by reading was the rise over the previous half century of a new literary practice—critical synthetic reviewing—and of review genres and sections of journals and entire review journals, such as *Schmidts Jahrbücher*, where Virchow's most extensive "white blood" reviews were abstracted for a wide audience but which also published original reviews.[48]

A review of a large literature, like that for "Hodgkin's disease" by the 1930s, could keep fifty-five disease names, one of which was "Hodgkin's," from being many, if not quite fifty-five, diseases.[49] Or, conversely, reviewing could enable one new case to recast others that had already been reported. And such reviewing happened not only in review articles and disease monographs with giant bibliographies, but also in singular case reports themselves, as we might predict from the fact that—two's a case—they were rarely singular. In writing his first "white blood" case report in 1845, for example, Virchow engaged in two kinds of library empiricism. On the one hand, he practiced reading as *compiling* similar cases. He searched the library for references to white blood, and found several, some of which had already been compiled in the eighteenth century by Albrecht von Haller. This was not just historical background. As we saw with Morgagni and epilepsy and as Heberden put into rules of interestingness, this practice of knowing the literature was how a phenomenon became notable (again, or differently). On the other hand, Virchow practiced reading as *differentiating* similar cases—that is, pulling apart phenomena that seemed similar. He discussed a case report published earlier that same year from the prosectorship of the renowned Vienna pathologist Carl Rokitansky, a case published as "general pyaemia" in a thirty-three-year-old locksmith. Rokitansky gave the postmortem only, no clinical history. Virchow ended his own case report by doubting that this "case belongs under the rubric pyaemia" and calling upon the *Wiener* to publish the clinical history to help decide this question—via more reading as secondhand observation.[50] What Virchow was doing here was reading the case literature as a form of critical review that was in fact equally a form of secondhand observation, reading medicine's written second nature. This shows the case library as object and apparatus of empirical research, but it shows more: the power of the library over bedside and postmortem. And the primacy of reading over observing—in the end. Rokitansky could say of this case: autopsia. Virchow could say of several: I read. And reading won.

Just as cases appearing in weekly medical chat and print were a form of news and thus subject to wider social histories, so too the medical review was not really medical, but belonged to a much wider emerging culture and practice of public life. Journals were changing. The old physicians' periodicals of the seventeenth century, such as the *Miscellanea curiosa* with which we began, were in effect *centuriae* by many people, rather than organs of review and forums of debate. In contrast, the new medical weeklies and monthlies of the late eighteenth and nineteenth centuries sported critical review, in line with a new generation of magazines and newspapers generally. The "review" was a generic manifestation and tool of the new periodical literature that arose from the late eighteenth century and characteristic of what Kant called the "age of criticism."[51] Thus did the new critical public sphere do much to create the new disease landscape and endless evaluative activity of modern medicine and, moreover, what it meant to be a disease category in this epoch, namely, that diseases read much more like a lively critical review literature than like the assemblage of symptoms and lesions under a heading in an encyclopedic compilation of *observationes*—or in a modern diagnostic manual.

LOOKING BACK AND FORWARD

Around 1950, as the antibiotics era dawned and history seemed to end for medicine, a bold-thinking historian divided medicine into four successive types. The most recent of these he called laboratory medicine (from around 1850), then, moving back in time, hospital medicine (from around 1800), bedside medicine (ancient and early modern), and library medicine (medieval). Only the last of these no longer existed. Consonant with narratives of scientific revolution, the story was that library medicine had been left behind when physicians turned their attention from books to bedside and autopsy table. This typology and periodization has structured much teaching and research in history of medicine and science.[52] In this essay, I have suggested instead that modern medicine arose, from the time of printing and in a new way since the late eighteenth century, as library medicine. This could be a trivial thesis. Any area of knowing involves the written word and its preservation, and dynamic information storage goes back to genres—doxa, problemata—in Antiquity.[53] But I am arguing that production of empirical novelty, generation of new knowledge, improving the art, what we usually attribute to getting out of the library, happened equally in and through the library, because it was a library not only of knowledge, but of the unknown.

All well and good until the twentieth century, one might think, when clinical knowledge became quantitative and experimental. Of course medicine has changed through the application of statistical method and controlled trial procedures. Yet medical practice and knowing continue to rely on literatures of cases. Case reports tend to contain more quantitative information than in the past and more of the formalized information that comes from monitoring, testing, and imaging technology. But these data are not often themselves aggregated and analyzed outside of the cases to which they first belonged, except in epidemiological research projects, which have to extract data from the patient-by-patient records and from the published case form in which medicine preserves it.[54] Meanwhile, because the data generated by clinical trials is specific to answering efficacy and safety questions about particular drugs, it is a less general and open-ended than the library of cases. That library has only grown, keeping pace with the rest of the medical literature. Since the 1970s the number of case reports listed in Medline has hovered at around 10 percent of all publications, rising from some 15,000 to over 50,000 annually today. Medical journals have spawned separate case journals. *BMJ Case Reports* began publication in 2008, years after the BMJ had launched its journal *EBM: Evidence-Based Medicine* in 1995, using the slogan of the movement that claims that, until now, medicine was not evidence based.[55] Digital, electronic medicine so far looks much like the old paper library medicine. In fact, evidence-based medicine *is* library medicine, using methods recognizable to early modern historians of scholarship and "information" since the Renaissance and used in medicine since that time (indeed partly invented or formalized by early modern physicians like Conrad Gesner, creator of the first catalogue of all printed books: *Bibliotheca universalis*, 1545–49).[56] Method today enjoys more powerful techniques of data synthesis and more rigorous mechanisms of comparative evaluation, though even these are probably continuous with the past rather than worlds apart from it, as could be tested by more detailed historical study of how physicians read and write their published literatures than has been possible in this essay. The title of a leading introductory EBM textbook is *How to Read a Paper*.[57] And in the late twentieth and early twenty-first centuries, the most important specific way of moving toward a more "scientific" procedure of informing medical decisions and making generalizations in medicine has been—back to the future—a better form of literature review. Systematic Cochrane reviews now make up a "database" of some 5,000 reviews. There seems to be little or no awareness of the centuries of bibliographic and reviewing information practice in and beyond medicine. Historical amnesia has left only the idea that "narrative reviews of healthcare research have existed for decades, but are often not systematic."[58] *Sic*. And, finally, computing has yet to change the structures laid out in this essay. As with chess, it will, or may, only beat the champion diagnosticians at their own game.[59]

Medicine of the past half millenium—once we see its library empiricism—bears striking similarities to today's data worlds, practices, and ambitions. I have indicated some of these similarities along the way. But the point may be the other way around. It is no coincidence that medicine provided a prototype of today's online digital knowledge indices and electronic search and mining tools. PubMed and MEDLINE continue indexing begun in the 1870s in the Library of the Surgeon-General's Office, United States Army, which by 1895 was the largest medical library in the world and held material dating back to the 1400s.[60] The *Surgeon-General's Catalogue* around 1900 is much larger than the *Royal Society Catalogue of Scientific Papers*, which was compiled at that same time (and the latter includes the medical science journals).[61] Published medicine has always been much bigger than published science. Moreover, the Royal Society's catalogue is an alphabetical author list only, whereas the Surgeon-General's is an elaborate index of subjects and topics, cross-referenced and nested in many layers—making both knowledge and the written unknown as minable as could be before the digital age. Through its long unbroken history, its size and global geography, its inseparability from both scientific and everyday life, its inclusion of so many areas of documented private and public life, medicine may be paradigmatic of third nature. As the greater library of natural and human reality grows and as inquiry and other activities become ever more data-productive, the rest of the documented, investigable world becomes more like medicine's library of the unknown, forever on its way to being known and used in new ways.

NOTES

1. On data mining, see Jones, chapter 12, this volume. On medicine and the renaissance of observation: Gianna Pomata, "Observation Rising: Birth of an Epistemic Genre, 1500–1650," in *Histories of Scientific Observation*, ed. Lorraine Daston and Elizabeth Lunbeck (Chicago: University of Chicago Press, 2011), 45–80.

2. Special issue on "Early Modern Information Overload," *Journal of the History of Ideas* 64, no. 1 (2003); Lars Behrisch, "Zu viele Informationen! Die Aggregierung des Wissens in der frühen Neuzeit," in *Information in der frühen Neuzeit: Status, Bestände, Strategien*, ed. Arndt Brendecke, Markus Friedrich, and Susanne Friedrich (Berlin: LIT Verlag, 2008), 455–73; Ann Blair, *Too Much to Know: Managing Scholarly Information before the Modern Age* (New Haven: Yale University Press, 2010).

3. For a succinct account with these details, see Lloyd G. Stevenson, "Exemplary Disease: The Typhoid Pattern," *Journal of the History of Medicine and Allied Sciences* 37 (1982): 159–81.

4. On archival practices of these fields, see Hsia (chapter 1), Sepkoski (chapter 2), Gere (chapter 8), and Lemov (chapter 10), this volume.

5. I am preparing a separate study on these points.

6. Kathryn M. Hunter, *Doctors' Stories: The Narrative Structure of Medical Knowledge* (Princeton: Princeton University Press, 1993); Brian Hurwitz, "Form and Representation in Clinical Case Reports," *Literature and Medicine* 25, no. 2 (2006): 216–40; Michael Stolberg, "Formen und Funktionen medizinischer Fallberichte in der frühen Neuzeit (1500–1800)," in *Fallstudien: Theorie—Geschichte—Methode*, eds. Johannes Süßmann, Susanne Scholz, and Gisela Engel (Berlin: trafo, 2007), 81–95; Steve Sturdy, "Knowing Cases: Biomedicine in Edinburgh, 1887–1920," *Social Studies of Science* 37 (2007): 659–89; Gianna Pomata, "Sharing Cases: The *Observationes* in Early Modern Medicine," *Early Science and Medicine* 15 (2010): 193–236; Rudolf Behrens and Carsten Zelle, eds., *Der ärztliche Fallbericht: epistemische Grundlagen und textuelle Strukturen dargestellter Beobachtung* (Wiesbaden: Harrassowitz Verlag, 2012).

7. Though research cited in note 6 discusses the overall function of case reports in medicine, few studies explore actual use of cases: Volker Hess and J. Andrew Mendelsohn, "Case and Series: Medical Knowledge and Paper Technology, 1600–1900," *History of Science* 48 (2010): 287–314; Rachel A. Ankeny, "Using Cases to Establish Novel Diagnoses: Creating Generic Facts by Making Particular Facts Travel Together," in *How Well Do Facts Travel? The Dissemination of Reliable Knowledge*, ed. Peter Howlett and Mary S. Morgan (Cambridge: Cambridge University Press, 2011), 252–72.

8. See most recently Warwick Anderson, "The Case of the Archive," *Critical Inquiry* 39, no. 3 (2013): 532–47, which seems to me to conflate case report and patient record and wrongly to construe patient files as medicine's case archive.

9. M. Gerbezius, "Pulsus mira inconstantia," *Miscellanea curiosa, sive Ephemeridum medico-physicarum Germanicae Academiae Caesareo-Leopoldinae Naturae* 10 (1692): 115–18.

10. Giovanni Battista Morgagni, *The Seats and Causes of Diseases Investigated by Anatomy*, trans. Benjamin Alexander, 3 vols. (London: Printed for A. Millar, T. Cadell, and Johnson and Payne, 1769), vol. 1, book I, letter 9, article 7, 192, 195.

11. William Burnett, "Case of Epilepsy Attended with Remarkable Slowness of the Pulse," *Medical-Chirurgical Transactions* 13 (1825): 202–11.

12. Alexandre Gérard, *Des perforations spontanées de l'estomac* (Paris: Impr. de Gillé fils, 1803). To reexamine Paris medicine in this light will require a separate study of the sources and their interpretations by Michel Foucault, Erwin Ackerknecht, and many other historians since the 1960s.

13. On the lively library, see Anthony Grafton, "Libraries and Lecture Halls," in *The Cambridge History of Science*, vol. 3: *Early Modern Science*, ed. Katharine Park and Lorraine Daston (Cambridge: Cambridge University Press, 2006), 238–50.

14. Volker Hess and J. Andrew Mendelsohn, "Sauvages' Paperwork: How Disease Classification Arose from Scholarly Note-Taking," *Early Science and Medicine* 19 (2014): 471–503.

15. Stanley J. Reiser, *Medicine and the Reign of Technology* (Cambridge: Cambridge University Press, 1978).

16. Robert Adams, "Cases of Diseases of the Heart, Accompanied with Pathological Observations," *Dublin Hospital Reports* 4 (1827): 353–453, on 396–99; William Stokes, "Observations on Some Cases of Permanently Slow Pulse," *Dublin Quarterly Journal of Medical Science* 2 (1846): 73–85.

17. Herbert Mayo, "An Account of Some Cases of Slowness of the Pulse," *London Medical Gazette* 22 (1838): 232–38; John R. Gibson, "Case of Fits with Very Slow Pulse" (letter to the editor), *London Medical Gazette* 23 (1839): 123–26, 155–60; J. H. Holberton, "Case of Slow Pulse with Fainting Fits . . . ," and discussion, Royal Medical and Surgical Society, *Lancet* 35, no. 916 (1841): 892; W. C. Worthington, Sr., "Remarkable Slowness of the Pulse," *Lancet* 36, no. 926 (1841): 336–37.

18. Thomas Spens, "History of a Case in Which There Took Place a Remarkable Slowness of the Pulse," *Medical and Philosophical Commentaries* 7 (1792): 458–65.

19. Kathryn M. Hunter, "'There Was This One Guy . . .': The Uses of Anecdotes in Medicine," *Perspectives in Biology and Medicine* 29, no. 4 (1986): 619–30.

20. John Ogle, "Fibrinous Deposit Infiltrated, and in Masses within the Substance of the Walls of the Heart. Tendency to the Formation of an Aneurismal Pouch. Peculiarities in the Pulse," *Transactions of the Pathological Society of London* 8 (1857): 118–19, and "Remarks," 119–21.

21. Daniel Webster Prentiss, "Report of Three Cases of Remarkably Slow Pulse, to which Is Appended Brief Abstracts of Ninety-Three Cases of Slow Pulse Found Recorded in Medical Journals in the Library of the Surgeon-General's Office, Washington, DC," *Transactions of the Association of American Physicians* 4 (1889): 120–59.

22. Henri Huchard, *Maladies du coeur et des vaisseaux* (Paris: Doin, 1889), lesson 13, type 5, 255–65.

23. Hubert Sattler, *Die Basedow'sche Krankheit* (Leipzig: W. Engelmann, 1909–10).

24. David Craigie, "Case I. Case of Disease of the Spleen, in Which Death Took Place in Consequence of the Presence of Purulent Matter in the Blood," *Edinburgh Medical and Surgical Journal* 64 (1845): 400–413, quotation on 411; John Hughes Bennett, "Case II. Case of Hypertrophy of the Spleen and Liver in Which Death Took Place from Suppuration of the Blood," *Edinburgh Medical and Surgical Journal* 64 (1845): 413–23; William R. Gowers, "Splenic Leucocythaemia," in *A System of Medicine*, ed. John Russell Reynolds, 5 vols. (Philadelphia: J. B. Lippincott, 1868–79), 5, no. 216–305, quotation on 219n.

25. See page 91 below on leukemia.

26. A. Wallhauser, "Hodgkin's Disease," *Archives of Pathology* 16 (1933): 522–62, 672–712.

27. Middlesex Hospital, "Cases of Bronzed Skin, etc.," communicated, with Remarks, by S. W. Sibley, *Medical Times & Gazette* 1 (1856), 188–89, on 188.

28. Jonathan Hutchinson, "Series Illustrating the Connexion between Bronzed Skin and Disease of the Supra-Renal Capsules," *Medical Times & Gazette* (1855), 2: 593–94, on 593.

29. Quoted in Jones, chapter 12, this volume.

30. See Daston, chapter 6, this volume.

31. Pomata, "Sharing Cases."

32. See page 89 above; using similar symptom categories and head-to-toe anatomical organization, see Théophile Bonet, *Sepulchretum sive Anatomia practica ex cadaveribus morbo denatis* (Geneva: Chouet, 1679); Joseph Lieutaud and Antoine Portal, *Historia anatomico-medico* (Paris: Vincent, 1767).

33. On *Verzettelung* and decontextualization, see Christoph Meinel, "Enzyklopädie der Welt und Verzettelung des Wissens: Aporien der Empirie bei Joachim Jungius," in *Enzyklopädien der Frühen Neuzeit: Beiträge zu ihrer Forschung*, ed. Franz M. Eybl et al. (Tübingen: Max Niemeyer, 1995), 162–87.

34. Jean-Jacques Leroux des Tillets, *Commission de l'Instruction publique. Académie de Paris: Faculté de Médecine—Clinique interne: Société d'Instruction médical: règlement* (Paris: Migneret, 1818).

35. William Heberden, *Commentaries on the History and Cure of Diseases* (Boston: Wells and Lilly, 1818), quotation from preface, viii.

36. Heberden, "A Sketch of a Preface, designed for the Medical Transactions" (1767), in Heberden, *Commentaries*, appendix, 393–401.

37. See most recently Mary Terrall, *Catching Nature in the Act: Réaumur and the Practice of Natural History in the Eighteenth Century* (Chicago: University of Chicago Press, 2014), emphasizing the domestic and public worlds that sustained such practices and virtues.

38. Jean Senebier, *L'art d'observer*, 2 vols. (Geneva: Chez Cl. Philibert & Bart. Chirol, 1775).

39. Clinique de la faculté de médecine de Paris: Hôtel-Dieu; Professeur: M. Trousseau. "Du goître exophthalmique," with case reports by attending physicians, *Union médicale* 8 (1860): 434–39, on 434.

40. Caleb Hillier Parry, *An Inquiry into the Symptoms and Causes of the Syncope Anginosa Commonly Called Angina Pectoris; Illustrated by Dissections* (London: Cadell and Davies, 1799), 3–4.

41. Pomata, "Sharing Cases," 225–26.

42. Samuel Wilks, "Cases of Enlargement of the Lymphatic Glands and Spleen (or, Hodgkin's Disease) with Remarks," *Guy's Hospital Reports*, 3rd ser., 11 (1865): 56–67, on 57; Herbert Fox, "Remarks on the Presentation of Microscopical Preparations Made from Some of the Original Tissue Described by Thomas Hodgkin, 1832," *Annals of Medical History* 8 (1926): 370–74.

43. Hess and Mendelsohn, "Case and Series," 296–300.

44. Charles E. Rosenberg. "The Tyranny of Diagnosis: Specific Entities and Individual Experience," *Milbank Quarterly* 80, no. 2 (2002): 237–60.

45. As in the example above, n20.

46. On the economic and business forces behind the data "deluge," see Strasser, chapter 7 in this volume.

47. Alex Csiszar, "Seriality and the Search for Order: Scientific Print and Its Problems during the Late Nineteenth Century," *History of Science* 48 (2010): 399–434.

48. "Weisses Blut und Milztumoren," *Schmidts Jahrbücher der in- und ausländischen Gesammten Medicin* 57 (1848): 181–88, abstracting two articles by Virchow that appeared in the weekly newspaper of the Prussian Medical Society in 1846–47.

49. Wallhauser, "Hodgkin's Disease."

50. Rudolf Virchow, "Weisses Blut," *Neue Notizen aus dem Gebiete der Natur- und Heilkunde* 36 (1845): cols. 150–56, quotation in col. 155.

51. Quoted in Thomas H. Broman, *The Transformation of German Academic Medicine, 1750–1820* (Cambridge: Cambridge University Press, 1996), 128, who situates physicians in this new critical literary public sphere; see also Broman, "J. C. Reil and the 'Journalization' of Physiology," in *The Literary Structure of Scientific Argument: Historical Studies*, ed. Peter Dear (Philadelphia: University of Pennsylvania Press, 1991), 13–42.

52. Usage of these terms is ubiquitous in scholarly and public discourse on medicine; see Erwin H. Ackerknecht, *A Short History of Medicine* (New York: Ronald, 1955), 146, 170; Nicholas D. Jewson, "The Disappearance of the Sick-Man from Medical Cosmology," *Sociology* 10 (1974): 369–85; John V. Pickstone, *Ways of Knowing: A New History of Science, Technology, and Medicine* (Chicago: University of Chicago Press, 2001).

53. See Taub, chapter 4 in this volume.

54. Leonard T. Kurland and Craig A. Molgaard, "The Patient Record in Epidemiology," *Scientific American* 245, no. 4 (1981): 46–55.

55. See http://casereports.bmj.com/, accessed June 2013.

56. See Helmut Zedelmaier, *Bibliotheca Universalis und Bibliotheca Selecta: Das Problem der Ordnung des gelehrten Wissens in der Frühen Neuzeit* (Cologne: Böhlau, 1992); Blair, *Too Much to Know.*

57. Trisha Greenhalgh, *How To Read a Paper: The Basics of Evidence-Based Medicine*, 5th ed. (Chichester: Wiley-Blackwell, 2014).

58. See http://www.cochrane.org/cochrane-reviews, accessed June 2013.

59. Katie Hafner, "Doctor vs. Computer for the Right Diagnosis," *International Herald Tribune*, 5 December 2012; Eta S. Berner, ed. *Clinical Decision Support Systems* (New York: Springer, 2007); Amit X. Garg et al., "Effects of Computerized Clinical Decision Support Systems on Practitioner Performance and Patient Outcomes: A Systematic Review," *Journal of the American Medical Association* 293 (2005): 1223–38.

60. Stephen J. Greenberg and Patricia E. Gallagher, "The Great Contribution: Index Medicus, Index-Catalogue, and IndexCat," *Journal of the Medical Library Association* 97 (2009): 108–13; M. H. Coletti and H. L. Bleich, "Medical Subject Headings Used to Search the Biomedical Literature," *Journal of the American Medical Informatics Association* 8 (2001): 317–23.

61. On the Royal Society Catalogue, see Csiszar, "Seriality and the Search for Order."

PART II

Spanning the Centuries: Archives from Ancient to Modern

FOUR

Archiving Scientific Ideas in Greco-Roman Antiquity

Liba Taub

As Lorraine Daston has highlighted in her Introduction to this volume, some of the natural and human sciences are dependent on collections of data and/or objects in order to pursue research, and to ensure the possibility of research in the future. Ancient Greeks and Romans confronted the need to select information from a huge corpus and developed various means to store this selected information with a view to subsequent retrieval. Indeed, there were a number of seemingly special textual formats that were used to store, organize, and permit eventual retrieval of various sorts of data and information. Here, I argue that for Greeks and Romans interested in studying topics in a range of fields—including physics, astronomy, mathematics, and medicine—certain types of texts provided an archival function, allowing the accumulation, organization, and use of data, information, and ideas. These texts provided a means for storing information with a view to future retrieval for later use.

For example, we have the records of astronomical data preserved in Ptolemy's *Handy Tables*.[1] Pliny the Elder presented another sort of "table," a table of contents, as the *summarium* in book 1 of the *Natural History* can be understood and used. There, at the beginning of his work, Pliny exhaustively lists the topics covered.[2] Both "tables"—in very different formats—provided the means to preserve, organize, transmit, and retrieve information important for doing scientific work (astronomy, astrology, and agriculture, as well as other endeavors). The types of texts these "tables" represent persist into our own time. Indeed, they are regarded as crucial components of many sorts of texts, not only the scientific.

The *Oxford English Dictionary* describes the noun "archive" as coming from the French *archif* or *archive*, which in turn derives from late Latin *archīum*, *archīvum*, itself coming from the Greek ἀρχεῖον, referring to magisterial residence or public office (where ἀρχή may refer—among other things—to "government"). Nevertheless, in this chapter, I am deliberately not using the term "archive" to refer to anything related to a political entity.[3] But the range of meanings recognized in the *OED* is now somewhat dated, as Oxford University Press itself acknowledges on the website for the online *OED*: "This entry has not yet been fully updated (first published 1989)." This is significant, because in the past thirty-odd years, the use of the word "archiving" as a name for an activity (recognized as a noun by the *OED*) has gained currency in popular as well as specialist circles; the earliest citation of "archiving" as an activity is in 1978 in *Nature* 328, no. 1 (November 23): "Most of these data . . . should be provided by satellites within the next few years but their processing and archiving will require considerable additional effort."[4] For the types of texts considered here, the archiving activity—storage with the view of retrieval for reuse and possible reconfiguration—is key; as distinct types of texts, their formats significantly relate to and contribute to their functionality.[5] That the information that has been archived in these texts is intended for later use is crucial: the point of this archiving is not simply to preserve and accumulate, but to provide resources that can be accessed and used for future intellectual work.

Archiving, retrieving, reconfiguring, transformation: these are the activities facilitated by the types of texts I consider here. I am concerned with the different ways that the same information (including data and ideas) can be reconfigured and transformed, through actively working with the contents of the archival text and subsequently contributing to the creation of a new text, providing a new expression of the material previously held in the archiving text. I am not here focusing on different interpretations of some material but on different, distinctive uses of that material. So how did ancient Greeks and Romans archive intellectual material for future use?

Aristotle has been credited as having assembled "the first large private library."[6] This library was, presumably, based on the acquisition of a great number of papyrus rolls. We know from our knowledge of his scientific work (for example, *History of Animals* and *Meteorology*) that he was adept at collecting data, a task in which he made use of the contributions of others. For example, together with members of his school, the Lyceum, he was responsible for assembling a collection of the constitutions of city-states, which informed his work on politics, the *Constitution of Athens* (*Athenaion*

Politeia). Aristotle was not collecting simply for its own sake, but with the intention of using what had been collected. This is an important clue that he valued the preservation of information with an eye to its future retrieval. In addition to accumulating information, he wanted to exert intellectual control over it, and to use it.

Some types of texts appear to have been intended to make it relatively easy to preserve, organize, and access specific information. The types of texts or genres[7] that may have served archiving functions include *hypomnemata* (notes, memoranda, or aides-mémoire) and question-and-answer texts (such as *problemata* [problems] as well as *zetemata*).[8] Each of these genres was used to collate and organize information on literary and ethical subjects, as well as scientific and mathematical topics. Another type of text, the so-called doxographical texts that preserve opinions (*doxai*) on various topics, will be the primary type of archiving text discussed here. Often, but not always, the opinions were attributed to particular individuals, or groups (for example, the Stoics); in some cases, the opinion is anonymized. The collection of opinions was one of the major ways that ancient Greeks and Romans created archives of ideas. The mining metaphor used today with regard to data mining is apt to describe the purposes of these ancient archiving genres; their intended use is for the retrieval and recovery of ideas, to be contemplated and further used. Of course, text mining can occur even when the text being mined is not in an archival genre, but it is not as easy to accomplish; the desire for easy retrieval—facilitated by different formats and organizing principles—will have contributed to the development of types of texts that facilitated such use. With the development and use of these archiving genres, there is here an implicit view that retrieval of the contributions of others and from the past is a worthwhile activity; these genres act as tools for pointing to topics on which attention should be focused.[9]

In what follows, I explore differences in the ways in which information was stored or archived by looking at some examples of texts produced by authors who were especially interested in gathering the opinions of others. I will be offering three key examples of such gatherers and users of others' opinions on a range of topics, Aristotle, Theophrastus, and Galen. I argue that doxographical texts served as archives for storing opinions and for making these available to other users for their own, sometimes divergent, purposes.

In considering ancient written formats that provided different ways of organizing information, Aristotle himself made it clear that he was working within a culture of the "book," a culture of both reading and writing. He suggests (*Topics* I.14) that "we should select also from the written handbooks of argument, and should draw up sketch-lists [*diagraphai*; Smith trans.=

"tables"; cf. *Eudemian Ethics* II.2, 1228a28] of them upon each several kind of subject, putting them down under separate headings." In other words, the reader produces his own text, in a format different from the one he is reading; however, the subsidiary text is based in crucial ways on the one being read. We can imagine ordered lists, or tables, as Robin Smith suggests, organized by headings, such as "good" or "animal." Aristotle recommends also that "in the margin, too, one should indicate also the opinions of individual thinkers, e.g. that Empedocles said that the elements of bodies were four." The lists or tables allow the classification of various opinions by subject, thereby allowing ease of reference when required, for example, to be used as a starting point in dialectic. This listing of opinions in tables, organized by subject, can be understood as archiving, enabling the accumulation, reconfiguration, and reuse of information—in this case, what others thought about a number of specific topics.[10]

The sort of sketch-list or table recommended by Aristotle is only one of several types of texts that appear to have been developed and used by ancient scholars to help organize and preserve information for future use in natural philosophy, medicine, and other subjects. In ancient texts we find references to and traces of what might be regarded as several different genres—which might be thought of as technologies—for organizing and storing knowledge, ideas, research questions, data, and various bits of information, as well as observations. Examples of some of these genres survive as extant texts; in other cases, we have only traces. A sort of archaeology of texts can help explore the types and genres of texts used for archival purposes in Greco-Roman Antiquity; the focus here will be on those concerned with natural philosophical and medical subjects. A somewhat similar archaeological approach was advocated by Jocelyn Penny Small, in her *Wax Tablets of the Mind: Cognitive Studies in Memory and Literacy in Classical Antiquity*. She describes her approach as being to "look at what was produced and try to figure out why it is the way it is and what that implies for tasks that it was used for."[11] Here, I will consider why and for what purpose ancient doxographical texts—lists or collections of opinions—were useful.

COLLECTING OPINIONS

Aristotle, for example, did not collect the views of others simply to preserve them; he used his collection of opinions to order and systematize these views as a first step in articulating his own. Other ancient authors, including Herodotus, seem to have had a somewhat similar aim, that is, listing the opinions of others' "wrong" ideas to set the stage for offering his own "correct" view.[12] But both Herodotus and Aristotle may have had historical motives as well, as did, for example, Diogenes Laertius. Galen also presented the views of predecessors, but wished to compare them, aiming to show their agreement. He regarded his demonstration of the similarity of the ideas of his great predecessors as being part of his own contribution to learning. Some ancient thinkers, including Theophrastus as well as several unknown authors, produced texts that are essentially a listing of the opinions of natural philosophers and physicians, typically organized by topic, but sometimes by individual or school.

In Antiquity, a variety of writings incorporated the opinions (*doxai*) of others. For example, accounts of the lives of the ancient philosophers usually reported their ideas and opinions; Diogenes Laertius' *Lives of the Eminent Philosophers* (3rd c. CE) is an important example. Collecting and recording the opinions of earlier thinkers may have begun in the fifth century BCE. In his discussion of the cause of the flooding of the Nile, Herodotus (5th c. BCE; book 2, 20–26) details the views of three unnamed but identifiable Greek thinkers, who he said wished to be regarded as clever.[13] Herodotus criticizes and dismisses their views, then offers his own. There is some evidence that Hippias of Elis (late 5th c. BCE) recorded the views of philosophers and organized them with a view to comparing them, indicating disagreements.[14] Plato's dialogue *Theaetetus* (152e, 180e) shows traces of such organization, contrasting the views of Heraclitus (535–475 BCE) and Parmenides (active earlier part of 5th c. BCE). Nevertheless, David Runia has argued that "what takes place in the fourth century indubitably involves a quantum leap. For the first time the earlier history of thought is systematically integrated into the practice of philosophy and science."[15] One motivation for collecting opinions appears to have been to produce intellectual histories, in some cases to enable authors and their readers to identify with other thinkers as well as particular philosophical schools. Furthermore, the reporting of opinions as part of an intellectual history is not limited to a single genre. Diogenes Laertius, in his *Lives of Eminent Philosophers*, reported the views of his subjects, seeing them as central to his biographical accounts of individuals.[16] He provided a record and account of ideas as part of the compilation of his intellectual history. To some extent, Seneca (4 BCE–65 CE), in an earlier period, may have been doing something similar in his *Natural Questions*, a text with a completely different format.[17] Another possible motivation for collecting opinions, which some scholars trace back to Aristotle, is for use as starting points in dialectic. Whatever the eventual use of opinions by ancient philosophers, historians, or biographers, scholars have identified a group of texts which listed—archived—opinions for later mining and use, in a variety of ways for a variety of purposes. These texts organized and archived opinions and ideas, but not other sorts of information, such as the biographical details, which were sometimes included in other texts, such as Diogenes Laertius' *Lives*. The embeddedness of the reporting of *doxai* within the broader tradition of ancient philosophical (including natural philosophical) literature is an important characteristic of that literature. Of course, there are texts that report the opinions of others, such as Plato's dialogues and Aristotle's various works, that are clearly not designed for easy mining. Those archiving texts which aim to preserve and specially organize opinions for later use may be understood to be "doxographies"; the texts that use the opinions in various ways may be described as "doxographical." However, the terminology (which is itself modern) is not always consistently applied by scholars.

A body of writings that were collections of opinions had traditionally been described by classicists as the "*placita* literature" (*doxai* or *placita* = "what it pleases someone to think"); one prominent example of the "genre" was the second-century CE work falsely attributed to Plutarch known as *Five Books on the Placita of the Philosophers Concerning Physical Doctrines* (*Placita philosophorum*). It is in the last decades of the nineteenth century that the terms "doxography" and "doxographer" appear in the scholarly literature, following the publication in 1879 of *Doxographi Graeci* (Greek doxographers) by the German classicist and historian of ancient philosophy and technology Hermann Diels (1848–1922).[18] There were no such terms in Antiquity, nor does the term "doxography" refer to a category of writing explicitly referred to by the ancients. It is not entirely clear how Diels intended the notion of doxography to extend beyond this literature.[19]

Runia has argued that "in coining the word [doxography] Diels appeared to suggest that it was possible to identify a group of writers who applied themselves to a particular kind of writing, just as *historiographi* and *comoediographi* . . . did"; he pointed to the adoption of another neologism, "biographer," as analogous.[20] But even though the term "doxographer" was not in use in Antiquity, the term *doxa* (opinion) certainly was, and there are numerous writings that are evidence of a deliberate and self-conscious aim to collect and report (and sometimes criticize) the *doxai* of others. The collection of *doxai* was undertaken by writers interested in natural philosophy and medicine, as well as (perhaps to a lesser degree) ethics.[21] Aristotle often prefaced the presentation of his own ideas with a synopsis of the views of his predecessors.[22] Theophrastus, presumably, dedicated an entire work to an account of the ideas of Aristotle's predecessors.[23]

Indeed, Diels argued, in the prolegomena to the *Doxographi Graeci*, that it was a lost work by Theophrastus that provided the material contained in the *placita* texts; in other words, according to Diels, the *placita* texts are excerpts from a work by Theophrastus that we no longer have. The exact nature of this work has been a topic of extended debate. In his *Lives of Eminent Philosophers*, Diogenes Laertius preserves the title as *Physikôn Doxôn*, in sixteen books. This title has been variously interpreted to mean either "Tenets of the Natural Philosophers" (*Physikōn Doxōn*) or "Tenets in Natural Philosophy" (*Physikai Doxai*). Diels preferred the first meaning, emphasizing the role of individual thinkers, but some scholars now opt for the second meaning, focusing on the tenets, ideas, or opinions themselves.[24]

In the *Doxographi Graeci*, Diels set out to reconstruct another lost work, a handbook of opinions (*doxai* or *placita*) that he ascribed to Aëtius (first or second century CE);[25] the fifth-century Christian writer Theodoret is the only extant source to mention him by name.[26] Diels further argued that Aëtius had not drawn directly from Theophrastus' collection of *doxai*, but from an intermediate epitome that he named the *Vetusta Placita*, traces of which may be found in other authors, including Cicero (*Academica priora* II, 37, 118); modern scholars have tended to question this latter hypothesis.[27]

IS DOXOGRAPHY A GENRE?

In recent years, a number of scholars have been reconsidering Diels' work. There is no unanimity among historians in the application of "doxography" and "doxographical text." Indeed, "doxographical" is widely used today by historians of ancient philosophy, and application of the term varies a good deal, being attached to texts that do not always share the same format. Jaap Mansfeld has argued that Diels' original sense of doxography applied only to texts that dealt with opinions regarding physical philosophy, noting that historians now use the term to include ethical philosophy. He also suggests that we may distinguish a narrow sense of doxography, relating to the *placita* literature, from one that is more broad, reflected in various philosophical texts; however, even his own classification of specific texts is not hard and fast.[28]

While the general sense is that doxographical texts record the opinions (*doxai*) of others, there are many types of texts (including, for example, biographies) that record opinions. Doxographical texts, broadly understood, do not constitute a separate genre.[29] If we define genres according to form as well as function, we must acknowledge that a variety of textual formats have been described as serving a doxographical function.[30] Furthermore, there are arguably not many extant examples of "doxography" in the narrow sense, if in fact it can be thought properly to be a genre in itself. Nevertheless, there are many ancient writings on scientific and medical subjects that are "doxographical," that is, that deliberately incorporate and, in some cases, discuss or elaborate on the opinions of others. The varied character of doxographical literature, broadly understood, with regard both to form as well as to content if not function, must be emphasized. Reports and collections of the opinions of philosophers appeared in a number of formats. Furthermore, in some cases it is possible to find multiple versions of *doxai*, some shorter, others more lengthy.[31] Philip van der Eijk has pointed out that doxography may have served a "variety of purposes, especially when it is incorporated into a wider framework and exploited . . . for literary or rhetorical purposes."[32]

Mansfeld has suggested that it may be useful to classify the overviews in Cicero (such as that found in *On the Nature of the Gods*) and Diogenes Laërtius' *Lives of Eminent Philosophers* as "belonging with an ancient genre which we may view as a sub-species of doxography, namely the (largely lost) literature *Peri Haireseôn* ('On Schools'), which deals with philosophical, or medical, schools and eventually may include arguments contra the position of a particular *hairesis*."[33] Another possible subgenre may be bio-doxographical writings, organized by individuals. Without attempting a taxonomy of writings that might be considered to be doxographical, we must nevertheless recognize that the label "doxographical" is applied to a variety of works, some of which may be regarded as separate genres, or subgenres. Indeed, the overlap and fluidity of such genre distinctions cannot be overemphasized. As has already been mentioned, Diogenes Laertius' *Lives* may be regarded as a collection of biographies,[34] as well as a contribution to intellectual history. This, too, signals the multiple functions of many of these texts. That "doxography" and "biography" are both neologisms should signal the potential pitfalls in attempting too rigid a classification.[35] Crucially for our concerns, the creation of "doxographies" (understood narrowly as lists of opinions) is a practice that has several possible uses, all of which rely on and demonstrate the "archival" nature of such texts. Understanding doxography collections as a form of archive makes it easier to understand how these works could have been exploited for various purposes.

THE FUNCTIONS OF DOXOGRAPHY: DOXOGRAPHY AS ARCHIVE

The method of dialectic in Aristotelian philosophy has been understood by some scholars to have been an important link to ancient doxographical texts. Others have argued that doxographical texts should be seen as part of a broader historical tradition, which includes biographical accounts.[36] By either reading, the function of doxographical texts to preserve and permit the retrieval of the opinions of predecessors, colleagues, and others—for any number of purposes, including doing philosophy as well as intellectual history—appears to be undisputed. We may then understand doxographical texts as themselves serving an archival function. Doxographical texts of various sorts (which offer opinions of others) may have themselves been composed with reference to other doxographical texts; this would serve as an example of the archive being reconfigured to create another archive, perhaps even being composed in a different format, such as the list or table recommended by Aristotle (*Topics* I.14, see above), but one that nevertheless serves as a means to store and retrieve information for later usage.

Within the framework of natural philosophical thought, doxography has been understood by some as an important method rather than a genre, related to other philosophical methods (and genres). Recognizing that doxographical writings (or writings that incorporate *doxai*) took many forms, it may be more helpful to focus on doxography as a tool or technique, rather than as a genre; it may even have been a project or end in itself.[37] Doxography enabled the archiving (that is, preservation, organization, storage, and retrieval) of information in the form of opinions on a variety of topics, including natural philosophy, mathematics, and medicine.

Even in Antiquity, authors relied on secondary sources, including summaries of the works of earlier thinkers, in handbooks and epitomes produced by others. So, for example, Diogenes Laertius proudly names his sources—well over two hundred; Pliny the Elder also boasted about the number of sources he had consulted.[38] And both Diogenes Laertius and Pliny produced secondary sources. Doxographies served their ancient, as well as their modern, readers as a sort of secondary literature; as such they often served as a tool or archive to be used to aid further work.[39]

USERS OF DOXOGRAPHY

With the notion of "doxography as archive" in mind, it will be useful to consider a number of examples of ancient authors who took the preservation, organization, and critical transmission of others' ideas very seriously, including Aristotle (384–322 BCE) and his colleague Theophrastus (371–287 BCE). A work that may be related to works of Aristotle and Theophrastus, the pseudo-Plutarchian *Placita Philosophorum* (*Opinions of the Philosophers*), will also be considered briefly. The Alexandrian physician Galen (second century CE) produced a work known as *On the Doctrines of Hippocrates and Plato* (also known as the *De placitis Hippocratis et Platonis* [*PHP*]). He used the views of great ancient thinkers (whom he referred to "the ancients" = *hoi palaioi*) as a starting point for critical discussion and investigation. His work indicates his reliance on some of the earlier *placita* literature, including that of an earlier doxographer particularly concerned with medical matters. Further, his work indicates an interesting divergence from Aristotle.[40]

Aristotle and Doxography

In many of his philosophical treatises, Aristotle deliberately relates his ideas to those of other thinkers; it was an essential part of his philosophical method to survey the opinions of others. Aristotle's focus on opinions might well be described as doxographical, but the reporting of opinions was not Aristotle's primary aim. Rather, doxography was a tool to be used in philosophical investigation. This might be in contrast, but not in conflict, with the reporting of opinions as part of the job of the intellectual historian; Aristotle was operating as a historian at least some of the time.[41]

The consideration of the opinions of others was an essential part of Aristotle's philosophical methodology. Aristotle emphasized the use of reputable opinion (*endoxa*) as a starting point for research, as part of his dialectical method, one of the types of deductive reasoning he advocated.[42] At the very beginning of the *Topics*, Aristotle defined dialectic as "the method by which we shall be able to reason from generally accepted opinions (*endoxa*) about any problem."[43] For Aristotle, dialectic, which proceeds from the starting point of *endoxa*, is to be contrasted with the strict demonstrations of the *Posterior Analytics*, which proceed from self-evident, indemonstrable premises and aim at demonstrative certainty. For this reason, knowledge and control of *endoxa* are crucial.

At the beginning of his philosophical investigations, Aristotle characteristically reports and discusses the opinions (*doxai*) of people in general and of experts (often philosophers) in particular, concerning issues in metaphysics and physics and other areas of interest.[44] Aristotle makes it clear (*Topics* I.2) that dialectic is particularly useful for certain purposes: intellectual training, casual encounters, and the philosophical sciences. He emphasizes the usefulness of learning to deal with others' opinions as part of ordinary life (101a30–34):

> For purposes of casual encounters, it is useful because when we have counted up the opinions held by most people, we shall meet them on the ground not of other people's convictions but of their own, shifting the ground of any argument that they appear to us to state unsoundly.

Within the philosophical sciences, however, he explains that the starting point will be working specifically from opinions that are reputable (*endoxa*), for "it is through reputable opinions about [the principles of the philosophical sciences] that these have to be discussed" (101a36–101b3).[45]

Those views regarded as worth considering are known as *endoxa*, defined in the *Topics* (100b21–24) as "what seems so to all or the majority or the wise." The plural noun *endoxa* can be understood as "reputable opinions"; the adjectival form *endoxos* may carry the meaning "reputable" or "respectable." Robin Smith translates *endoxos* as "acceptable," emphasizing that the appellation is a relative term: "a proposition is *endoxos* with respect to some definite group of persons, whether it be the public generally, or the community of experts, or someone famous."[46] The use of "acceptable premises" as a starting point for dialectic has a philosophical and logical function for Aristotle; furthermore, the sometimes silent or anonymous attribution of "acceptable premises" by Aristotle provides an active link to other thinkers, including his predecessors.

In this way, doing dialectic forces one to engage with other, even older, ideas. This philosophical method was, for Aristotle, to some extent grounded in a familiarity with earlier thinkers and their ideas. One important way of engaging with these thinkers and ideas was through their writings. In much of his discussion of dialectic, Aristotle seems to be operating within the framework of oral argument, especially when one remembers that dialectical argument may be useful for *casual* encounters. But we know that he was working within, and made significant contributions in many ways—not only through writing, but also in assembling a library—to the growing culture of the book.[47]

It should also be noted that Aristotle's philosophical method uses *endoxa*; doxography is based on *doxai*, which are not necessarily *endoxa*. Zhmud has suggested that, for this reason, we should not assume a dialectical origin for doxography. He suggests, instead, that the origin may be historical, and points to the example of Herodotus' discussion of the Nile, written at least eighty years before the *Topics*. Aristotle reports the same views as did Herodotus (including Herodotus' own) in his *On the Rising of the Nile* (*De inundation Nili*). Was Herodotus the source Aristotle used for these views of his predecessors?[48] Furthermore, just as Herodotus and even Aristotle report some *doxai* that they regard as wrong, silly, or even childish, so do other authors who relate the opinions of others, often with a view of refuting them (consider, for example, Seneca in the *Natural Questions* and Ptolemy in the opening chapters of the *Almagest*). At times Seneca explains that he is recording certain opinions for posterity, but both Seneca and Ptolemy present the views of others against whom they argue, in part, it seems, to gain a rhetorical advantage.

A fundamental tactic of argument in Aristotle's works on physics, including the *Physics*, *On the Heavens*, and the *Meteorology*, is to survey the views of others, especially "the wise." Cynthia Freeland argues persuasively that Aristotle's discussion of the *endoxa* presents a picture of natural philosophy as a "problem-solving activity," in which important questions and problems are highlighted. Furthermore, questions and problems arise from the process of critical examination of the *endoxa*. So, for example, Aristotle (*Meteorology* 348a14–30) rejects Anaxagoras' explanation of hail, in which clouds are forced by the summer heat to the upper region, where the water in them freezes. He cites evidence, based on observations, to refute the explanation: Aristotle points out that hail falls from clouds close to the earth, rather than from far above, so contradicting Anaxagoras' claim. Additionally, careful consideration of *endoxa* may actually motivate a program of observation and data collection, in order to test and, if necessary, contest the views of others.[49] On some level, certainly, *doxai* can be distinguished from data, but sometimes opinions almost serve as proxies for data, when none is available.[50]

When observations and other data are available—along with the *endoxa* themselves—they need to be "archived" in order to be accessible for interpretation and theory-building. How have these observations been collected and archived? To some extent, through doxographical philosophical and/or historical texts that record and report—and in some cases deliberately gather and organize—this information, in some cases in lists and tables; the information may then, in turn, be used in composing philosophical texts. For Aristotle, the intellectual archive of doxography—that is, of doxographical texts—was meant to be used to think with.

Theophrastus and Doxography

Theophrastus' now-lost collection of *doxai* may have served more than one function, for Theophrastus himself and for his colleagues, students. and later readers. Runia has noted that some evidence suggests that in his work Theophrastus "appears to have added his own critical judgements, evaluating the views he recorded in relation to the norm of Peripatetic doctrine."[51] As Zhmud has pointed out, this is exactly what Herodotus had done earlier, in presenting others' views on the Nile, followed by his own. This may suggest that Theophrastus was following the model of Aristotle, giving the opinions of others, and then concluding with his own "correct," considered view; such a model has been associated by some historians with Peripatetic philosophical practices. Zhmud has suggested that the collection of *doxai* "was mainly of historical interest, showing the difficult path to the truth that was finally revealed in Aristotle's physical teaching."[52]

Mansfeld has argued that Theophrastus' work "was a huge collection of materials to be used in dialectical and/or scientific discussion."[53] Such a collection can be regarded as an archive for doing natural philosophy. Mansfeld's understanding of the work reinforces this interpretation, as he suggests that "the best working hypothesis available at the moment is that Theophrastus set about collecting the materials in a systematic and complete way and made this collection available to colleagues and pupils." He sees this as having been rather different from the practice of Aristotle, whom he regards as having "compiled an overview of tenets whenever the need arose in the course of a scientific discussion (he may of course have made such overviews for his private use . . .)."[54] By this reading, Theophrastus' collection of material would qualify as an archive, deliberately assembled and preserved for use by others, for a number of possible purposes.

The texts referred to as the *placita* literature may have had a derivative relationship to Theophrastus' (and even Aristotle's) work.[55] Some scholars have understood them to be concerned with identifying topics for use in the dialectic, including Mansfeld, who has also suggested that they were used more widely. By his account, collections of *doxai*, of the type attributed to Aëtius, "offered a frame of reference and enabled philosophers or scientists to provide an overview of and arguments against those views they wanted to discuss." Moreover, as he noted, "various motives could be involved: rejection, appropriation, revision, supplementation or complete replacement." In other words, collections of *doxai* were employed in different ways, and for different ends.

As noted above, the pseudo-Plutarchian *Placita Philosophorum* (*Opinions of the Philosophers*) is normally regarded as an epitome of an earlier doxographical work, produced by Aëtius; at least, that was Diels' working hypothesis. It is organized by subject matter or question and shares a good deal in terms of organization with question-and-answer texts, such as the pseudo-Aristotelian *Problems*. A brief look at the pseudo-Plutarchian work gives an impression of how such *placita* texts were organized. Following a short introduction, chapter 1 addressed the question "what is nature?," chapter 2 "what is the difference between a principle and an element?," chapter 3 "of principles and what they are," and chapter 4 "how was this world composed in what order and after what manner is it?"[56] To provide a further glimpse of the organization, here is chapter 9, dealing with the question "of matter":

Matter is that first being which is substrate for generation, corruption, and all other alterations.

The disciples of Thales and Pythagoras, with the Stoics, are of opinion that matter is changeable, mutable, convertible, and sliding through all things.

The followers of Democritus aver that the vacuum, the atom, and the incorporeal substance are the first beings, and not obnoxious to passions.

Aristotle and Plato affirm that matter is of that species which is corporeal, void of any form, species, figure, and quality, but apt to receive all forms, that she may be the nurse, the mother, and origin of all other beings. But they that do say that water, earth, air, and fire are matter do likewise say that matter cannot be without form, but conclude it is a body; but they that say that individual particles and atoms are matter do say that matter is without form.[57]

Notably, and in contrast to the question-and-answer texts, the opinions (unlike the "answers") normally name individuals associated with specific views. Each section treats a particular topic, giving the opinions of different thinkers. Such collections of *doxai* would have been useful for readers with various needs, including the historical as well as the philosophical. Indeed, there is evidence—including a title attributed again to Theophrastus by Diogenes Laertius (5.48)—that collections of *doxai* were further epitomized. So, Theophrastus is credited with having produced a work entitled *Physical Opinions* in sixteen books and another entitled *Epitome of Physical Opinions* in a single book. That readers made use of such collections of *doxai* in presenting and defending their own points of view is clear from the range of ancient writings that reproduced the *doxai* in various formats.

Galen and Doxography

Galen's *On the Doctrines of Hippocrates and Plato* is a rich example of a doxographical text. And, as Mario Vegetti has pointed out, the work has been more often mined for doxographical information than studied in its own right.[58] That Galen was concerned with the ideas and opinions of his predecessors is clear from the title of the work, which is found in other writings by Galen that make specific reference to his *On the Doctrines of Hippocrates and Plato*.[59] His purpose in the work is to examine his predecessors' ideas and provide a basis for these *doxai* to serve in argument. This is, on the face of it, a similar purpose to the use of *endoxa* in Aristotelian dialectic. However, strictly speaking, Galen differs from Aristotle in his methodological view of the role of *doxai* in argument, in that he regards them as rhetorical rather than dialectical. However, in considering the archival function of doxographical texts, this difference is not significant. Vegetti suggests that traditional doxography serves as "a repository of dialectical and rhetorical arguments that can be used in discussion and in the encounter between rival theories." This doxographical "repository" serves as an archive for further work in a number of fields, including natural philosophy and medicine.[60]

R. J. Hankinson has argued that Galen, unlike some ancient thinkers, thought it possible to make progress in science, particularly medicine. Galen believed that the doctrines of past masters, particularly Hippocrates, Plato, and Aristotle, should be taken as a starting point, examined, and then built upon.[61] He treats the agreement of opinions as a conceptual starting point—a purified kind of *endoxa* marked by clarity and obviousness.[62] For example, in book VIII, he presents an account of the views of Hippocrates and Plato on the elements, quoting extensively from both and pointing to their agreement.[63]

Galen's ambition may have been to transform the *doxai* of the great ancients, Hippocrates, Plato, and Aristotle, into syllogisms, as appropriate. In *On the Doctrines of Hippocrates and Plato* he certainly focuses on opinions as starting points. He seeks to restate and clarify the views of the ancients, as he understands them. And he is at pains to argue that he does understand the views of Hippocrates and Plato better than others do.[64] He is offering an "enriched" archive.

The comparison of the ideas of Hippocrates and Plato provides a starting point and a rationale for *On the Doctrines of Hippocrates and Plato* as a whole, but the status of these *doxai* within Galen's work is somewhat different from that of *doxai* reported, on the one hand, by Aristotle and, on the other, by the *placita* literature. Galen makes it clear that he does not expect readers to simply read *On the Doctrines of Hippocrates and Plato* and be done; it is only one of several resources available. He indicates that *PHP* is not meant to be read in isolation.[65] While part of the rationale for quoting at length from Plato is Galen's wish to make life easier for his readers, he also refers them to his other works, including his work *On Medical Terms*, in which he discussed the term "element."[66] Here we get a strong sense of the culture of the book, in which doxographical works stand beside other types of texts, including commentaries. Like the *placita* literature, *On the Doctrines of Hippocrates and Plato* is not meant to be used entirely on its own; it forms part of a larger body of discourse, and was intended to be used in conjunction with other works, including others that were in the process of composition.

CONCLUDING REMARKS

Doxography (the reporting, preserving, and "archiving" of opinions for use or "mining" by others) has links to other genres of Greek and Roman discourse, particularly the dialogue and the so-called problem texts. The links may be, to some extent, related to argumentative and investigative methods.[67] For example, the questions and answers of the *problemata* texts are in some ways reminiscent of the dialogic form, minus the literary setting. Certainly, they hint at oral exchange. They also share something of the character of the written lists associated with doxographies (particularly in the narrow sense). Arguably, some of the question texts are intended to be doxographical. For example, the presentation of Plutarch's *Platonic Questions* is similar to Aristotle's handling of the opinions of others. One function of the question-and-answer text may have been to serve as an archive of active and interesting research problems.[68] Furthermore, it is easy to imagine that the questions posed in these texts might well reflect opinions, *endoxa*, and so be related to doxographies, which also had archival functions.

Certain types of ancient texts (for example, doxographies, or collections of the views or opinions of philosophers) provided an archival function for Greeks and Romans studying topics and issues in a range of scientific fields, including natural philosophy and medicine. These types of texts provide special windows on how information was processed, controlled, and mobilized by those engaged in natural philosophy and medicine. Considering the functions of these and related texts can helpfully shape our comprehension of ancient Greek and Roman scholarly and investigative methods.

A particular feature of the doxographical texts is their archiving of already-produced opinions about nature, rather than raw data. As we have already seen, in some cases, as Aristotle and others noted, we have to work with opinions because data is not available. The ancient doxographies—lists of opinions of various people on specific topics—seem to qualify as second natures: the compiler of the doxography has does preliminary sorting and sifting, categorizing and arranging of, presumably, a vast range of possible candidates to be included. What was "indigestible first nature" has become "intelligible second nature," as Daston describes it, even though—in this case—first nature was not empirical data. In any case—and this is what is significant about the second-nature status of doxography-as-archive—the aim of the doxography, as an archiving text, is to allow the philosophical and scientific work of hypothesizing and explaining to proceed. The creation and formation of scientific archives does not follow a set pattern; issues regarding the second-nature character of what is being archived today emerge in others chapters in this volume, including those by Vladimir Janković on the US National Climate Program (chapter 9) and Bruno Strasser on the Protein Data Bank (chapter 7). Notably, it is perhaps only in later periods that ancient doxography becomes a third-nature entity, a repository of second-nature findings that have been selected to endure as archives of philosophical and scientific enterprises, rather than as material to be mined for active scientific projects.[69] Future work might consider the legacy and intellectual consequences of the development and use of ancient archiving texts.

ACKNOWLEDGMENTS

I am grateful to other members of the Working Group, particularly Lorraine Daston and Suzanne Marchand, for their careful reading of and useful suggestions for my work.

NOTES

1. Such texts are being addressed by others, and are not considered here.

2. Aude Doody, *Pliny's Encyclopedia: The Reception of the Natural History* (Cambridge: Cambridge University Press, 2010), chapter 3 (92–131), and Jocelyn Penny Small, *Wax Tablets of the Mind: Cognitive Studies of Memory and Literacy in Classical Antiquity* (London: Routledge, 1997), 16–18, have interesting things to say about Pliny's table of contents.

3. *OED* online accessed 21 September 2014: http://www.oed.com/view/Entry/10416?rskey=WRzjLa&result=1&isAdvanced=false#eid.

4. *OED* online http://www.oed.com/view/Entry/10417?redirectedFrom=archiving&; accessed 17 October 2014.

5. In a stop list, the very process of retrieval is reconfiguration; see Daniel Rosenberg, chapter 11 in this volume.

6. Rudolph Pfeiffer, *History of Classical Scholarship from the Beginning to the End of the Hellenistic Age* (Oxford: Clarendon Press, [1968] 1978), 67; Small, *Wax Tablets*, 43.

7. E. Werlich, *A Text Grammar of English* (Heidelberg: Quelle & Meyer, 1976); I. Taavitsainen, "Changing Conventions of Writing: The Dynamics of Genres, Text Types, and Text Traditions," *European Journal of English Studies* 5, no. 2 (2001): 139–50.

8. See Liba Taub, "'Problematising' the *Problems*: The *Problemata* in Relation to Other Question-and-Answer Texts," in *The Aristotelian* Problemata Physica: *Philosophical and Scientific Investigations*, ed. Robert Mayhew (Leiden: Brill, 2015).

9. How this reflects views regarding invention and novelty is not always clear. On the valorization of discovery in Greek Antiquity, see Leonid Zhmud, *The Origin of the History of Science in Classical Antiquity*, trans. Alexander Chernoglazov (Berlin: De Gruyter, 2006).

10. Robin Smith, *Aristotle Topics,* books 1 and 8, with excerpts from related texts, translated with a commentary (Oxford: Clarendon Press, 1997), xxiii–xxiv. Aristotle, *Topics* 105b12–16, trans. W. A. Pickard-Cambridge, in *The Complete Works of Aristotle,* ed. J. Barnes, 2 vols. (Princeton: Princeton University Press, 1984). Aristotle's instructions indicate the desirability of active involvement by the reader with the text.

11. Small, *Wax Tablets,* 240. I am not certain whether my approach is an example of what Small herself refers to as "anticipatory plagiarism," referring to "those situations when a scholar thought of something totally independently, only to discover later that someone else had already thought of the same thing" (Small, *Wax Tablets,* 245, preface, n1).

12. On nineteenth-century reading of Herodotus, see Suzanne Marchand, chapter 5 in this volume.

13. Cf. Leonid Zhmud, "Revising Doxography: Hermann Diels and His Critics" *Philologus* 145 (2001): 219–43, on 242.

14. I am grateful to Thomas Buchheim for reminding me of Hippias' work in this context. Cf. Jochen Althoff, "Aristoteles als Medizindoxograph," in *Ancient Histories of Medicine: Essays in Medical Doxography and Historiography in Classical Antiquity,* ed. Philip J. van der Eijk (Leiden: Brill, 1999), 57. On Hippias, see J. Mansfeld, "Aristotle, Plato, and the Preplatonic Doxography and Chronography," in *Storiografia e dossografia nella filosofia antica,* ed. G. Cambiano (Turin: Tirrenia Stampatori, 1986, 1ff. (reprinted in J. Mansfeld, *Studies in the Historiography of Greek Philosophy* [Assen: Van Gorcum, 1990], 1ff, 22ff); A. Patzer, *Der Sophist Hippias als Philosophiehistoriker* (Freiburg: Alber, 1986).

15. David T. Runia, "What Is Doxography," in *Ancient Histories of Medicine: Essays in Medical Doxography and Historiography in Classical Antiquity,* ed. Philip J. van der Eijk (Leiden: Brill, 1999), 33–55, 46.

16. Diogenes Laertius tends to provide information about the views of the subjects of his *bioi.* He was particularly interested in recounting the ideas of his subjects, as part of the aim of his work was to provide an intellectual history, a history organized around philosophical "schools" or "successions" (Cf. Liba Taub, "Presenting a 'Life' as a Guide to Living: Ancient Accounts of the Life of Pythagoras," in *The History and Poetics of Scientific Biography,* ed. Thomas Söderqvist (Aldershot: Ashgate, 2007), 17–36.).

17. Cf. Liba Taub, *Ancient Meteorology* (London: Routledge, 2003), 143, 159–61.

18. In preparing his work, Diels had collected and compared evidence found in a number of extant later writings, including the pseudo-Plutarchian text. Diels began his work on the *placita* literature while he was a third-year student working under Hermann Usener at the University of Bonn. He started by investigating the relationship between collections of *placita* including the pseudo-Plutarch *Placita,* along with an anthology compiled by Johannes Stobaeus in the early fifth century CE, and a text attributed to Galen (now pseudo-Galen), the *History of Philosophy* (*De historia philosophica*). In 1877 the Berlin Academy awarded a prize for the best work on pseudo-Plutarch's *Placita* to Diels for his manuscript of *Doxographi Graeci.* Eckart E. Schütrumpf, "Hermann Diels," in *Classical Scholarship: A Biographical Encyclopedia,* ed. Ward W. Briggs and William C. M. Calder III (New York: Garland, 1990), 52–60, 53.

19. Hermann Diels, *Doxographi Graeci* (Berlin: G. Reimer, 1879). See J. Mejer, *Diogenes Laertius and His Hellenistic Background* (Wiesbaden: Franz Steiner Verlag,1978), 81–82, and

J. Mansfeld and D. T. Runia, *Aëtiana: The Method and Intellectual Context of a Doxographer* (Leiden: E. J. Brill, 1997),101ff., on Diels and the invention of these terms. See Mansfeld, "Sources," on the use of the term "doxography," 17–19. See also Jaap Mansfeld, "Physikai doxai and Problemata physica from Aristotle to Aëtius (and Beyond)," in *Theophrastus: His Psychological, Doxographical, and Scientific Writings*, ed. W. W. Fortenbaugh and D. Gutas (New Brunswick: Transaction Publishers, 1992), 63–111; 36; Zhmud, "Revising Doxography."

20. Runia, "What Is Doxography," 35. Runia notes that these latter are ancient terms, while *paroemiographi* (proverb writers/collectors) and *epistolographi* (letter writers) were used by the earlier nineteenth-century scholars von Leutsch-Schneidewin and Hercher; cf. Mansfeld and Runia, *Aetiana*, 102.

Even though doxography was not an ancient category, as Philip J. van der Eijk, "Historical Awareness, Historiography, and Doxography in Greek and Roman Medicine," in *Ancient Histories of Medicine: Essays in Medical Doxography and Historiography in Classical Antiquity*, ed. Philip J. van der Eijk (Leiden: Brill, 1999), 1–33, on 29, has pointed out, "this is not to say that ancient doxographical writers were not applying a set of genre conventions, but these have to be extracted from the evidence first before being used in turn as explanatory factors."

21. But Mansfeld argues that "a doxographical literature in the field of ethics, which as to scale and taxonomy would be even remotely comparable to physical doxography, never existed. Yet one occasionally encounters short lists and overviews of ethical tenets, in some later authors. It is therefore possible that modest doxographical collections of ethical views did circulate, and we may have some evidence concerned with the circulation (and adaptation) of a diaeretical overview of tenets about the End, or Highest Good"; Jaap Mansfeld, "Doxography of Ancient Philosophy," *The Stanford Encyclopedia of Philosophy* (Spring 2004), ed. Edward N. Zalta., http://plato.stanford.edu /archives/sum2012/entries/doxography-ancient/, accessed June 2013. Cf. K. A. Algra, "Chrysippus, Carneades, Cicero: The Ethical *Divisiones* in Cicero's *Lucullus*," in *Assent and Argument: Studies in Cicero's Academic Books: Proceedings of the 7th Symposium Hellenisticum* (Utrecht, 21–25 August 1995), ed. Brad Inwood and Jaap Mansfeld (Leiden: Brill, 1997), 112–13; Zhmud, "Revising Doxography."

22. Mansfeld, "Doxography" (2004). For example, see Aristotle, *Meteorology* 342b25–343a22, where he reviews some of his predecessors' explanations of comets; he then goes on to criticize these views.

23. Cf. Diogenes Laertius, *Lives*, 5: 48. The title is disputed, as is its meaning. Cf. Zhmud, "Revising Doxography."

24. See Zhmud, "Revising Doxography," for a discussion of various interpretations. Mansfeld, "Physikai doxai," esp. 63–67.

25. See also Runia, "Doxography," 35–36.

26. David T. Runia, "The Placita Ascribed to Doctors in Aëtius' Doxography on Physics," in *Ancient Histories of Medicine: Essays in Medical Doxography and Historiography in Classical Antiquity*, ed. Ph. J. van der Eijk (Leiden: Brill, 1999), 189–250, on. 189–96.

27. Diels, *Doxographi Graeci*; John Burnet, *Early Greek Philosophy* (London: A & C Black, 1920). See also G. S. Kirk, "The Sources for Presocratic Philosophy" in *The Presocratic Philosophers*, 2nd ed., ed. G. S. Kirk, J. E. Raven, and M. Schofield (Cambridge:

Cambridge University Press, 1983), 5. Cf. Jaap Mansfeld, "Doxography of Ancient Philosophy," in *The Stanford Encyclopedia of Philosophy* (Summer 2012), ed. Edward N. Zalta., http://plato.stanford.edu/archives/sum2012/entries/doxography-ancient/.

28. Mansfeld, "Doxography" (2012).

29. Cf. Jaap Mansfeld, "Sources," in *The Cambridge History of Hellenistic Philosophy*, ed. Keimpe Algra, Jonathan Barnes, Jaap Mansfeld, and Malcolm Schofield (Cambridge: Cambridge University Press, 1999), 3–30; Zhmud, "Revising Doxography." It is not entirely clear whether doxography is a method (as Mansfeld thinks) or a project.

30. Zhmud, "Revising Doxography," emphasizes that Mansfeld doesn't think that doxography is a genre, but a method. Mansfeld, "Sources," 19, protests against what he regards as misuses of "doxographical."

31. Mansfeld, "Doxography" (2012), sect. 7. While doxographies served to preserve ideas that otherwise might have been lost, it is difficult to judge how accurately authors quoted and reported the views of others. It is clear that those who presented opinions and views of other thinkers had their own motives for doing so. On the doxographical tradition, see Kirk, "Sources," 4–6; see Mansfeld, "Sources," 17–19, for a reappraisal of the use of "doxography," and also Mansfeld and Runia, *Aetiana*, xiii–xiv.

32. Van der Eijk, "Historical Awareness," 23.

33. Mansfeld, "Doxography" (2012), sect. 6. In his view, works by Philodemus and Arius Didymus (on ethics) should also be included in such a broader grouping.

34. We can understand biography to be focused on giving a report of the *bios*. The *bios* ("life"; plural = *bioi*) as an account or celebration of a particular life can be found in a range of ancient writings and may also include a discussion of the opinions (or *doxai*) of the individual being described; such accounts often carried an ethical or religious message.

35. Cf. Taub, "Presenting a 'Life.'"

36. See the work by Mansfeld, Runia, Balthussen, and Zhmud cited *passim*. I have argued elsewhere (Taub, "Presenting a 'Life'") that the one of the functions of ancient biographies of philosophers was to establish intellectual histories and ancestries, useful for self-identification, as well as the categorization of others.

37. Cf. Zhmud, "Revising Doxography," 230; Zhmud has criticized Mansfeld for focusing on doxography as a philosophical method.

38. R. D. Hicks, *Diogenes Laertius Lives of Eminent Philosophers*, 2 vols (Cambridge, MA: Harvard University Press, [1925] 1972), I: xix, refers to Richard Hope's count of about 250 authors; cf. Richard Hope, *The Book of Diogenes Laertius, Its Spirit and Its Method* (New York: Columbia University Press, 1930), 59–60.

39. Mansfeld, "Doxography" (2012), sect. 7; Cf. van der Eijk,"Historical Awareness," 9. The philosophical use by Aristotle has already been noted and will be discussed in further detail below.

40. For more on doxography of views related to medicine, including the doxography credited to the Peripatetic Meno, see *Ancient Histories of Medicine: Essays in Medical Doxography and Historiography in Classical Antiquity*, ed. Philip J. van der Eijk (Leiden: Brill, 1999), in particular the chapters by David Runia, and Daniela Manetti.

41. Mansfeld, "Doxography" (2012); Taub, *Ancient Meteorology,* 93–96. Zhmud, "Revising Doxography," has argued that Aristotle may have done some of his doxographical work in historian mode.

42. Han Baltussen, "Peripatetic Dialectic," in *Theophrastus: His Psychological, Doxo-*

graphical, and Scientific Writings, ed. W. W. Fortenbaugh and D. Gutas (New Brunswick, NJ: Transaction Publishers, 1992), 6 ff.; cf. also Daniela Manetti, "The Role of Doxography," in *Ancient Histories of Medicine: Essays in Medical Doxography and Historiography in Classical Antiquity*, ed. Philip J. van der Eijk (Leiden: Brill, 1999), 95–141, 116n53.

43. See Aristotle, *Topics*, 100a20ff. See also G. E. R. Lloyd, *Aristotelian Explorations* (Cambridge: Cambridge University Press, 1996), 218.

44. Mansfeld, "Doxography" (2012).

45. Trans. W. A. Pickard-Cambridge, in *The Complete Works of Aristotle*, ed. J. Barnes, 2 vols. (Princeton: Princeton University Press, 1984), 1: 168.

46. Smith, *Aristotle* Topics, 42. The translation of this term has been the focus of some debate. See J. Barnes, "Aristotle and the Methods of Ethics," *Revue Internationale de Philosophie* 133–34 (1980): 490–511; Myles Burnyeat, "The Origins of Non-deductive Inference," in *Science and Speculation: Studies in Hellenistic Theory and Practice*, ed. J. Barnes, J. Brunschwig, M. Burnyeat, and M. Schofield (Cambridge: Cambridge University Press, 1982), 197n11, on the importance of translating *endoxos* as "reputable" and "respectable." On the "factual" status of *endoxa*, see K. Pritzl, "Endoxa as Appearances,"*Ancient Philosophy* 14 (1994): 41–51, 44; cf. C. A. Freeland, "Scientific Explanation and Empirical Data in Aristotle's *Meteorology*," *Oxford Studies in Ancient Philosophy* 8 (1990): 67–102, 94 on "reputable facts."

47. Leslie Kurke, *Aesopic Conversations: Popular Tradition, Cultural Dialogue, and the Invention of Greek Prose* (Princeton: Princeton University Press, 2010), for example 1–50, emphasizes the fluidity and permeability of both oral and written traditions within ancient Greek cultural practices valorizing wisdom (*sophia*), including philosophy. While it is outside of the scope of the current study, it would be interesting to examine the practices of those who believed that philosophy should not be confined to writing.

48. Cf. Zhmud, "Revising Doxography," 242. J. Balty-Fontaine, "Pour une edition nouvelle du 'Liber Aristotelis de Inundatione Nili,'" *Chronique d'Egypte* 34 (1959): 95–102; D. Bonneau, "Liber Aristotelis De inundatione Nili," *Etudes de Papyrologie* 9 (1971): 1–33.

49. Aristotle's use of *endoxa* has been much discussed by modern philosophers and historians of philosophy. There is no agreement on the role of *endoxa* for Aristotle; this is, however, a bigger debate than can be addressed here. Freeland, "Scientific Explanation," 76–78, provides a helpful discussion; at 78 she suggests that Aristotle "regards it as important to see where an existing scientific theory fails, because it makes a false prediction about something that one might otherwise not have considered relevant, so might not otherwise have observed."

50. Aristotle makes it clear that on some subjects we really have no data available, nor can we have demonstrable knowledge; he points to the question of whether the universe is eternal as an example (*Topics* I.11). The view that we have insufficient data on some subjects and must therefore resort to opinions or even multiple possible explanations of phenomena was accepted by others, including Theophrastus, as well as Epicurus and Lucretius; cf. Liba Taub, "Cosmology and Meteorology," in *Cambridge Companion to Epicureanism*, ed. James Warren (Cambridge University Press, 2009), 105–24.

The view that opinion may not represent the truth, but may still worth considering goes back at least to Parmenides (c. 515–460 BCE), whose poem, following the introductory proem, was organized into two major sections, the way of truth (*alēthia*), contrasted with the way of opinion (*doxa*). Surely Parmenides' views on *doxa* lie importantly at the root of the background to ancient uses of *doxai*. However, this is too big a topic to contemplate here.

51. Runia, "Doxography," 46, with reference to Diels' comment, *DG*, 103.

52. Zhmud, "Revising Doxography," and Leonid Zhmud, "The Historiographical Project of the Lyceum: The Peripatetic History of Science, Philosophy and Medicine," *Antike Naturwissenschaft und ihre Rezeption* 13 (2003): 109–26, p. 120; see also p. 125.

53. Mansfeld, "Physikai doxai," 67.

54. Mansfeld, "Physikai doxai," 68. Cf. Runia, "Doxography," 46n54. Runia cautions against being too enthusiastic in calling these doxographies.

On Theophrastus and topics, see T. L. Tieleman, *Galen and Chrysippus on the Soul* (Leiden: Brill, 1996) 123ff.; Johannes M. van Ophuijsen, "Where Have All the Topics Gone?," in *Peripatetic Rhetoric after Aristotle*, ed. W. W. Fortenbaugh and D. Mirhady (New Brunswick: Transaction Publishers, 1994), 138.

55. In some cases, all we have is a list of very brief statements and attributions, e.g., *Plutarchi Moralia*, vol. V, fasc. 2, part 1, ed. J. Mau (Leipzig: Teubner, 1971), bk. I, p. 67, bk. II, p. 81. See also Tryggve Göransson, *Albinus, Alcinous, Arius Didymus*, Acta Universitatis Gothoburgensis (Göteborg: 1995), 51–52, on authorship.

56. I don't know when these chapter headings were inserted; it would be worth considering how many were posed as questions.

57. *Plutarch's Morals*, translated from the Greek by several hands, corrected and revised by William W. Goodwin (Boston: Little, Brown, and Company; Cambridge: Press of John Wilson and Son, 1874), 3, Perseus Project, http://data.perseus.org/citations/urn: cts:greekLit:tlg0094.tlg003.perseus-eng1:1.9, accessed June 2, 2013.

58. Mario Vegetti, "Tradition and Truth: Forms of Philosophical-Scientific Historiography in Galen's *De placitis*," in *Ancient Histories of Medicine: Essays in Medical Doxography and Historiography in Classical Antiquity*, ed. Philip J. van der Eijk (Leiden: Brill, 1999), 333–57, 333–34n1, points to the relative lack of interest in this work.

59. Phillip De Lacy, ed. and trans., Galen, *On the Doctrines of Hippocrates and Plato* (Berlin: Akademie-Verlag, 1984), 39.

60. Vegetti, "Tradition," 356–57. He adds that "there is something more to Galen's effort to make tradition 'speak the truth,' an effort that does not neglect any of the most sophisticated tools of textual and conceptual exegesis. To prove the truth of tradition means to cancel the chronological distance between past and present in the timeless homogeneity of theory; it means having Hippocrates, Plato and Aristotle close to oneself, or even going back among them."

61. R. J. Hankinson, "Galen's Concept of Scientific Progress," *Aufstieg und Niedergang der Römischen Welt (ANRW)*, 2, no. 2 (1994): 1775–89, 1779; cf. his n14. Galen thought that one should not accept the doctrines of the greats uncritically, recognizing that not all of them were sound. Further, even in those cases in which they were sound, it is important to know why they were, and to be able to offer a demonstration. Hankinson has noted that "the *antiqui* are presumably Hippocrates, Plato and Aristotle, people whose intelligence and probity enabled them to make genuine advances in science."

62. Tieleman, *Galen*, 125.

63. Galen, *On the Doctrines*, 50.

64. Yet a reading of the *On the Doctrines of Hippocrates and Plato* suggests that, in practice, the transformation of *doxai* into syllogisms was not a simple trajectory; Galen's

treatment of the ideas of Hippocrates and Plato involves a good deal of discussion, exegetical work, and interpretation on his part. As Vegetti has noted, for Galen, development (or progress) "consists in an explanation and development of what was implied. Vegetti, "*Tradition and Truth,* 355; cf. Galen, *De method medendi* 9.8 (10.633 K.). Vegetti adds (355–56n101) that "in Galen the idea of a systematic restoration of the knowledge of the past often takes on the aspect of clarification or explanation of what had remained implicit, abbreviated or obscure." On this topic, cf. J. Mansfeld, *Prolegomena: Questions to Be Settled before the Study of an Author, or a Text* (Leiden: Brill, 1994), 148 ff.

65. De Lacy in Galen, *On the Doctrines*, 48–49, points to two testimonies to the lost portion of book I, indicating that Galen's intention was to include all of the points of agreement between Hippocrates and Plato. Cf. Galen, *On the Doctrines*, I.1, 65 (my emendations):

> But it is not my purpose in this work [treatise] either to assemble all the passages from these men or to interpret them, since I intend to present my interpretations separately in other treatises. My present purpose, as I stated at the outset, is to examine and judge their views only in respect to the question whether or not they agree with each other on every point.

See Katharina Fischer, "Der Begriff der '(wissenschaftlichen) Abhandlung' in der griechischen Antike—eine Untersuchung des Wortes πραγματεία," in *Antike Naturwissenschaft und ihre Rezeption* 23 (Trier: Wissenschaftlicher Verlag, 2013), 93–114, on the use of the word "pragmateia," which I here read as "work."

66. Galen states that he is quoting from Plato at a number of points: , Galen, *On the Doctrines*, 492–93, 494–95 ("Plato himself will explain this to you"), 496–97 ("I shall quote the passages for you"). He makes it clear that "there is no need for me to include proofs of these (elements) here, since I have already written elsewhere a commentary *On the Elements According to Hippocrate*s." If readers desire a full account of the nature of the elements, the *PHP* will not be sufficient on its own to provide that. , Galen, *On the Doctrines*, VIII 10, 493. Galen explains what he set out to do there: "My work *On the Elements according to Hippocrates* is an explication of the [Hippocratic] work *On the Nature of Man*. It does not explicate every word, however, as writers of explications commonly do; rather, it comments only on those statements which give continuity to the doctrine, along the pertinent scientific proofs." Galen, *On the Doctrines*, VIII 12–13, 493. Once again, he advises his readers: "If you wish to learn what they [the proofs] are, I refer you to that book." Galen, *On the Doctrines*, VIII 13, 493.

67. Socrates, as portrayed in Plato's dialogues, had earlier used opinions as the starting points for argument and investigation. Cf. Smith, *Aristotle Topics*, xiii–xiv. In his treatise on dialectic, the *Topics*, Aristotle emphasized the value of opinions as the starting points for argument.

68. Taub, "Problematising."

69. But see Zhmud, "The Historiographical Project," 120, 125.

FIVE

Ancient History in the Age of Archival Research

Suzanne Marchand

One of the rarely remarked-upon oddities of our modern university system is that historians of the ancient world are very often found outside of history departments. Housed in departments of classics or Near Eastern or East Asian studies, they are presumed to have more in common with linguists or archaeologists than with historians proper—and their historiographical traditions and practices are thought to be entirely different from those beloved by garden-variety historians. Ancient historians readily and repeatedly confess that their evidence is more difficult to assess accurately and their conclusions necessarily more tentative than is the case for modern historians; for all of the recent hand-wringing among historians about the impossibility of objectivity, it is still clearly the case that ancient historians foreground the limits of their knowledge much more than do students of the postclassical age (medievalists, I would submit, often fall between the ancient and the modern, with many also admitting forthrightly the limits of their ability to know "the facts"). We have come to see ancient history as in some way less "true" than modern history—but the question is, of course, in what sense it is less true, and when and why we learned to think this way.

It is certainly not true that in the later eighteenth century, most European scholars believed that to be an ancient historian was professionally and intellectually something different from being a modern historian In fact, many of the most popular writers of history in the eighteenth century—including Charles Rollin, Edward Gibbon, William Robertson, and A. H. L. Heeren—focused chiefly on ancient history, and felt quite comfortable crossing the ancient/modern divide; this was, of course, also true of ecclesiastical historians, such as Jacques-Bénigne Bossuet, whose 1681 *Discourse on Universal History* continued to be read (by both admirers and detractors) throughout the next century. Leopold von Ranke himself wrote his (subsequently lost) dissertation on Thucydides, and spent his first few years as a professor lecturing on "general world history" (*allgemeine Weltgeschichte*). It is worth noting, too, that these historians often included in their histories discussions of the Orient as well as the Occident, even though they could read no "oriental" languages apart from (in some cases) rudimentary Hebrew. All of these writers believed that there were lessons to be learned, for the moderns, from studying ancient history, and its long-admired historians; all knew, too, that impartiality and truth were rare commodities in any history, ancient or modern. For none of these men, even for those who saw the nineteenth century dawn, was it self-evident that modern history should be deemed the most "scientific" and the most relevant form of historical study, or that archival use would become the acid test for what made a history most compelling and complete.

Previous historians have regularly, and reasonably, argued that the rise of liberal nationalism after 1789 led scholars and students to devote more attention to modern, national histories. But in this essay I offer an additional reason for modern history's rising prestige and prominence, at least for scholars, and that has to do with the development of new sets of expectations for historical truth, and practices for obtaining it, that fell unevenly on ancient and modern, as well as on European and non-European histories. Leopold von Ranke, in particular, would make the case for the superior truthfulness of modern and national histories over ancient and universal ones in large part *because* the former could be written on the basis of a particular kind of archive, subjected to a particular set of critical skills and criteria for establishing truth. Ranke's championing of the archive also emerged at a time in which the models provided by classical historians began to lose their grip,[1] though we will see throughout the paper the ways in which Herodotus and Thucydides, in particular, were tested against modern findings, and the former, in particular, cast as a naïve child, by contrast to the mature history of the moderns. In the first part of the essay, I revisit historiographical practices and discussions in ancient and universal history before the age of Ranke, lavishing attention on changes underway in the later eighteenth and early nineteenth centuries. I will then seek to understand the rise of archivally driven modern history, and the opening of the gulf between the "ancients" and the "moderns" that today marks our humanist institutions.

As this volume shows, there are many different sorts of archives, and many understandings of what it means to collect, curate, and query them; and in the widest sense of an archive—as a collection of data brought together to resist its being lost to memory—archives seem always to have been part of human culture and statecraft. It would be folly to try to pinpoint the first history written from archival sources, as this would depend entirely on one's definitions of a history and of an archive, as well as on the inferences the interpreter made—with respect to eras before modern footnoting—about how much of the narrative depended only on archived sources. What we can say, however, thanks to the excellent recent work of Markus Friedrich, is that in the wake of the Council of Trent, we begin to see something like institutional and state archives develop out of what had been something more like private archives, used by bureaucrats, lawyers, clergymen, and nobles as part of everyday life in a society ruled by the quest to protect or extend one's traditional privileges. Not surprisingly, then, writers of history—many of them royal officials, clergymen, or nobles— also increasingly drew on these archives to write their stories. Already by the seventeenth century, Friedrich argues, contemporaries had begun to think that using archives improved the quality of historical research.[2] Heavily archival histories and biographies, such as Johannes Müller's history of Switzerland and Angelo Fabroni's life of Lorenzo di Medici, entered the lists long before Ranke's birth in 1795.[3] As the work of Anthony Grafton, Arnaldo Momigliano, Donald Kelley, and Ulrich Muhlack has shown, many of the text-critical tools—including classical and oriental philology, chronology, and the antiquarian study of artifacts such as coins, gemstones, and inscriptions—central to modern "scientific" history had also been pioneered as early as the Italian Renaissance.[4] There was nothing new then in Ranke's commitment to archival sources, and the methods he applied to those sources also owed much more to early modern, and non-German, scholarship than his nineteenth-century fan club was willing to admit. Neither archival work nor source criticism alone made it possible for Ranke to be hailed the "father of scientific history," or in the telling encomium lavished on him by his contemporaries, "the modern Thucydides."

What Ranke does have some claim to have pioneered, however, is the idea that writing history based on archival documents was the *only* way to fully implement Thucydides' injunction (in 1.22), in Ranke's famous paraphrase, to describe history "as it actually happened"; and the corollary that only in modern history would it be possible to carry out such an endeavor. In part this was the case because few people in the later eighteenth century would have thought that modern history could be the exemplary form of the genre. Kaspar Eskildsen even suggests that archival use lost some of its respect in the eighteenth century, as it was seen merely to offer knowledge about particulars, rather than about universals, the true objects of a proper science.[5] Despite its lack of an archive in the early modern European sense, it was still ancient history that provided the richest and most universally valid lessons, perhaps especially in the age of philosophical history, which by definition had to begin at the beginning (either at Adam's creation, or, for secularists, in the "state of nature") and take in the sweep of human events. Sharp-eyed and critical though he was, according to an admiring early nineteenth-century editor, the great eighteenth-century universalist Charles Rollin did not waste his time with the annals of modern history, "so much less fecund in noble remembrances; [Rollin] . . . showed us the human species as it leaves nature's hands, and flourishes under the influence of a nascent civilization." Modern societies, wrote the commentator, exhibit "corruption without politeness and barbarism without virtues. The history of Antiquity, on the other hand, offers us two great subjects of study: institutions and men. The ancients are our teachers about liberty, and that education is not their least title to our recognition."[6] Written in 1818, this tribute to Rollin reinforces Grafton's point that even as history ceased to be what Felix Gilbert described as "a branch of rhetoric," historians continued to dedicate themselves to the proposition that history should be the teacher of life, rather than an unbiased account of past events.[7] And it illustrates my claim that it was to ancient history more than modern history that readers and writers chiefly looked for "great subjects" of study.[8] But they also looked to it for something else: the confirmation or negation of biblical events. That made ancient history even more critical as well as controversial for Europeans, down at least to the nineteenth century's end.

One might rightly call the eighteenth century the great age of universal history. But that scarcely prepares us to appreciate the sheer ambition and diversity of the historical projects of this era. In a period predating the existence of historical seminars or history departments, virtually anyone could and did write things called "histories"—but that does not mean that it was an easy genre to master, or one spared scathing criticism. Many enlightened histories were enormous undertakings; even those authors who did not wish to narrate all of human history after the Flood often bit off very large chunks of time and space, as did Voltaire, A. L. Schlözer, J. C. Gatterer, and the Abbé Raynal. After the collapse of the Four Monarchies model in which, as Voltaire quipped, everything that happened in the world, no matter how irrelevant or trivial, was explained as having happened the sake of the Jewish nation,[9] how was one to write it, especially as the materials for doing so were multiplying at a rapid rate? If Bossuet was able to write his own universal history, the British *Universal History from the Earliest Account of Time to the Present* (1736–65) required a team of authors—and ran to forty-four volumes![10] Nor were the skeptics getting any easier to refute; chronology continued to be, in Herder's words, little more than a "shouting match,"[11] and periods before the Persian Wars to be particularly problematic. Undaunted, or believing they could overcome objections with new evidence, authors sought to fill in these gaps with geographical, ethnographic, or etymological materials extracted from travel accounts. But many, too, continued to rely on secondhand Greek accounts—including Hellenistic texts assumed to be deeply ancient, such as the Chaldean Oracles, or the works of Hermes Trismegistus.

Perhaps because religion was such a major subject of contention in this century, and because the Old and New Testaments themselves averred that the most ancient religions lay in the Orient, quarrels about ancient Near (and to a lesser extent, Far) Eastern chronology and history proved especially vehement, and paper consuming. The argumentative terrain here stretched far and wide, and admitted of no conquerors. The vast tangle that was pre-Roman chronology simply couldn't be unpicked, not even by Newton, the great hero of the natural scientists, though he spent the last thirty years of his life working on the subject.[12] The orthodox churchmen could not succeed in defending miracles, nor did any of the modern Euhemerists provide convincing analyses of the human or natural origins of the world's religions. Voltaire and, in the Germanies, P. E. Jablonski exulted in the prospect of demonstrating that texts earlier than the Old Testament had already revealed the great ethical ideas to mankind—and travelers such as J. Z. Holwell and Hyacinthe Anquetil-Duperron claimed to have found them; but skeptical voices were at least as loud. In the wake of the huge popular successes of Montesquieu's *Persian Letters* and Galland's *1,001 Nights*, large numbers of exoticizing fantasies, parodies, or purported "translations" flooded the market—to say nothing of the hybrid Chinese, Indian, or Persian "documents" created by Jesuit fathers.[13] Methodologically too, there was no agreement about what constituted a probable historical truth: while the rationalists insisted on the universality and consistency of natural phenomena and human behavior, Herodotean heirs—including, for example, J. G. Herder—believed it wrong to dismiss reports of strange practices or beliefs just because they did not meet modern standards of "vraisemblance."

But as the century wore on, more Europeans traveled or learned Asian languages, and radical forms of Orientophilia came under fire, both from enlightened rationalists seeking to defend the superiority of Western progress and from enlightened Christians seeking to save the truth of the Scriptures. Writing in 1769, Voltaire, responding to the hyper-Egyptophiles, declared that virtually everything supposedly known about Egyptian history had been written with a phoenix feather, and that history proper really only commenced with Herodotus' account of Xerxes's preparations for battle with the Greeks.[14]

In the 1770s, critics attacked the very early Indian dates proposed by Holwell and N. B. Halheld, and William Jones and William Robertson declared Anquetil Duperron's *Zend Avesta* a forgery. By 1781, it had become clear that the Veda that Voltaire hoped was deeply ancient had been written by a modern Jesuit missionary. While radicals such as C. R. Dupuis speculated about the origins of all religions in ancient Egyptian agricultural practices and star worship (subsequently turned into dogmas and rituals by priestly conspiracies), scholars—usually those in universities, or close to the circles of the érudits—increasingly abandoned investigations of the earliest eras of human history (always presumed to be "oriental") where polemical and religious controversies raged most fiercely. Gibbon, once intrigued by chronological debates, now gave up writing about prehistorical periods.[15] In his own version of universal history (*Ideas for a Philosophical Historical of Mankind*), written in 1784–85, J. G. Herder registered the damage to the credibility of Near Eastern ancient history:

> The eastern part of Asia has become known to us only recently through religious or political parties, and in the hands of scholars in Europe has become so confused in parts that we still see great stretches of it as a fairytale land. In the Near East and in neighboring Egypt everything from all periods appears to us as a ruin or a vanished dream; what we know from written sources we know only from the mouths of passing Greeks, who were partly too young and partly of too foreign a way of thinking to understand the deep antiquity of these states; they were only able to grasp what interested them. The archives of Babylon, Phoenicia, and Carthage are no more: Egypt was in its decline, almost before a single Greek visited its interior. Everything has been shrunk down to a few faded pages, containing fables of fables, fragments of history, a dream of the prehistorical world.[16]

Already in Herder's lament we can pick up a longing for lost archives, or at least for direct testimony from the Egyptians and Carthaginians, to resolve the "shouting match" in which his contemporaries remained embroiled. Oriental prehistory has become a fable, one whose riddles Herder has almost given up trying to solve. He follows the passage just cited with a noteworthy announcement of the subject of his next chapter, and a change in tone: "With Greece," he writes, "the morning breaks, and we joyfully sail forth to meet it."[17]

In fact, in the fields of Greek and Roman history, proving one's claims had also become increasingly challenging as skeptics such as Pierre Bayle challenged their accounts and travelers, collectors, and antiquarians began to check their "facts," not against one another but against geographical, ethnographic, or archaeological *Realien*. Thucydides, valued already by Jean Bodin and Thomas Hobbes as a firsthand (and military) participant who had *seen* rather than merely *heard* what had transpired,[18] and as someone who said almost nothing about religious matters, came through this process rather well, as did Gibbon's hero Tacitus. On the other hand, it became increasingly difficult to trust Livy or Plutarch, two of the early modern world's favorite ancient authors. "The nonsense of Xenophon," wrote Thomas Macaulay in 1828, "is that of a dotard."[19] Herodotus, who remained much beloved by readers for his stories and for what John Gillies, the author of one of the first histories of Greece, called his gaiety, in contrast to Thucydides' gloom,[20] was subjected to a barrage of fact checking—exemplified by Pierre-Henri Larcher's translation and commentary on the *Histories* (1786–1802), which ran to nine volumes, and by James Rennell's 800-page *The Geographical System of Herodotus Examined; And Explained by a Comparison with Those of Other Ancient Authors, and with Modern Geography* (1800). The consensus opinion was that this author could be trusted only with respect to what he had himself seen, but not with respect to what he heard from strangers.[21] This application of Bodin's principle did not actually solve the question of Herodotus' credibility, as there was no consensus then (and there is barely one now) on where Herodotus had actually traveled. But it became an endlessly repeated mantra, and it had, for our purposes, a significant effect, throwing into question Herodotus' evidence on the period of Greek-Oriental interaction, and making his work truly historical only insofar as it treated war and things Greek. The only real lesson of Herodotus' histories, Voltaire concluded, was to demonstrate "The superiority of this generous, little nation, free while all of Asia was enslaved."[22]

We cannot discuss in detail here the veritable industry that arose to check Herodotus' facts—a gargantuan, multidisciplinary enterprise that still operates on a much smaller scale today.[23] But it produced some remarkable historiographical inquiries, including one composed by Göttingen professor (and universal historian) Christoph Meiners in 1775, and dedicated to hauling Herodotus' book 2 (on Egypt) before what Meiners called "the judgment seat of criticism as powerful as it is careful." Meiners did, by and large, ratify Herodotus' testimony, but he did so in a new and telling way. "Up until today," he wrote, "opinions are just as free and unfettered in their judgment of [Herodotus'] credibility (*Glaubwürdigkeit*) as they were more than 1,000 years ago."[24] Instead of juxtaposing ancient authors, or testing them by modern experiments, scholars, in his view, ought to inspect each ancient author's source usage, and to understand the facts and fictions of texts as products of their own era. To be a discriminating reader of a historian like Herodotus, one had to understand not just the errors he made, but *why* he made those specific errors. "The facts themselves, which [Herodotus] describes, the era in which they occurred, and the sources which he drew on for advice, give each part of his history another measuring stick for his credibility and produce a different judgment," Meiners wrote.[25] In the case of Herodotus' accounts of Greek and Egyptian religious similarities and the deep antiquity of Egyptian gods, for example, Herodotus had erred not because he was a liar, but because he was writing during a period in which Egyptian religion had already been "corrupted" by borrowings from Greek mythology—and because he had credited the accounts of the Egyptian priests, who were no more than "fantastic chatterboxes." In the end, Meiners suggested, it might be possible to know only what the Greeks had *thought* about Egyptian religion—at least until one could read the hieroglyphs for oneself.

Meiners' critique, I argue, can be seen as a first step across the threshold that marked the opening of a new age of historical skepticism, one in which ancient historians would be treated less as contemporaries with whom one could argue and more as repositories from which to build better and more enduring histories. Scholars now became attentive in a new way to the rhetoric of ancient sources, no longer in the interests of imitating their style or refuting their claims, but as a means to identify the distance between the methods and intentions of ancient composition and the actual events ancient writers purported to describe. They now believed—as natural historians since Francis Bacon had believed—that they could obtain a clearer understanding than the ancients about ancient history itself, perhaps even one that would resolve the chronological and theological polemics of the last generations—though this would require undertaking a painstaking, even pseudo-Kantian, process of delimiting what could actually be known.

One might describe the evolution of this conviction (by no means one achieved overnight) as the victory of the moderns over the ancients in the field of historiography; or, alternatively, one might term it the birth of historicism from the desire to salvage some sort of historical truth. One might also, and rightly, chalk up some of the new attitude to Romantic melancholy, which emphasized the pastness of the past, and the impossibilities of perfect translation, lamenting—to use Friedrich Schiller's dichotomy of 1795—the loss of the "naïvete" of the ancients while also highlighting the moderns' superior wisdom and capacity to understand the agonies of "sentimental" adulthood.

In any event, this new form of melancholic skepticism prompted the development of a specific set of textual practices that came to be called

"*Quellenkritik*," or source criticism. As we can see clearly now, thanks to the work of Grafton, Kelley, Momigliano, and Peter Miller, source criticism did not spring fully-formed even from the erudite skulls of the faculty of the University of Göttingen in the 1750s and 60s. German scholars such as J. D. Michaelis, C. G. Heyne, and J. S. Semler perfected, rather than invented, the careful philological methods that allowed them to produce their daring interpretations of classical myths and biblical scriptures.[26] But it was chiefly German scholars who were credited by their contemporaries for developing a modern form of historicizing *Quellenkritik*, beginning about 1790, whose distinguishing feature is its relentless insistence on using philological tools to strip away all historical accretions and later additions from each text, and on applying the most rigorous, rational, and historicizing suspicion to each claim made in the original, in order to understand the author's identity (if in question), motivations, and limitations.

Given the passionate debates about the Bible and the origins of the world's religions in the eighteenth century, it is not terribly surprising that first treatises to self-consciously employ *Quellenkritik* were works treating the Old Testament, nor that methods developed by J. G. Eichhorn on texts attributed to Moses were soon adapted by F. A. Wolf to the study of that alternative secular mythmaker, Homer.[27] Wolf's study of the Homeric poems was published in 1790, the same year that saw the printing of Johannes Voss's assault on previous interpretations of mythology (*Mythologische Briefe*). Both Wolf and Voss, we should note, had been trained in the interpretation and translation of literary texts, and in their view only scholars who possessed excellent linguistic skills could be trusted to provide reliable accounts of the ancient world. Many of their philological descendants would later celebrate their accomplishments, but for contemporary universalists such as Philipp Buttmann, their critiques seemed to apply a philological Occam's razor both to ancient texts and to older inquiries, "and precisely those results which one had grown accustomed to seeing as completely credible (*ganz unbezweifelt*), have [now] been explained as totally false." Voss in particular, Buttmann wrote, in applying "the frigidity of painstaking research," had thrown away the whole of Euhemerism, arguing that the ancients were too simple to have been capable of any sort of allegory.[28] Completing a process begun much earlier, under the pressure of the philological expertise of Wolf, Voss, and Eichhorn, Homer and Moses could be seen as, at best, naïve and distant "poets" rather than conveyors of truths.

These new tactics of using formal features to date and distance one from ancient texts came into currency quite quickly, aided by Romantic sensibilities and the critical new attitude toward myth and toward oriental chronologies. In 1798, this critical approach would be turned on the earliest historians as well by another young German philologist, Friedrich Creuzer. Creuzer's *Herodotus and Thucydides: An Attempt at a Closer View of Their Historical Principles with References to Lucian's Text, How Must One Write History* (1798) plotted the same catastrophic development in historiography that Schiller had plotted for poetry, making his protagonists distinctly ancient men, with aims and objects completely different from those of their readers in the present. A much more poetic and primitive Herodotus, Creuzer wrote, stood behind Thucydides, who—as countless authors would repeat for the remainder of the century—had brought *ancient* history writing to its pinnacle. But both Herodotus and Thucydides, Creuzer continued, had failed to achieve the higher perfection of the new, peculiarly German form of source-critical writing. "Historical criticism in its narrow sense, requires a wholly other form of understanding than was available to the ancients," Creuzer wrote; but his next line demonstrated his ambivalence about this recent development, and perhaps its contingency: "it is a young, tender plant, which only in the most recent times and chiefly, indeed perhaps only, under German skies, can be tended."[29]

But tended historical source criticism was in the Germanies, perhaps above all by the philologically trained and politically active B. G. Niebuhr, who made himself exemplary by taking no aspect of Livy's Roman history for granted, and by claiming he could critically sift ancient poetry for fragments of historical truth.[30] After Niebuhr—along with Wolf—commenced lecturing at the newly founded University of Berlin in 1810, his reputation soared, and imitation of his methods spread. The British historian George Grote, an admirer of German scholarship and equally confident in the rational powers of the modern individual to pick apart the past, described Niebuhr's method as "recomposing the ancient world by just deduction from small fragments of history, like the inferences of Cuvier from the bones of fossil animals."[31] And just as geologists grew increasingly confident that they could create a full and credible fossil record,[32] humanists and radical theologians began to believe that source-critical scholarship, especially as applied to newly discovered documents and materials, would lay the foundations for a new and more convincing science of Antiquity.

Even more eager than Niebuhr to establish the virtuosity and autonomy of the classical philologist, Friedrich Wolf in 1807 produced a manifesto for his vision of a new science of Antiquity in his *Darstellung der Altertumswissenschaft*. Here, Wolf gave a new name to a trend in history-writing underway at least since Voltaire's *Essai sur les moeurs* (1756); in his hands, *Altertumswissenschaft* was to include a wider set of source materials, the humble and prosaic along with the more conventional classical texts. His efforts to create a "total" science of the nations of Greece and Rome were also enabled by the much older humanist and antiquarian traditions of collecting materials relating to customs and political institutions. The result, for our purposes, was the forwarding of projects of collecting together, in national-linguistic groupings, materials formerly thought to be of purely antiquarian interest, such as inscriptions, coins, and vases, and the lopping off of speculative biblical chronologies and natural histories of the human race. For in Wolf's hands, to make his inquiries *wissenschaftlich* meant that both prehistorical speculation and the Orient—inferior in "civilization" and not directly accessible to the nonorientalist—would be left out. "Nature in the abstract—the rationalist standard—was replaced by the image of a specific historical culture," as Anthony La Vopa beautifully summarizes the change from universal and conjectural to critical, national history.[33] And so too did Wolf neatly excise the cultures inaccessible to the expert classical philologist—those of the ancient Near East—from the project of writing "real" history.

There is a clear relationship here between Wolf's *Altertumswissenschaft* and the kinds of archival projects Lorraine Daston describes so beautifully in her essay in this volume (chapter 6). Among those projects, many of them related directly to the history of the ancient world, including the *Corpus Inscriptionum Graecarum*, launched by Wolf's student Philipp August Boeckh in 1815, the *Corpus Vasorum Antiquarum* (also edited by Boeckh, vol. 1, 1828), and the *Corpus Scriptorum Byzantinae Historiae* (revived from an early modern French model by Niebuhr and the Greek philologist Immanuel Bekker in the 1820s). All of this activity amounted to the creation of archives for ancient history, and the desperate attempt to provide a repository of independent facts and fragments by which the ancient narratives could be judged, dismantled, and reconstructed for perpetuity by modern compositors.

Already by the 1820s—the era in which Ranke launched his career—there are signs that *Quellenkritik* and its associated skepticisms about the ancient Orient, myth, and prehistorical events were having a corrosive effect on the beloved genre of universal history. In this decade, than Johannes Voss and his Graecophile friends launched a vicious attack on Creuzer's attempt to plumb the oriental origins of religious ideas in his *Symbolik und Mythologie der alten Völker* (1810–12), assaulting the universal historian for depending on indirect sources (including Herodotus and the Hellenistic Greeks), and exulting when they could show that his documents—such as one purporting to demonstrate the Indian origins of Dionysos—rested on forgeries.[34] Creuzer's friends, many of them orientalists or universal historians, tried to defend him, but the Creuzer Streit proved to be a warning to many to narrow their field of vision to the more directly knowable post-archaic Greeks and Romans. K. O. Müller, for example, wrote his Greek histories in ways that allowed him to avoid discussing the Greeks' debts to the Orient, a move Martin Bernal has denounced as racially motivated, but is more likely to have been a self-protective reaction to the Creuzer Streit.[35] The next generation of ancient historians—including Ernst Curtius, George Grote, and Theodor Mommsen—would find that focusing on one national history would be a sufficient scientific endeavor and left the writing of universal histories to those increasingly regarded in academic circles as dilettantes.

Even for those retreating to national ancient histories, however, the repeated narrowing of the source base by philologists wielding ever-sharper source-critical hatchets, and the periodic revolutionizing of the field by archaeological finds, made for ongoing uncertainty and for more modest claims about what really could be known about the past. Over time, *Quellenkritik* and the cutting off of prehistory diminished the authenticity of many ancient texts, and the reliability of their direct testimony, even as the collection of inscriptions and material evidence made the ancient world richer in prosaic, but not self-conscious, forms of testimony. George Grote, for one, applauded the new austerity imposed by the advent of stricter methods, writing in 1843 in *The Westminster Review*:

> Estimated by a poetical standard, the loss has been serious indeed: but it has been far more than compensated by the acquisition of lasting and substantial benefits. We have obtained in exchange an ascertained, methodical, and constantly increasing body of authentic truth: and we have obtained it, let us remark, not by transforming and refining the imperfect ancient physics themselves, but by following cautiously the track and respecting the limits, of positive evidence.[36]

In the preface to his *History of Greece* (1846), Grote again praised the new German critical scholarship, and celebrated progress in the field since William Mitford's first attempts to write Greek history. "The general picture of the Grecian world," Grote argued, "may now be conceived with a degree of fidelity, which, considering our imperfect materials, it is curious to contemplate." And yet, two pages on, Grote agonized about the trustworthiness and incompleteness of his sources, and imagined his reader's reaction to so much authorial uncertainty:

> The question of credibility is perpetually obtruding itself, and requiring a decision, which, whether favourable or unfavourable, always introduces more or less controversy; and gives to those outlines, which the interest of the picture

requires to be straight and vigorous, a faint and faltering character. Expressions of qualified and hesitating affirmation are repeated until the reader is sickened; while the writer himself, to whom this restraint is more painful still, is frequently tempted to break loose from the unseen spell by which a conscientious criticism binds him down—to screw up the possible and probable into certainty, to suppress counterbalancing considerations, and to substitute a pleasing romance in place of half-known and perplexing realities.[37]

As if in answer to Grote's longing, in 1855 the colonial bureaucrat James Talboys Wheeler, author of a respectable *Geography of Herodotus* (1854), would revolt against the frustrations of missing evidence and the oppressive restrictions of *Quellenkritik* to pen a two-volume imaginary biography of Herodotus. "In a word," Wheeler explained, "the author has sought to clear antiquity from the dust of the schools, and teach it in shady playgrounds and flowery gardens."[38] Not surprisingly, Wheeler's fictional biography proved an academic flop—and a popular success.

I have quoted Grote extensively and invoked Wheeler here to capture what I believe to be a larger sentiment among historians of the period, namely, the expanding sense that ancient history, even when written on the basis of careful *Quellenkritik*, could never be as full and rich a story as modern history could be. Theodor Mommsen recognized even more clearly than did Niebuhr the dangers of ranging backward into Roman prehistory, and self-consciously set himself the task of piecing together fragmentary and problematic sources legal sources to obtain a persuasively "scientific" history of Rome.[39] After the publication of Curtius' *Greek History* (1857–67) and the first volumes of Mommsen's *Roman History* (1854–56), academic ancient historians largely refrained from writing full national histories, preferring to wait for the specialized *Quellenkritik* to fill in at least some of the gaps with proper sources. Writing the fifth edition of his history of Greece before the death of Pericles in 1888, Max Duncker criticized his contemporaries for restricting themselves to "micrology," but admitted that more than forty years after the original edition of the volume, he was no more confident now than before that he had solved the evidentiary problems, and that there was still considerable guesswork involved.[40]

If we look at the teaching profiles of Prussian historians—those most directly exposed to *Quellenkritik*, as well as to nationalist pressures—from the 1820s forward, we can track a striking gravitation away from universal and even ancient history proper and toward national and modern subjects. F. C. Dahlmann, trained as a classical philologist by Wolf himself, began teaching universal history—including the history of the Persians, Chinese, and Ottomans—at the University of Kiel in 1812, and continued down to

1827, increasingly, however, focusing on the peoples whose languages he had mastered, namely the Greeks and Romans. Shortly after the Napoleonic wars, Dahlmann began to teach a course entitled "Vaterländische Geschichte" (essentially Danish history). Encouraged by rising student demand, he increasingly devoted himself to this course, and to his ambitious source-collection project, the *Quellenkunde der deutschen Geschichte*.[41] F. C. Schlosser, author of a hugely popular universal history in 1815, made his next project a history of the eighteenth century; after the disappointing reception of his *Geschichte des Hellenismus* in 1834, J. G. Droysen took to teaching Prussian history and historiography. And Ranke himself after 1834 gave up lecturing on the first half of world history (except for one series in 1848). In 1867–68, he explained why he and so many of his colleagues no longer taught the course. It was, he said, impossible to teach it properly. In the preceding decades, all periods of history had been subjected to source-critical critiques, but "not always has research arrived at results which can be presented as such. Historical science is just not mature enough for the reconstruction of universal history on new foundations."[42] Ranke would prove this himself, inadvertently, when his final publication—a *Weltgeschichte* based on his old lecture materials—would be fiercely criticized for its obsolete findings and secondhand sources, above all by orientalists, enraged that Ranke had ignored advances in their field.[43]

Fully instructed in the techniques of *Quellenkritik* and well aware of the controversies in ancient historiography, Ranke turned his attention to modern history in the early 1820s, during the height of the Creuzer Streit. His 1824 *Geschichte der romanischen und germanischen Völker von 1494 bis 1514*—a book far more chronologically circumscribed than Creuzer's *Symbolik*—was based on printed sources; but in volume two of this work, subtitled *Zur Kritik neuerer Geschichtsschreiber*, he subjected his sources to a thorough critique and comparison, and worried openly about the fact that modern historians, too, had borrowed from one another, and then embroidered upon their borrowings.[44] Even modern historians, faced with modern documents, resembled the man who found himself entering

> a great collection of ancient things [*Alterthumern*], in which the authentic and the inauthentic, the beautiful and the repulsive, the impressive and the inconspicuous, from many nations and eras, lie together without any order. . . . They speak to us in a thousand voices; they display the most divergent natures; they are arrayed in all colors. Some treat things with ceremony; they want to illustrate; this seems to be the path of the ancients, which they follow. Others want to draw lessons for the present from the past. Many want to defend or to attack; not a few are careful to report the facts for deeper reasons, resulting from their dispositions

or passions. Then there are a few who only have the goal of transmitting what has happened [*was geschehen ist*]: to these we can add the reports of confirmed eye-witnesses. The participants themselves get the word; documents, apparent and real, are available in large numbers.[45]

Of course, Ranke concludes, we must listen to those in the latter categories, those disposed only to the reporting of facts, and the eyewitnesses—something easily done, he reiterated, for the modernist, if only he would put his shoulder to the archival wheel.[46]

Again, we should reiterate that Ranke was by no means the first historian to comment on the superior number of sources available to the modernist—no one who had visited Rome or who had turned the pages of Eusebius could doubt that much had been lost. Nor was he even the first to suggest that the public documents held in archives could be used to lay the foundations for more solid and believable historical accounts; Voltaire had said as much in praising the work of Thomas Rymer, royal historiographer and editor of English diplomatic documents dating back to the twelfth century.[47] But Ranke developed and broadcast the advantages of his method at an opportune time, just as *Quellenkritik* had so complicated the lives of ancient and universal historians. And Ranke did not fail to advertise the great advantages of studying modern history. In lectures given in 1847, Ranke described the bountiful firsthand memoirs, "which relate from their standpoint what has been lived through and experienced," available to the moderns. Of such sources, Ranke argued,

> antiquity and even the Middle Ages knew little. Written commerce between parties has been drawn into archives; often it concerns unimportant things, often also the most important; we see in any case the movement of the minds of the persons involved, the manuscripts produced by their activity, where then what is wholly uncontestable appears: talent, bigotry, truthfulness and lying stand before us as in everyday life. [In modern history], we have reports of those who stand near the facts and who have had conveyed upon them the official duty to report the truth, the reports of those, who in the moment of the event have knowledge about those who are involved in it. Countless representations of the personalities that in their contrast illuminate one another. Thus is the historical material endlessly enlarged by every nail that one pounds into the wall as time goes on, presuming that one knows how to find it [the material; the nail] and is in the position to use it.[48]

The archive could provide the means by which a modern historian could avoid being tricked by the rhetorical redescriptions, forgeries, and lacunae that dogged ancient history. Instead of sifting backward, as had the proponents of *Quellenkritik*, one could, he announced excitedly to his students in 1830, "put together a modern history based exclusive on original and certain . . . materials."[49]

In 1856, Ranke fleshed out these ideas. In ancient history, he wrote, we can still find our best models (again he is thinking of Thucydides, and not Herodotus).

> Unfortunately we lack the monuments that permit us direct knowledge [*freie Erkenntnis*] and could make possible to go beyond their interpretations. . . . [by contrast] for the modern period we have significant and contemporaneous writers of history. [We have] rich sources in the archives, which as of recently have been published and clearly represent the inner lives of states. . . . The study of history has been, thereby, enlivened and made fruitful. Its goal is the reaching of objective truth in all things. In ancient and medieval history one has to do without this, but in the modern it is possible and the path has been charted. . . . Historical research must give each its due. One extracts the earlier object and treats it according the conditions under which it occurred, that is the goal of historical research. We are able to do more now than the ancients could. The goal is, to embrace the general. The more embracive history is as a whole and the more secure it is in particulars, the greater is its service.[50]

"In antiquity," he repeated a few pages later, "Thucydides is the unrivaled master of contemporary historical writing. In modern times much [has been written from] personal standpoints. To this we can add another element: the exchange of written documents held in archives, which allows for the testing of these personal accounts. It is not everywhere possible to arrive at objective truth, but lies can be fended off. Therefore, the character of modern history is more believable."[51] By 1867–68, Ranke was exulting in success of his archival method, and the new forms of objectivity this had yielded:

> The era, which I prophesied thirty years ago has really arrived. One writes modern history no longer from tradition, which shaped earlier writers, and which they then continued, but from unmediated documents of previous centuries, which can be found in archives; dispatches, correspondence, documentary fragments of different kinds; this is the way opposing minds worked; and although [in this process] the parties clashed powerfully, thus can truth stretch a wider arc.[52]

Although Ranke and his younger contemporary J. G. Droysen disagreed about many things, and Droysen thought Ranke overly obsessed with unpublished sources, Droysen could not help but admire Ranke's archival practices. "How excellent is Ranke's interpretation and representation," he exulted in his own lectures on historical method, drafted in 1857 and given to large crowds at the University of Berlin in the 1860s and 1870s. "If one wanted to treat anything having to do with this period [the Reformation era] in an orderly way, one could oneself move from him to the archives that he used; though one cannot disassemble his account in such a way that the individual archival bits of data are freed from the form and connections in which he has put them; one could take apart the mosaic that he has composed and use the individual pieces to create, cleanly and easily, a new composition."[53] Unlike Herodotus' "jumble"—which for Droysen could not be picked apart and reassembled—Ranke's histories represented the right kind of historical artistry, the preservation of truths like tesserae, eternally recoverable from an orderly archive for the next mosaic maker. Archival evidence gave Ranke the ability—or at least the appearance of the ability—to enter into the composition of a narrative without a creator of another narrative to contend with, and to open a hermeneutical frame that could be securely built upon without fearing that the sources themselves had been corrupted. Something of the same technique could be pursued by Mommsen for certain periods of Roman history; but as Grote regretted, the question of credibility was constantly intruding itself, and forcing modesty and acknowledgments of incompleteness upon scholars of Antiquity. Thus the modern descendants of Ranke alone seemed to be authentic practitioners of historical *Wissenschaftlichkeit*. Ranke had certainly displaced the first "father of history," Herodotus, whom Thomas Macaulay now described as a "delightful child."[54] Droysen, like Ranke, maintained that Thucydides ought to be hailed as "the greatest historian of all time;"[55] but in some respects, he and his contemporaries admitted, Ranke had gone Thucydides one better, as he could stand among the archival sources as a contemporary, and thus reconstruct from *direct* evidence much more than the ancient historian could extract from his inquiries and experience. Ranke was well aware that one could find many stories in the archive; but he convinced himself, and the students in his famous seminar, that this testimony was credible in a way that predigested or fragmentary ancient history could never be.

What made Ranke singular and famous, argues Wolfgang Hardtwig, was the combination of his method of *Quellenkritik* with the power of his modern narrative.[56] For me, that still does not really explain how Ranke attained such prominence, and modern history so quickly displaced ancient and universal histories as the exemplary form of the genre. My view is much closer to that of Kaspar Eskildsen, who emphasizes the ways in which Ranke, in the 1830s, "turned the status relationship between archive and armchair upside down. He convinced readers, colleagues, and students not only that archival work was independently purposeful, but also that proper history could be written only from within an archive."[57] I would not argue that he made ancient history, right away, seem a field inferior in importance or in scholarly virtuosity; his generation, and the next, teemed with highly esteemed ancient historians. But he did succeed in making the sorts of archives he used in the writing of modern history seem the exemplary type, in part by capitalizing on the insecurities of universal and ancient historians introduced by enlightened rationalism, by Romanticism, and by *Quellenkritik*.

Hailing Ranke as the modern father of history signified the opening of a methodological divide that has, until today, remained in force. We are now much more savvy about the kinds of materials modern archives leave out—including much of the history of women, of the underclasses, and of the colonized—and the ways in which they themselves can deceive. But those who seek to pioneer new territory without a conventional archive behind them ever run the risk of being cashiered for failing to offer "positive" evidence, or having *Quellenkritik* deployed against them. Perhaps this explains why, in part, ancient historians today are still so often found in departments of classics, where both those who work on "facts" and those who work on "fictions" are presumed to share a corpus of materials, while modern historians are presumed to base their work on archivally based facts.[58] We are no longer Rankeans, but we live in the shadows of a professional history constructed around a kind of archival research that makes modernists and ancient historians seem to have little in common, and that, it might be added, still makes universal history seem amateurish and the ancient Near East appear to be "a vanished dream." Perhaps reconstructing the process by which Ranke's archival practices made specialized modern history safe—until the later twentieth century—from the dangers of *Quellenkritik* will help us understand a divergence of ancients and the moderns that continues today to prevent us from understanding the common origins and challenges we share.

NOTES

1. See here Arnaldo Momigliano, "The Place of Ancient History in Modern Historiography," in idem, *Settimo contributo alla storia degli studi classici e del mondo antico* (Rome: Edizioni di Storia e Letteratura, 1984), 13–36.

2. Markus Friedrich, *Die Geburt des Archivs: Eine Wissensgeschichte* (Munich: Olden-

bourg Verlag, 2013), 231–33, 256. Also Donald R. Kelley, *Faces of History: Historical Inquiry from Herodotus to Herder* (New Haven, CT: Yale University Press, 1998), 207–8.

3. Anthony Grafton, *The Footnote: A Curious History* (Cambridge, MA: Harvard University Press, 1997), 81–82.

4. See Grafton, *What Was History? The Art of History in Early Modern Europe* (New York: Cambridge University Press, 2007), and idem, *The Footnote*; Kelley, *Faces of History*, and idem, *Fortunes of History: Historical Inquiry from Herder to Huizinga* (New Haven, CT: Yale University Press, 2003); Arnaldo Momigliano, "Ancient History and the Antiquarian," *Journal of the Warburg and Courtauld Institutes* 13 (1950): 285–315, and other essays; Ulrich Muhlack, *Geschichtswissenschaft im Humanismus und in der Aufklärung* (Munich: C. H. Beck, 1991).

5. Kaspar Risbjerg Eskildsen, "Leopold Ranke's Archival Turn: Location and Evidence in Modern Historiography," *Modern Intellectual History* 5, no. 3 (2008): 431.

6. Saint-Albin Berville, "Éloge de Rollin," in *Oeuvres complètes de Rollin, nouvelle edition, accompagnée d'observations et d'éclaircissements historiques, par M. Letronne*: vol. 1: *Histoire ancienne* (Paris: Firmin Didot Freres, 1820), xii.

7. Grafton, *What Was History?*, 28.

8. For a fascinating account of which ancients were most read in the early modern period, see Peter Burke, "A Survey of the Popularity of Ancient Historians, 1450–1700," in *History and Theory* 5, no. 2 (1966): 135–52.

9. Voltaire, "Le Pyrrhonisme de l'Histoire" [1768], in *Oeuvres Complètes de Voltaire*, vol. 14 (Stuttgart: Frères Hartmann à la Haye, 1829), 26–27.

10. This compendium was, nonetheless, rapidly translated into German, French, Italian, and Dutch. Guido Abbatista, "The English *Universal History*: Publishing, Historiography, and Authorship in an European Project (1736–1790)," *Storia della storiografica* 39 (2001): 101n2.

11. Suzanne Marchand, "Where Does History Begin? J. G. Herder and the Problem of Near Eastern Chronology in the Age of Enlightenment," *Eighteenth-Century Studies* 47, no. 2 (2014): 157–75, here 166.

12. Jed Z. Buchwald and Mordechai Feingold, *Newton and the Origins of Civilization* (Princeton: Princeton University Press, 2012), 216–36.

13. See here Urs App, *The Birth of Orientalism* (College Station: University of Pennsylvania Press, 2010).

14. Voltaire, "Le Pyrrhonisme," 18, 24. On the academy, see Kelley, *Faces of History*, 210.

15. Frank E. Manuel, *Isaac Newton: Historian* (New York: Belknap Press, 1963), 184–85.

16. J. G. Herder, *Ideen zur Philosophie der Geschichte der Menschheit*, in *Herders Sämmtliche Werke*, ed. Bernhard Suphan (Berlin: Weidmannsche Verlag, 1883): 14: 90.

17. Herder, *Ideen*, 92.

18. This prioritization, as Glenn Most has shown, was already embedded in Herodotus' *Histories* and in the Gospel of John. Glenn Most, *Doubting Thomas* (Cambridge, MA: Harvard University Press, 2007), on Herodotus 6. On Bodin, see Kelley, *Faces of History*, 199.

19. Thomas Macaulay, "History," in *The Works of Lord Macaulay: Essays and Biographies*, vol. 1 (London: Longmans, 1896), 184.

20. John Gillies, *The History of Ancient Greece*, vol. 3 (of 4) (Basil: J. J. Tourneisen and J. L. Grand, 1790), 299, 303.

21. Voltaire, "Le Pyrrhonisme," 23. This principle was reiterated by Robert Wood, in his attempt to use Herodotus to validate the geographical accuracy of the Homeric poems the very next year. See Wood, *An Essay on the Original Genius and Writings of Homer* [1769] (London: H. Hughs, 1775), 184.

22. Voltaire, "Le Pyrrhonisme," 26.

23. The most recent major challenge to Herodotus' credibility as a historian came from Detlev Fehling in his hyperskeptical *Die Quellenangaben bei Herodot: Studien zur Erzählkunst Herodots* (Berlin: Walter de Gruyter, 1971).

24. Christoph Meiners, *Versuch über die Religionsgeschichte der älteste Völker besonders der Egyptier* (Göttingen: Johann Christian Dieterich, 1775), 5, 72, 81, 82.

25. Meiners, *Versuch*, 83.

26. For a careful study of developments at Göttingen, see Michael C. Legaspi, *The Death of Scripture and the Rise of Biblical Studies* (Oxford: Oxford University Press, 2010).

27. Anthony Grafton, Glenn W. Most, and James E. G. Zetzel, introduction to Friedrich August Wolf, *Prolegomena to Homer, 1795*, ed. and trans. Grafton, Most, and Zetzel (Princeton: Princeton University Press, 1985), 18–26.

28. Philipp Buttmann, "Ueber die philosophische Deutung der griechischen Gottheiten, inbesondere von Apollon und Artemis" (1803), in idem, *Mythologus, oder gesammelte Abhandlungen über die Sagen des Alterthums*, vol. 1 (Berlin: Mylius'schen Buchhandlung, 1828), 2–3.

29. Friedrich Creuzer, *Herodot und Thucydides: Versuch einer nähern Würdigung ihrer historischen Grundsätze mit Rücksicht auf Lucians Schrift Wie man Geschichte schreiben müsse* (Leipzig: Neue Akademische Buchhandlung Marburg, 1803), 88. For a fine interpretation of this text see Ulrich Muhlack, "Herodotus and Thucydides in the View of Nineteenth-Century German Historians," in *The Western Time of Ancient History*, ed. Alexandra Lianeri (Cambridge: Cambridge University Press, 2011), 179–209.

30. Barthold C. Witte, *Der preussische Tacitus: Aufstieg, Ruhm und Ende des Historikers Barthold Georg Niebuhr, 1776–1831* (Düsseldorf: Droste Verlag, 1979), 83.

31. George Grote, "Grecian Legends and Early History" (1843), in Alexander Bain, ed., *The Minor Works of George Grote* (London: John Murray, 1873), 75.

32. See David Seposki, chapter 2 in this volume.

33. Anthony J. La Vopa, *Grace, Talent, and Merit: Poor Students, Clerical Careers, and Professional Ideology in Eighteenth-Century Germany* (New York: Cambridge University Press, 1988), 323.

34. Anonymous, "Triumph der Symbolik," in *Hermes, oder kritisches Jahrbuch der Literatur* 26 (1826): 344–55. On the Creuzer Streit, see George S. Williamson, *The Longing for Myth in Germany: Religion and Aesthetic Culture in Germany from Romanticism to Nietzsche* (Chicago: University of Chicago Press, 2004), 137–50; and Josine Blok, "Quest for a Scientific Mythology: F. Creuzer and K. O. Müller on History and Myth," *History and Theory*, suppl. 33 (1994): 26–52.

35. Martin Bernal, *Black Athena: The Afroasiatic Roots of Classical Civilization*, vol. 1: *The Fabrication of Ancient Greece, 1785–1985* (New Brunswick, NJ: Rutgers University Press, 1987), 306–16; cf. Blok, "Quest for a Scientific Mythology."

36. Grote, "Grecian Legends and Early History," 133.

37. George Grote, *A History of Greece*, 4th ed., vol. 1 (Bristol: Thoemmes Press, [1872] 2000), iv, vi.

38. J. Talboys Wheeler, *The Life and Travels of Herodotus in the Fifth Century before Christ: An Imaginary Biography Founded on Fact, Illustrative of the History, Manners, Religion, Literature, Art and Social Condition of the Greeks, Egyptians, Persians, Babylonians, Hebrews, Scythians and Other Ancient Nations in the Days of Pericles and Nehemiah*, vol. 1 (London: Longman, 1855), vi.

39. Kelley, *Fortunes of History*, 125–28, quotation 125.

40. Max Duncker, *Griechische Geschichte bis zum Tode des Perikles*, 5th ed., vol. 1 (Leipzig: Duncker & Humblot, 1888), v–vii.

41. Wilhelm Bleek, *Friedrich Christoph Dahlmann: Eine Biographie* (Munich: CH Beck, 2010), 47–53.

42. Ranke, "Neuere Geschichte seit dem Anfang des 17. Jahrhunderts" (lecture notes, 1867–68), in Leopold von Ranke, *Aus Werk und Nachlass*, vol. 4, *Vorlesungs-Einleitungen*, ed. Volker Dotterweich and Walter Peter Fuchs (Munich: R. Oldenbourg Verlag, 1975), 411.

43. See Suzanne Marchand, *German Orientalism in the Age of Empire: Religion, Race, and Scholarship* (New York: Cambridge University Press, 2009), 208.

44. See Grafton's excellent reading of this text in *The Footnote*, 62–93.

45. Leopold von Ranke, *Zur Kritik neuerer Geschichtsschreiber* (Leipzig: G. Reimer, 1824), iv–v.

46. Indeed, the conclusion of Ranke's text berated German historians, in particular, for failing in their archival duties. See Ranke, *Zur Kritik*, 173–81.

47. Voltaire, "Le Pyrrhonisme," 38.

48. Ranke, "Neuere Geschichte, seit dem Westfälischen Frieden" [April 1847], in idem, *Aus Werk und Nachlass* 4: 189–91.

49. Ranke, quoted in Walter Peter Fuchs, "Einleitung," in Ranke, *Aus Werk und Nachlass* 4: 18.

50. Ranke, "Neuere Geschichte" (1856), in idem, *Aus Werk und Nachlass* 4: 265.

51. Ranke, "Entstehen und Ausbildung der Großmächte" (1856–57), in idem, *Aus Werk und Nachlass* 4: 269. Note that quoted here are a student's notes, not Ranke's lecture itself.

52. Ranke, "Neuere Geschichte seit dem Anfang des 17. Jahrhunderts (1867/68), 414.

53. Droysen, *Historik*, vol. 1: *Rekonstruktion der ersten Fassung der Vorlesungen* (1857) (Stuttgart: Friedrich Fromann Verlag, 1977), 155.

54. Macaulay, "History," 168.

55. Droysen, *Historik*, 96.

56. Wolfgang Hardtwig, *Deutsche Geschichtskultur im 19. und 20. Jahrhundert* (Munich: Oldenbourg Verlag, 2013), 65–66.

57. Eskildsen, "Leopold Ranke's Archival Turn," 433.

58. This is still a presumption we share, I think, despite the advent of cultural history.

SIX

The Immortal Archive:
Nineteenth-Century Science Imagines the Future

Lorraine Daston

MONUMENTS OF THE MODERN AGE

Two million stars, 180,000 Latin inscriptions; 20,000 photographic plates of the night sky and 254 volumes of published data, 20,000 paper squeezes from ancient stones and 30 volumes of transcriptions and supplementary material. Projects conceived on the scale of a decade extended to a century or more; budgets endlessly revised, always upward. Careers dedicated to undertakings that outlasted individual lifetimes and subordinated individual ambitions to the needs of the disciplinary collective—and not the actual collective of one's here-and-now colleagues but the envisioned collective of future astronomers and classical philologists. The proponents of the Corpus Inscriptionum Latinarum (officially begun in 1853 by the Prussian Academy of Sciences in Berlin) and the Carte du Ciel (launched in 1887 by an international congress organized by the Observatoire de Paris) described their projects as "monuments," the modern age's answer to ancient pyramids and medieval cathedrals. But the form these nineteenth-century monuments took was not architectural but archival: published compendia and mechanical copies of the working materials that nineteenth-century scholars and scientists imagined would enable their successors to conduct research for centuries (if not millennia) to come.

Nineteenth-century archival projects like the Corpus Inscriptionum Latinarum ("collection of Latin inscriptions," known as the CIL to scholars of all aspects of Roman Antiquity) and the Carte du Ciel ("sky map") defined the objects, methods, organization, and scope of scientific inquiry in their respective disciplines in the latter half of the nineteenth century. Massive investments of capital, labor, and time on an international scale in the present would, it was promised, secure future scientific progress, whether in understanding the development of the Latin language or tracking the movement of the so-called fixed stars. These two towering archival monuments were simply the most grandiose expression of a much broader archival movement in nineteenth-century science and scholarship. Just as the collections of the great national museums and libraries from Paris to London to Berlin created canons of everything from paintings to plants,[1] so the compendia of the philologists and the astronomers canonized epigraphs and stars—and moved the site of research from the field to the study.[2] Supporters of the Carte du Ciel enthused that astronomy would finally become a day job by transporting "the image and the study of the heavens into the cabinet."[3] Advocates of the paper squeezes and other mechanical reproductions that played so prominent a role in the CIL hoped that future epigraphers could learn their craft from such collections without wetting their feet in a muddy Italian field or risking pestilential fevers in the Levant.[4] Glass photographic plates and paper squeezes were (and still are) the archives behind the published compendia that standardized objects of inquiry and made them portable and, so the pious hope, permanent.

This was the first wave of Big Science, and the human sciences were very much in the van.[5] It was the moving spirit behind the CIL, ancient historian Theodor Mommsen, who coined the very term "Big Science" (*Großwissenschaft*).[6] Addressing the Berlin Academy of Sciences in 1890, he observed that "science [*Wissenschaft*, embracing both the natural and human sciences] also has its social problem; as in the big city and big industry, big science cannot be achieved by the lone individual, although it can be directed by one, a necessary element of our cultural development and one whose proper bearers are or should be the academies."[7] Although Mommsen's view about the academies as the natural site of Big Science was not universally shared—other projects like the *Patrologia Latina* (1844–64), the *Atlas International des nuages* (1896), or the Internationales Gradmessungsprojekt (begun 1867) were variously launched by individuals, learned societies, and international consortiums—, some institutional backing and often state funding were inevitably involved in projects that spanned generations and continents.

On the mid nineteenth-century diplomatic stage, cultural and scientific prestige counted, especially for countries such as Prussia (and after 1871, the Second German Empire) and France, which saw themselves as losing out in imperial competition with Great Britain and sought compensatory recognition. The organizers of the CIL and the Carte du Ciel never tired of reminding their respective ministries that national glory was at stake whenever they lobbied for more funding. International scientific congresses were treated as state occasions; expeditions to bring back antiquities for national museums enjoyed lavish government patronage; the first proofs of the Carte du Ciel photographs were displayed as part of the French exhibition at the Paris Exposition Universelle in 1900.[8] Geopolitics mingled with academic politics. France had barely conquered Algeria when the minister for public instruction (and noted literary scholar) Abel-François Villemain constituted a commission to compile a French version of the CIL, on the grounds that it would now be possible to collect Roman inscriptions from North Africa, thanks to "the vigilance of the [French] military administration."[9] Coverage of the stars at southern latitudes for the Carte du Ciel would have been impossible without the cooperation of observatories in European colonial outposts such as Algiers, the Cape of Good Hope, and Sydney.[10] For the scholars and scientists, the nation-state was not only a source of material and moral support but also the guarantor of the continuity of their long-term ventures. Mommsen equated Big Science with organization, but he admitted that organization alone would not suffice for scientific projects that surpassed a human generation, especially for those that established the archives of the sciences: "above all the fundamental work of the collection and selection of scientific materials must be assumed by the state."[11] Shaping and shaped by the expanded spatial and temporal horizons of nineteenth-century nation-states, vast scientific archives proudly took their place among the other monuments, including temple-like national museums and libraries, erected to prove the superiority of the self-proclaimed *nations civilisées* or *Kulturnationen* to other cultures far away and long ago.

Like all Big Science, these nineteenth-century archival projects devoured labor, money, and time. But unlike much contemporary Big Science, they did not aim first and foremost at new results—the latest particle detected by the Large Hadron Collider or exoplanet discovered by the Hubble Space Telescope—but rather at the creation of an archive. The CIL and the Carte du Ciel were understood as the preconditions for discovery, the firm foundation upon which *future* edifices of knowledge would be erected. "It is the foundation of historical science," Mommsen preached to his fellow academicians, "that the archive of the past be put in order."[12] Future research, whether about making sense of the Roman legal system or identifying a new comet, would be made possible by the discipline's carefully assembled compendium of Latin epigraphs or star positions. The participants in the Carte du Ciel hoped to "will [their astrophotographs] to the astronomers of year

3000 at least"—the qualification in deference not to the fragility of human civilizations, much less to any fears that astronomers might eventually become extinct, but only to the possible deterioration of the photographic plates over the millennia, "inasmuch as today we do not know whether the chemical deposited on the glass will remain eternally unalterable."[13]

The Corpus Inscriptionum Latinarum and the Carte du Ciel were arguably the grandest of these monuments and their many convergences underscore the danger of projecting current oppositions between the natural and human sciences backwards in time. Both were backed by massive state funding on the part of Prussia (later united Germany) and France. Both were international collaborations; both pioneered new methods, standards, and forms of labor organization (paid and volunteer); both required a high degree of standardization and therefore of consensus on matters of techniques, selection, format, and schedule; both took decades to finish, if they were ever finished at all. Both were conducted within a context of international competition and cooperation (often two sides of the same coin) framed by global imperial ambitions that loosened state purse strings and enlisted the sciences in new forms of cultural rivalry—first and foremost among contemporary European nations but also between them and other civilizations, past and present. Both marched under the banner of self-sacrifice and self-effacement. The organizers often likened their huge, laborious, protracted projects to medieval cathedrals: the work of many anonymous craftsmen who would never live to see the fruits of their labors. The pathos of progress and positivism saturated their manifestoes. As the pace of scientific advance accelerated, and theories and interpretations lived barely as long as mayflies, such compendia seemed to offer the one fixed point in a changing world—the archive of the future.

SLOW-RIPENING FRUIT: THE *CORPUS INSCRIPTIONUM LATINARUM*

In 1874 classical philologist Theodor Mommsen, member of the Prussian Academy of Sciences and professor at the University of Berlin, told his colleagues and students that it was time to forget about "German modesty": "We are most certainly not modest, and we do not want to be or be seen to be so. On the contrary, we want to continue in art and science, state and church, in all living and striving, to reach for all that is highest, all at once, everywhere, and absolutely." This vaulting ambition owed less to recent German military victories than to the triumphs of German science and scholarship. "Long before German weapons conquered on the battlefield, German research had won in its realm the same recognition and forced our neighbors against their will to learn our austere but by now essential language." But glory was achieved by inglorious methods. Woe betide the student who thought to escape hard work by taking refuge from mathematics or physiology in history—he would soon submit to the yoke of *Quellenforschung* and learn the "banausic patience of backbreaking labor."[14]

Neither soaring ambition nor inexhaustible diligence, however, sufficed for Mommsen. Without organization both were unavailing. Although he enjoyed a reputation for individual erudition and brilliance without peer among his contemporaries as a historian of ancient Rome and Roman law, and wrote with such verve that he was awarded the Nobel prize for literature in 1902, Mommsen thought that the chief task of modern science and scholarship was to organize research on the model of factory work. Newly elected to the Prussian Academy of Sciences to lead the *Corpus Inscriptionum Latinarum*, a published comprehensive collection of Latin epigraphs gathered from all over the lands once part of the ancient Roman empire, Mommsen told his fellow academicians in 1858 that only organization would save this project and indeed all the philological and historical disciplines from sinking into a sea of detail. To "copy, compare, examine without a plan for the whole work" was to dissipate energies, waste money, and postpone completion indefinitely.[15] Mommsen probably had the Prussian Academy's earlier publication project on Greek inscriptions, the *Corpus Inscriptionum Graecarum* (CIG), in mind here: begun in 1815 under the leadership of philologist August Böckh, the first volume appeared over a decade later in 1828 and the second a dawdling fifteen years after that, in 1843.[16] Mommsen had no intention of following Böckh's example. Four years after Mommsen addressed the Prussian Academy on the virtues of scientific organization, the first volume (on inscriptions relating to the death of Julius Caesar) appeared in 1862; eleven volumes had appeared by 1883. Like the CIG, the CIL was organized geographically by region, but there the resemblance ended: the Latin inscriptions would be transcribed after close, firsthand inspection, the collection would aspire to completeness (and therefore geographic comprehensiveness, potentially covering the entire ancient Roman Empire, including Britain, northern Europe, the Iberian Peninsula, the Near East, and northern Africa), and the volumes would come out with the regularity of manufactured goods. The CIL was the prototypical nineteenth-century compendium: more rigorous in its methods, more comprehensive in its scope, and more industrial in its production than earlier encyclopedic works of learning.[17]

Such innovations were essential, Mommsen believed, if the Prussian Academy's version of the CIL was not to share the fate of its predecessors. Since at least the seventeenth century, scholars had recognized the need for a collection of all known Latin inscriptions, which humanists and anti-quarians had by then been transcribing and publishing in dribs and drabs for over a century. Despite Olympian erudition and Herculean labors, the *Inscriptionum antiquae totius orbis Romanae in corpus absolutissimum redactae cum indicibus XXV* (1603 and later editions) of Jan Gruter, Marcus Welser, and Joseph Justus Scaliger was outdated almost as soon as it was published. The eighteenth-century efforts of the Veronese poet and polyhistor Scipio Maffei and of the French botanist and classicist Jean-François Séguier (in collaboration with his Modena colleague Lodovico Muratori) to supersede and supplement the Gruter collection were either judged incomplete or remained unfinished in manuscript form.[18] In the 1830s the young Danish philologist Olaus Kellermann, inspired by the work of the grand old man of Latin epigraphy, Count Bartolomeo Borghesi of San Marino,[19] and sup-ported by small stipends from the academies of Copenhagen and Berlin, declared that the time for a Corpus Inscriptionum Latinarum was ripe. But Kellermann died of cholera in Rome in 1837, and in 1843 the French initiated their own CIL, under the protection of Minister Villemain, with the justification that France, "which has preserved in its language, its cus-toms, its laws, its arts so many vestiges of Roman civilization, is particularly called to this work."[20] The French project in its turn faltered (and collapsed altogether when Villemain resigned during the Revolution of 1848), and it was Mommsen's teacher Otto Jahn, then professor of archaeology and philology at the University of Greifswald, who inherited Kellermann's epigraphic manuscripts and hatched plans for a Prussian-led CIL with Mommsen (then in Italy on a fellowship studying sources on Roman law and gathering inscriptions).[21] Their plans bore fruit in Mommsen's 1847 CIL proposal to the Prussian Academy, eventually approved in 1854, albeit only with determined backing from Friedrich Carl von Savigny, jurist and chancellor to King Friedrich Wilhelm IV, against stiff opposition within the academy, led by Böckh.[22]

Why did Mommsen think that he could succeed where so many before him had failed? In a word, or rather two: critique and organization. Raised in the new philological religion of textual criticism and tutored by Borghesi in its application to inscriptions, Mommsen believed that the new methods would detect false readings and forgeries propagated, whether wittingly or not, by his predecessors. He had harsh words for the negligence of Scaliger and Muratori, and branded others (e.g., the sixteenth-century antiquar-ian Pirro Ligorio) as outright forgers. So far, the methods of critique had mostly been applied only to published sources and, in some cases, manu-scripts (especially those held by the Vatican, though Mommsen despaired of a library that was open for only ninety days a year, and then for a mere

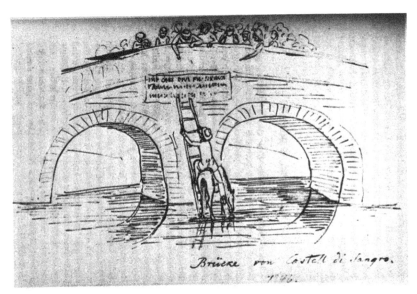

Fig. 6.1 Julius Friedländer, sketch of Mommsen climbing up to an inscription on a bridge at Castel di Sangro, 1846.

three hours a day). These were the methods and sources that had sufficed for Böckh in the preparation of the CIG—and for the abortive French CIL project, in explicit imitation of the CIG. But Mommsen insisted that it was necessary at some point to leave the library for the field, to return to the original inscriptions: "If the editor of Livy is prepared to travel to Florence and Paris [to consult manuscripts] for a critical edition, how can an editor of a complete critical edition of inscriptions fail to avail himself of the inspection of the stones themselves? All the more so, since he would otherwise sacrifice one of the chief advantages of epigraphy over other literature, the indubitable, unassailable certainty of the text?"[23] Back to the sources! This was the battle cry of critique, and Mommsen practiced what he preached. Whereas Böckh hadn't bothered to travel even as far as Paris museums to check Greek inscriptions (and regarded modern Greece, newly independent from the Ottomans, as almost as inaccessible as ancient Greece),[24] Mommsen woke up the curators of provincial Italian museums from their midday siestas with demands to inspect their antiquities and clambered up ladders to transcribe weathered inscriptions on bridges and buildings (fig. 6.1).[25]

But critique demanded more than autopsia, seeing with one's own eyes. If possible, the intrepid field epigrapher should also make a mechanical reproduction of the transcription. Easiest, cheapest, and most reliable was the paper squeeze (*Abklatsch* in German, *estampage* in French): after cleaning the stone, the epigrapher stretched over the inscription a sheet of ordinary paper, moistened it with a sponge dipped in water, then (the crucial part) banged it firmly to push the wet paper into all the incisions cut into the stone, let the squeeze dry in the sun, and finally rolled it up to be mailed or transported[26] (fig. 6.2). Other methods of making facsimiles were possible—Mommsen had made aluminum (*Stanniol*) impressions of a bronze edict by Claudius in Lyon[27]—and eventually photography became affordable and feasible under field conditions. Yet no other method was as portable, as versatile, and as faithful as the simple paper squeeze. It could be used everywhere where water and paper were available; it was as flat, lightweight, and shelvable as the stones were bulky, heavy, and cumbersome; it reproduced every detail, even those overlooked at first glance but in hindsight crucial, without the distortions of light and shadow that marred photographs; it allowed careful and repeated study back in Paris or Berlin, where the epigrapher could consult specialist libraries and escape the baking sun and flea-bitten inns of the Mediterranean lands where the inscriptions came from (a particular concern for Mommsen, who hated the heat and dreaded meeting a fate like Kellermann's if he lingered in Italy).[28] The squeeze could "not only replace the study of the original but even surpass it" in the opinion of its proponents:[29] autopsia in the comfort of the study, critique sharpened by the possibility of repeated consultation as knowledge of Latin paleography and Roman history advanced. Although squeezes had been made since the sixteenth century, their significance as archival sources, permanent records to which future scholars might return again and again with new questions and tools to challenge old readings, emerged first in the context of nineteenth-century Egyptology and Mommsen's CIL.[30] Backing up the published multivolume compendium of Latin inscriptions was the archive of paper squeezes, like gold reserves backing a paper currency.

"Organization" was Mommsen's second watchword. Although Mommsen often conjured up "Big Industry" as a model for "Big Science," described tasks like making squeezes as "mechanical," and demanded unprecedented sums as the "industrial capital [*Betriebskapital*]" for his projects at the Prussian Academy, the only form of organization he seems actually to have borrowed from industry was the division of labor. Mommsen's conception of scholarly organization was a curious mixture of the artisanal and the industrial, the chauvinist and the cosmopolitan, the modern and the medieval. He dispatched several generations of talented young philologists, not all of them German, to the farthest reaches of the ancient Roman Empire to find and transcribe Latin inscriptions of any and all kinds. Even the "mechanical" task of cutting up folio volumes of past publications of inscriptions would be, Mommsen asserted, better accomplished by Berlin

Fig. 6.2 Paper squeeze by Emil Hübner of Latin inscription from Écija (Roman Astigi), Spain. CIL II2 /5, 1176.

Gymnasium students than by Italian assistants, who were pricey and hard to come by.[31] Mommsen himself assumed editorial responsibility for the first few volumes, but later volumes often appeared under the aegis of one of his many lieutenants. These were more like trusted apprentices than nameless factory workers: among others, Otto Hirschfeld, Hermann Dessau, Emil Hübner, Elimar Klebs, Alfred von Domaszewski, Heinrich Dressel, and Eugen Bormann all started their distinguished academic careers in the service of the CIL.[32]

Their work was anything but mechanical, involving a combination of strenuous (and sometimes dangerous)[33] travel and the most exacting application of what Mommsen called "the rigorous philological method," which was as much ethos as technique: "simply the relentlessly honest search for the truth, with no effort spared in matters small and large, no doubt dodged, no gap in the transmission or in one's own knowledge glossed over, and always with full accountability to oneself and others."[34] In practice this meant scrupulous respect for the source as transmitted (to the point of not correcting what were obvious scribal errors, as was otherwise standard among his fellow philologists), familiarity with all the relevant published literature, excellent skills in Latin paleography (including the abbreviations and eccentric spellings often employed by Roman stonecutters), inspecting the inscriptions firsthand wherever they might be found (and if possible making squeezes for a still more exact record), and leaving out nothing, however apparently trivial or insignificant. Despite Mommsen's industrial rhetoric, these standards were hardly compatible with speed and efficiency.

He also oscillated between nationalism and internationalism in organizing his many projects. We have already heard his proud paean to German scholarship and science, which no longer needed to hide its light under a bushel basket. He was often deeply pessimistic about recruiting scholars of other nationalities to his vast archiving projects. In an 1874 lecture to the Prussian Academy once again calling for better organization in the "fundamental work of collecting and examination of scientific materials," he despaired of foreign allies: "That the English, the French, the Italians will help us bind the sheaves in this field is more to wish than to hope for; universalism in the realm of science is not native to these nations and Germany stands here, as always and in everything, on its own."[35] Yet he in fact not only recruited foreign scholars (many of them Italians, starting with the indispensable Borghesi) to help with the work of the CIL; he was an active supporter of any number of international projects, including the Royal Society's *International Catalogue of Scientific Papers*, the *Thesaurus Linguae Latinae* (which was to be undertaken by a "cartel" of academies), and plans for an international association of European and American academies that would coordinate large research projects. Although old and infirm, the eighty-three-year-old Mommsen made a point of attending the Paris congress of the latter organization in 1901.[36]

Finally, he was a thoroughly modern scholar of medieval sensibility. Although he was the doyen of academic entrepreneurs, mobilizing his network of contacts in the academy, the ministry, the universities, and even his own extended family (the great classical philologist Ulrich von Wilamowitz-Mollendorff was his son-in-law) to rake in money and powerful support for his ever larger projects to enrich the archives of *Altertumswissenschaft*, he never wavered in his conviction that no private person or entity could undertake such work. Only an academy, backed by the state, had the institutional longevity needed to see such *Langzeit Projekte* to their completion (and some were very long indeed, such as the interminable Leibniz edition, begun in 1901 and still as of this writing nowhere near completion). The scholars and scientists who amassed these compendia would die; others would replace them. "People come and go; science endures."[37] Like the medieval church, the academy would not only sustain but also preserve the fruits of these Herculean labors, the archive of the future.

Mommsen returned often to metaphors of the harvest: sowing seed and binding sheaves, but also crops lost to thorns and sudden storms. Like the farmer, the archivist for the future was at the mercy of fate. Mommsen warned that some, perhaps many of the academy's compendia projects might never bear fruit. And even if the crop eventually was reaped, patience was necessary, "for our fruits [. . .] in the best case ripen slowly."[38] So slowly that the farmer would never enjoy them, or even be able to imagine to what use they would be put. Compendia like the CIL were only preparation, *Vorarbeiten*, for the real work eventually to be done. This was why the archivist may omit nothing. Fusing the literal farmers' fields in which he had tracked down ancient stones with the metaphorical fields of inscriptions collected by the archivist, Mommsen pleaded for exhaustiveness: "the archivist does not ask now whether every piece that he preserves and must preserve is really worth preserving. When the broad field of Latin inscriptions is once surveyed, then the mute stone will lie without causing harm, [and] the truly fertile land will be plowed."[39] Compendia were compendious because the present could not choose for the future; it could only prepare the ground.

FOR THE ASTRONOMERS OF THE YEAR 3000: THE CARTE DU CIEL

"150 lightweight chairs, gilded and upholstered in red velvet"; "2 candelabra with seven lights each in the style of Louis XVI"; "36 cut-crystal bowls"; not to mention a small garden's worth of flower arrangements and ornamental plants.[40] It is April 1887, and the Paris Observatory awaits distinguished guests, the world's astronomical elite, who expected to be wined and dined in style. Nor were they disappointed: nine-course banquets and evening concerts leavened the long days of deliberations on whether reflecting or refracting telescopes were best suited to astrophotography, how to divide up the labor of photographing the whole sky among the eighteen participating observatories, and the merits of making a star catalogue as well as a map of the heavens.[41] Admiral Ernest Mouchez and subsequent directors of the Paris Observatory treated the Carte du Ciel from the outset as a grand state occasion, for which everything from Sèvres porcelain to Louis XIV armchairs could be requisitioned from the Mobilier National, and repeatedly reminded the French government that the success of the project was "a point of honor for France."[42]

In contrast to the CIL, planned and mostly executed by the Prussian Academy of Sciences, the Carte du Ciel was conceived from the start as an international undertaking, albeit one propelled by France's efforts to recoup prestige lost in military, imperial, and commercial defeats through cultural and scientific initiatives centered on Paris.[43] From the beginning, Mouchez collaborated with David Gill, director of the British-run observatory at the Cape of Good Hope in South Africa, and Otto Struve, director of the celebrated Russian observatory at Pulkova,[44] to organize an international project that "no astronomer could accomplish . . . in his lifetime" and that "should be undertaken on a carefully prepared plan, and be continued on a uniform system with the same instrument, persistently and continuously till the whole has been accomplished."[45] National competition was, however, perfectly compatible with this avowed spirit of international cooperation. Indeed, the astronomers counted on national rivalries to cajole more money from their governments as the gigantic project ran way over budget and schedule. From the outset, Mouchez, Gill, and Struve insisted that all invitations to astronomers in other countries be sent via diplomatic channels, despite the bureaucratic delays entailed,[46] in order that the Paris conferences to plan the Carte du Ciel might resemble diplomatic congresses to negotiate treaties, with all the pomp and circumstance necessary to solemnify the occasion and prod delegates into binding commitments. Not for nothing those gilt chairs, lavish flower arrangements, and *soirées musicales*.

In contrast, earlier collective star-mapping projects had been both more modest in scale and informal in organization. One had been funded by the Prussian Academy of Sciences at the initiative of Friedrich Bessel, director of the Königsberg observatory and member of the academy's Physical-Mathematical Class. In his proposal to the academy of 4 November 1824,[47] Bessel opined that "absolute completeness is here unattainable" and that a map of at least most stars down to the ninth or tenth magnitude would be more than sufficient for most purposes, such as spotting new comets and discovering heretofore unknown planets. He was also nonchalant about staffing and standardization of instruments and methods. Albeit time-consuming, the task of star mapping required "little equipment and still less knowledge"; a prospectus distributed in French would recruit volunteers, each of whom would then be assigned a part of the sky to chart, with a sample to guide them, and rewarded with an honorary medal of "20 to 25 ducats" when they delivered their assignment. Yet the exalted language of a "great and celebrated monument" willed to posterity had already crept into Bessel's decidedly uncompendious proposal, and it was to assume almost megalomaniac proportions in the rhetoric of the Carte du Ciel, advertised as a compendium of heaven itself.

From 16 through 25 April 1887 fifty-eight astronomers from sixteen countries plus three colonies met in Paris at the invitation of Mouchez, on behalf of the Paris Observatory, and the Paris Academy of Sciences to plan what one contemporary called "the greatest venture yet undertaken in astronomy,"[48] namely a complete photographic map of the sky, including all stars to the fourteenth magnitude.[49] Only the combined and prolonged efforts of almost a score of observatories in both the Northern and Southern Hemispheres could produce what promoters hailed as an "imperishable monument," a photographic record of "the authentic state of the universe visible from the earth at the close of the nineteenth century."[50] The proportions of the project were indeed monumental in every sense: eighteen observatories around the world, from Helsinki at +60.9 degrees latitude to Melbourne at –37.5[51] labored for decades—publication of the catalogue was not completed until 1964[52]—to amass charts projected already in 1912 to stack 32 feet high and weigh about 4,000 pounds.[53] Armed with this snapshot of the sky circa 1900, future astronomers would be able, it was hoped, to detect changes in the heavens that unfolded on too long a timescale to be perceptible within a short human lifetime—appearance of new stars, nebulae, and comets, the telltale motion of as yet undiscovered planets, the extended periods of variable stars, the incremental proper motions of the so-called fixed stars.[54] As the deliberations of the 1887 International Congress and of subsequent meetings (1889, 1891, 1896, 1900, 1909) of the Permanent Committee make clear, the intricate coordination of telescopes, photographic plates, micrometric measurements, and myriad other details to insure that the parts of the map would be commensurable required that participants relinquish control not only over instruments and methods but also over the choice of research area for years to come.[55] By uniting astronomers around the world and across generations, the Carte du Ciel aspired to nature's own Brobdingnagian scale.

Astrophotography, pioneered by the brothers Paul and Prosper Henry in Paris using a refracting telescope and by Andrew Ainslie Common and Isaac Roberts in Britain with reflecting telescopes,[56] made this immense project conceivable, and its supporters at times invoked the ideals of mechanical objectivity, of images made "by photography alone and without the intervention of any human errors."[57] As in the case of the paper squeezes made of inscriptions for the CIL, the mechanical and therefore deliberately unselective character of the reproductions was intended to guarantee the value of the Carte du Ciel as an archive of the future. Who could predict which apparently trivial detail might suddenly become significant in the context of as-yet-undreamed-of research programs?[58] As part of the published compendium of the Carte du Ciel, the photographic plates were to be printed at vast expense (estimated at about one million francs for the French observatories alone in 1898)[59] by a special photoengraving (*héliogravure*) technique (fig. 6.3). Although each star appeared on at least two plates,

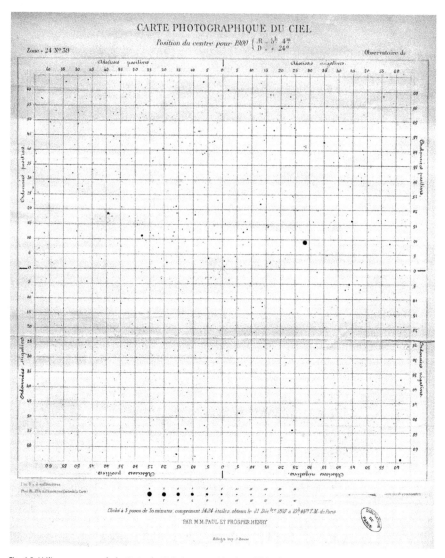

Fig. 6.3 Héliogravure proof of a Carte du Ciel photographic plate. Bibliothèque de l'Observatoire de Paris.

there were worries that the engraver might introduce errors and hence calls for meticulous proofreading of each engraving (which might contain upwards of 3,000 stars) by at least two people.[60]

This laborious process would squander all the advantages of speed gained by recording stars photographically rather than manually, and proofreading would inevitably insinuate further errors. Only the plates themselves could serve the astronomers of the future as an unimpeachable source for the number and position of stars observable from the Earth circa 1900, which is why they have been solicitously preserved even after the printing of the star map.[61]

Analogous to the paper squeezes behind the printed volumes of the CIL, the plates of the Carte du Ciel were the archive within the archive. The published compendia—transcriptions, maps, catalogues, even the images of the squeezes and photographic plates—served to disseminate and standardize the objects of inquiry for entire disciplines and for centuries to come. Philologists are still consulting and updating the CIL; astronomers have used the Carte du Ciel to detect stellar motion and dark matter.[62] But the material contained in the compendia was hardly raw data: the inscriptions must be reproduced (whether by squeeze, drawing, or photograph), transcribed, and interpreted; the plates must be checked, measured, and reduced by calculation.[63] The archives *were* the squeezes and the plates— not the rows of published volumes nor even the stones menaced by weather and railway lines and the ephemeral stellar positions. Stored in tailor-made cabinets and sleeves, the archives awaited a new query (or a new doubt) that no one could have anticipated and that the published compendia were never intended to resolve. The compendia consolidated disciplines around a shared, essential source; the archives behind the compendia were the discipline's insurance policy against an unpredictable future.

Like the CIL, the Carte du Ciel invoked industrial metaphors to capture the novelty of its undertaking—the grandeur of its scale (millions of stars instead of a few hundred thousand); its global scope; its remarkable success in the coordination of telescopes, film emulsion, measuring machines, and many other technical details, often requiring considerable sacrifice on the part of some participating observatories.[64] These sacrifices were particularly grave for the smaller observatories participating in the Carte du Ciel project: for example, the Australian observatories of Sydney, Perth, Melbourne, and Adelaide took eighty years to complete their three assigned zones of the sky (18 percent of the entire sky), at the price of missing out on much of twentieth-century astrophysics and spectroscopy.[65] The aim was standardization as thorough as any manufacturing process, in order to insure commensurable results: the parts of the great star map had to fit together seamlessly.

And as in the case of the CIL, the industrial rhetoric sometimes extended to the labor force required for the more "mechanical" aspects of the Carte du Ciel, such as the measurement of the star positions on the photographic plates using purpose-built machines outfitted with microscopic lenses and grids, as well as calculating the positions against a uniform coordinate system.[66] Just as Mommsen had wanted to enlist Gymnasium students for the more humdrum tasks of the CIL, Oxford astronomer H. H. Turner was by 1911 able to complete at least that part of the astrophotographic catalogue assigned by the International Congress to his observatory by resorting to the labor of poorly paid schoolboys and legions of volunteers, although the Oxford contribution to the photographic map was never completed.[67] The French observatories preferred to hire women at the cut-rate wage of 75 centimes an hour, and hoped that two or three working for three to four years at the Observatory of Paris might complete half the star catalogue, about a million stars.[68]

Yet as in the case of the CIL, upon closer inspection the division of labor in the Bureau of Measurements turned out to resemble the artisanal workshop as much as the factory assembly line. The French attempted to centralize measurements for the entire Carte du Ciel at the Paris Observatory—in part to underscore French leadership in the project but also to insure uniformly high quality of workmanship. Directed by Dorothea Klumpke, an American who received her doctorate in astronomy from the Sorbonne in 1893 and became the first woman to work at the Paris Observatory, the Paris Bureau of Measurements was deemed essential to the success of the Carte du Ciel. Klumpke did not eventually advance to a professorship, as Mommsen's lieutenants did, but she was awarded various honors by the Société Astronomique and the French Academy of Sciences and was in all likelihood the model for the woman calculator who warns the world of an impending collision with an asteroid in Camille Flammarion's science fiction novel, *La fin du monde* (1894). Reflecting in 1921 on the qualities required for the measurement of the plates, Klumpke modestly listed "patience, scientific conscientiousness, some intelligence, and much good will."[69] However, the fact that she was the only woman (besides the wife and daughter of Admiral Mouchez) to attend the gala dinner held for the delegates of the 1887 international congress suggests that she was hardly regarded as an anonymous cog in the machinery of the Carte du Ciel.[70]

Although the use of schoolboys and women paid a pittance (if they were paid at all) in the Carte du Ciel and of apprentice epigraphists in the CIL has often been described as the industrialization of science, the labor organization of these projects resembles an equally hierarchical but more ancient model of the division of labor still very much in evidence in nineteenth-century science: the patriarchal family. Just as Charles Darwin's entire household was mobilized in support of his scientific pursuits, the Mommsen children were trained to file Latin inscriptions as soon as they had mastered the alphabet.[71] Nineteenth-century German classicists made a habit of marrying into each other's families; the genealogy of terms like *Doktorvater* hardly needs belaboring.[72] Big Science nineteenth-century style still bore the imprint of the old-fashioned family business, despite its metaphors of newfangled industrial organization. The incessant appeals to duty and self-abnegation in the name of the discipline's future, made by both the CIL and the Carte du Ciel, would have rung hollow to workers in a gigantic factory. But they resonated in the context of a transgenerational family undertaking.

CONCLUSION: THE PATHOS OF POSITIVISM

Only the compendious ambitions of the Carte du Ciel to map forty to fifty million stars and to thereby produce "what will certainly be the most important scientific document willed by the nineteenth century to future centuries and the source of discoveries of the highest interest for knowledge of the universe" could have justified the sacrifices of especially the outlying observatories far from centers like Paris and Greenwich. For these collaborators in the Carte du Ciel, almost all of their limited resources were sunk into the project for generations. The Vatican Observatory was the only institution to mobilize nuns for the task of measuring the plates,[73] but the displaced language of religious commitment, of selfless und unstinting labor in the service of the hereafter was ubiquitous among the project's moving spirits. Conjuring up images of pyramids, cathedrals, and other architectural grandiosities that had enlisted armies of laborers, late nineteenth-century adherents to the faith of the archives appealed to that which endured after time had erased all other traces of civilizations past. Despite all the positivist bombast that surrounded these gigantic projects to compile every last Latin inscription or the positions of millions of stars, melancholy lurked at their heart.

Or rather, positivism in the latter half of the nineteenth century was intrinsically melancholy. In contrast to the swaggering optimism of firstwave, Comtean positivism of the 1830s, with its three-stage schemes and utopian promises of a social physics (later social physiology) just around the corner, the positivism of the 1880s and '90s was less exhilarated by the promise of scientific progress than disconcerted by its relentless pace. In 1830 Auguste Comte, self-declared prophet of positivism, had written confidently of the "fundamental revolution [that] will be necessarily accomplished in its full extent"[74]—nothing short of the happy end of the history of science. Some forty years later, Austrian physicist Ernst Mach wrote resignedly of perpetual revolutions in science, history and no end: "In fact, if one learned nothing more from history than the mutability of [scientific] views, then it would still be invaluable. [. . .] One gradually accustoms oneself [to the fact] that science is incomplete, mutable."[75] And mutable at allegretto tempo: new theories succeeded one another faster than Parisian fashions. Applied mathematician and physicist Henri Poincaré wrote elegiacally in 1902 of how ephemeral scientific theories had become, of "ruins piled upon ruins."[76] Positivism, which had trumpeted the final victories of science in the 1830s, ended the century intimidated by those very same victories—so many of them, in such swift and seemingly endless succession. Only modest but stable facts seemed to offer shelter from breakneck, never-ending scientific progress.

This was the pathos of positivism, and the key to the strangely mixed emotional register of nineteenth-century Big Science, at once megalomaniac (180,000 Latin inscriptions! 50 million stellar positions!! forever and ever!!!) and modest (stone placed upon stone by anonymous laborers, crops planted now but harvested only by future generations). Monuments and ruins vied for pride of place in the metaphorical imagination of Big Science entrepreneurs. What present science could secure was only the archive, no longer eternal truths. No philological commentary, no interpretation of ancient history, no astronomical theory from the present would endure even a lifetime, much less an eon. In his celebrated 1917 lecture "Science as Vocation," Max Weber drew the lessons of late nineteenth-century science with bitter clarity: scientists and scholars must be resigned to the inevitability that all they have done will be surpassed in ten or twenty years. A scientific career was written on water. This was the context in which Weber formulated his tough-minded, teeth-clenched view of modernity as the "disenchantment of the world," shadowed by the very personal disenchantment of the late nineteenth-century scientist with a seemingly solid science that melted into air.[77] And this was the context in which projects like the CIL and the Carte du Ciel promised a more sober enchantment of eternal truths replaced by eternal facts—all of them, always, safe in the archives of science.

ACKNOWLEDGMENTS

I am grateful to the archivists at the Berlin-Brandenburgische Akademie der Wissenschaften and the Bibliothèque de l'Observatoire de Paris for their expert help in locating relevant documents and to Philip Aubreville and Andréas Richier for invaluable research assistance. Unless otherwise noted, all translations are my own.

NOTES

1. See for example Jim Endersby, *Imperial Nature: Joseph Hooker and the Practices of Victorian Botany* (Chicago: University of Chicago Press, 2008), on the role of Kew in botany, and Charlotte Klonk, *Spaces of Experience: Art Gallery Interiors 1800–2000* (New Haven: Yale University Press, 2009), on the arrangement of key British and German collections.

2. For analogous developments in paleontology, see David Sepkoski, chapter 2 in this volume.

3. [Félix Tisserand], draft of letter to Ministère de l'Instruction Publique, early 1890s, Bibliothèque de l'Observatoire de Paris, MS 1060-IV-A-2-3, Boite 25.

4. Wilhelm Larfeld, *Handbuch der griechischen Epigraphik*, 2 vols. (Leipzig: O. R. Reisland, 1907), vol. 1, 254.

5. For lack of a better English term, I use the "human sciences" to cover what in English would be called the humanities (including some social sciences), in German the *Geisteswissenschaften*, and in French *les sciences humaines*. On the emergence of such terms in the nineteenth century see Ernst Rothacker, *Logik und Systematik der Geisteswissenschaften* (Bonn: H. Bouvier u. Co. Verlag, 1947), 4–16.

6. Rüdiger vom Bruch, "Mommsen und Harnack: Die Geburt von Big Science aus den Geisteswissenschaften," in *Theodor Mommsen: Wissenschaft und Politik im 19. Jahrhundert* (Berlin: Walter de Gruyter, 2005), ed. Alexander Demandt et al., 121–41; Stefan Rebenich, "'Unser Werk lobt keinen Meister': Theodor Mommsen und die Wissenschaft von Altertum," in *Theodor Mommsen: Gelehrter, Politiker und Literat*, ed. Josef Wiesehöfer (Stuttgart: Franz Steiner Verlag, 2005), 185–205. Derek J. de Solla Price seems to have reinvented the term independently in his *Little Science, Big Science* (New York: Columbia University Press, 1963).

7. Theodor Mommsen, "Antwort auf Harnack," 3 July 1890, in Theodor Mommsen, *Reden und Aufsätze*, 2nd ed. (Berlin: Weidmannsche Buchhandlung, 1905), 208–10, on 209.

8. Revue Germanique Internationale, *La fabrique internationale de la science: Les congrès scientifiques de 1865 à 1945* (Paris: CNRS Editions, 2010); Bärbel Holtz, "Preußens Kulturstaatlichkeit im langen 19. Jahrhundert im Fokus seines Kultusministeriums," in *Kulturstaat und Bürgergesellschaft: Preußen, Deutschland und Europa im 19. und frühen 20. Jahrhundert*, ed. Wolfgang Neugebauer and Bärbel Holtz (Berlin: Akademie Verlag, 2010), 55–77; Suzanne L. Marchand, *Down from Olympus: Archaeology and Philhellenism in Germany, 1750–1970* (Princeton: Princeton University Press, 1996), 62–65 (on the Prussian subsidies for Richard Lepsius' 1842–45 expedition to Egypt) et passim; "Rapport adressé à M. le Directeur de l'Enseignement Supérieur," 27 May 1899, Observatoire de Paris, MS IV-A-3, Boite 25.

9. [Ambroise Firmin Didot], *Projets et rapports relatifs d'un recueil général d'épigraphie latine* ([Paris?]: [Firmin Didot?], [1843?]), 3n1. This brochure seems to have been printed privately by the publisher Firmin Didot, scion of a famous French publishing dynasty and himself a classicist of note.

10. Ileana Chinnici, "La Carte du Ciel: Genèse, déroulement et issues," in *La Carte du Ciel: Histoire et actualité d'un projet scientifique international*, ed. Jérôme Lamy (Paris: Observatoire de Paris, 2008), 19–43.

11. Theodor Mommsen, "Rede am Leibnizschen Gedächtnistage," 2 July 1874, in idem, *Reden und Aufsätze*, 39–49, on 47.

12. Theodor Mommsen, "Antrittsrede," delivered 8 July 1858 to the Königliche Preussische Akademie der Wissenschaften, in idem, *Reden und Aufsätze*, 35–38, on 37.

13. Ernest Mouchez to David Gill, 30 April 1887, Bibliothèque de l'Observatoire de Paris, MS IV.A, "Comité international de la Carte du Ciel," carton 7.

14. Theodor Mommsen, "Rede bei Antritt des Rektorates," 15 October 1874, in idem, *Reden und Aufsätze*, 3–16, on 5, 7, 14. On the cult of *Quellenforschung* and other critical methods, see Suzanne Marchand's essay in this volume (chapter 5) and Glenn W. Most, "*Quellenforschung*," in *The Making of the Humanities*, ed. Rens Bod, Jaap Maat, and Thijs Weststeijn (Amsterdam: Amsterdam University Press, 2014), 207–17.

15. Theodor Mommsen, "Antrittsrede," 8 July 1858, in idem, *Reden und Aufsätze*, 35–38, on 35.

16. On other such archival projects relating to ancient history sponsored by the Prussian Academy of Sciences, see Suzanne Marchand, chapter 5 in this volume.

17. Bruch, "Mommsen und Harnack," 133.

18. Wilhelm Larfeld, *Handbuch der griechischen Epigraphik*, 2 vols. (Leipzig: O. R. Reisland, 1907), vol. 1, 39–67.

19. Both the French and Prussian CIL projects (as well as Kellermann) acknowledged that Borghesi's cooperation was an essential precondition for the success of their ventures: Archiv der Berlin-Brandenburgishe Akademie der Wissenschaften, *Akten der Preußischen Akademie der Wissenschaften, 1812–1945*, Titel CIL, nr. 1575 (copy of letter from Borghesi to Kellermann, 31 July 1835); Theodor Mommsen, *Tagebuch der französisch-italienischen Reise 1844/1845*, ed. Gerold and Brigitte Walser (Bern: Verlag Herbert Lang, 1976), 185; Firmin Didot, *Projets et rapports*, 2.

20. Firmin Didot, *Projets et rapports*, 3n1.

21. Lothar Wickert, ed., *Theodor Mommsen–Otto Jahn: Briefwechsel 1842–1868* (Frankfurt am Main: Vittorio Klostermann, 1962), 20–64; Thorsten Kahlert, "Theodor Mommsen, informelle Netzwerke und die Entstehung des Corpus Inscriptionum Latinarum um 1850," in *Geschichtsforschung in Deutschland und Österreich im 19. Jahrhundert: Ideen—Akteure—Institutionen*, ed. Christine Ottner and Klaus Ries (Stuttgart: Franz Steiner Verlag, 2014), 180–97.

22. Adolf Harnack, *Geschichte der Königlich Preussischen Akademie der Wissenschaften zu Berlin*, 3 vols. (Berlin: Reichsdrückerei, 1900), v. I.2, 896–914, provides a detailed account.

23. Theodor Mommsen, *Ueber Plan und Ausführung eines Corpus Inscriptionum Latinarum* (Berlin: A. W. Schade, 1847), 4–8, in *Acta der wissenschaftlichen Unternehmungen der philosophisch-historischen Klasse*, vol. 17,a: Sammlungen lateinischer Inschriften, 1836–48, Archiv der Berlin-Brandenburgische Akademie der Wissenschaften, PAW-II-VIII.96.

24. Larfeld, *Handbuch*, vol. 1, 76.

25. Mommsen, *Tagebuch*, 190 (Fermo); Tafel III (Castel di Sangro).

26. Emil Hübner, *Über mechanische Copien von Inschriften* (Berlin: Weidmannsche Buchhandlung, 1881), 5–10. Although the French CIL commission intended to work mostly from published sources and manuscripts, they encouraged correspondents to send them squeezes of inscriptions and included detailed instructions in their prospec-

tus: Joseph Tastu, "Instructions pour l'estampage des inscriptions," in Firmin Didot, *Projets et rapports*, 33–35.

27. Mommsen, *Tagebuch*, 61.

28. Mommsen, *Tagebuch*, 168.

29. Hübner, *Über mechanische Copien*, 5.

30. Manfred G. Schmidt, "Spiegelbilder römischer Lebenswelt: Inschrift-Clichés aus dem Archiv des Corpus Inscriptionum Latinarum ausgewählt und kommentiert," in idem, *150 Jahre Corpus Inscriptionum Latinarum* (Berlin: Walter de Gruyter, 2003), 3–32.

31. Mommsen, *Ueber Plan und Ausführung*, 29–30.

32. Rebenich, "Unser Werk lobt keinen Meister," 195.

33. When in November 1882 one of Mommsen's young epigraphists, Johannes Schmidt, went missing in southern Tunisia (then occupied by French troops) while in search of inscriptions, Mommsen fretted about the safety of both apprentice and project: Stefan Rebenich and Gisa Franke, eds., *Theodor Mommsen und Friedrich Althoff: Briefwechsel 1882–1903* (Munich: Oldenbourg Verlag, 2012), 6.

34. Theodor Mommsen's obituary for Otto Jahn, quoted in Rebenich, "Unser Werk lobt keinen Meister," 188.

35. Mommsen, "Rede am Leibnizschen Gedächtnistage," 2 July 1874, in idem, *Reden und Aufsätze*, 39–49, on 48.

36. Brigitte Schroeder-Gudehaus, "Die Akademie auf internationalem Parkett: Die Programmatik der internationalen Zusammenarbeit wissenschaftlicher Akademien und ihr Scheitern im Ersten Weltkrieg," in *Die Königliche Preussische Akademie der Wissenschaften*, ed. Jürgen Kocka (Berlin: Akademie Verlag, 1999), 175–95, on 177, 185; Rebenich, "Unser Werk lobt keinen Meister," 196.

37. Mommsen, "Ansprache am Leibnizschen Gedächtnistage," 3 June 1887, in idem, *Reden und Aufsätze*, 154–56, on 156.

38. Mommsen, "Rede zur Vorfeier des Geburtstags des Kaisers," 18 March 1880, in idem, *Reden und Aufsätze*, 89–103, on 102–3.

39. Mommsen, "Antrittsrede," delivered 8 July 1858 to the Königliche Preussische Akademie der Wissenschaften, in idem, *Reden und Aufsätze*, 35–38, on 38.

40. "Soirées: Dîners à l'occasion des réunions du Comité de la Carte du Ciel." Bibliothèque de l'Observatoire de Paris, MS 1060.IV-A-2-3, Carton 25.

41. On the Carte du Ciel see Théo Weimer, *Brève histoire de la Carte du Ciel en France* (Paris: Observatoire de Paris, 1987); Suzanne Débarat et al., eds., *Mapping the Sky: Past Heritage and Future Directions* (Dordrecht: Kluwer, 1988); and Lamy, *La Carte du Ciel*. Correspondence relating to the project is available in Ileana Chinnici, *La Carte du Ciel: Correspondence inédite conservée dans les Archives de l'Observatoire de Paris* (Paris: Observatoire de Paris,1999).

42. Admiral Mouchez, Directeur de l'Observatoire de Paris, au Ministre de l'Instruction Publique, 25 April 1891, Bibliothèque de l'Observatoire de Paris, MS 1060. IV-A-2, Carton 24.

43. Anne Rasmussen, "L'Internationale scientifique, 1890–1914," thesis, Ecole des Hautes Etudes en Sciences Sociales, 1995.

44. Simon Werrett, "The Astronomical Capital of the World: Pulkova Observatory in the Russia of Tsar Nicholas," in *The Heavens on Earth: Observatories and Astronomy*

in Nineteenth-century Science and Culture, ed. David Aubin, Charlotte Bigg, and H. Otto Sibum (Durham: Duke University Press, 2010), 33–57.

45. D. Gill to E. Mouchez, Royal Observatory of the Cape of Good Hope, 23 February 1885, Observatoire de Paris, MS IV.A, Carton 7.

46. E. Mouchez to D. Gill, Observatoire de Paris, 18 October 1886. Bibliothèque de l'Observatoire de Paris, MS IV.A, Carton 7, Folder 1.

47. Friedrich Bessel, Akademische Sternkarte (dated 15 October 1824, read 4 November 1824), *Acta der wissenschaftlichen Unternehmungen der Mathematischen Klasse*, vol. 1 (1824–37), Archiv der Berlin-Brandenburgische Akademie der Wissenschaften, sig. II–VII, 3, 3r.–4v.; 47v. (suggested standardization by following the sample included in the prospectus).

48. Julius Scheiner, *Die Photographie der Gestirne* (Leipzig: Wilhelm Engelmann, 1897), 311.

49. A star catalogue down to the eleventh magnitude was also planned as part of the Carte du Ciel project. On astrophotography see John Lankford, "The Impact of Photography on Astronomy," in *Astrophysics and Twentieth-Century Astronomy to 1950*, ed. Owen Gingerich (Cambridge: Cambridge University Press, 1984), 16–39.

50. Camille Flammarion, "La photographie céleste à l'Observatoire de Paris," *Revue d'astronomie populaire* 5 (1886): 42–57, on 55. A list of the participants in the 1887 congress is given in Chinnici, "La Carte du Ciel," 23–24.

51. For the zone assignments of individual observatories see Nathy P. O'Hora, "Astrographic Catalogues of British Observatories," in Débarat et al., *Mapping*, 135–38, on 136. For a tabulation of the final contributions, see Lankford, "Impact," 30.

52. Commission 23 of the International Astronomical Union, established in 1919 to oversee the Carte du Ciel, was dissolved in 1970: Théo Weimer, "Naissance et développement de la Carte du Ciel," in Débarat et al., *Mapping*, 29–32, on 30.

53. Herbert H. Turner, *The Great Star Map* (New York: E. P. Dutton, 1912), 145. The project ultimately produced some 22,000 plates: Lankford, "Impact," 30.

54. Ernest B. Mouchez, *La photographie astronomique à l'Observatoire de Paris et la Carte du Ciel* (Paris: Gauthier-Villars, 1887).

55. Institut de France–Académie des Sciences, *Congrès astrophotographique international tenu à l'Observatoire de Paris pour le levé de la Carte du Ciel* (Paris: Gauthier-Villars, 1887); Albert G. Winterhalter, *The International Astrophotographical Congress and A Visit to Certain European Observatories and Other Institutions: Report to the Superintendent* (Washington: Government Printing Office, 1889); and the irregularly published *Bulletin du Comité Permanent International pour l'Exécution Photographique de la Carte du Ciel*.

56. On the Henry brothers and the French telescopes, see Françoise Le Guet Tully, Jérôme De la Noë, and Hamid Sadsaoud, "L'Opération de la Carte du Ciel dans le contexte institutionnel et technique de l'astronomie française à la fin du XIXe siècle," in Lamy, *La Carte du Ciel*, 69–107; on Common's and Roberts' photographic work with reflectors, see John Lankford, "Amateurs and Astrophysics: A Neglected Aspect in the Development of a Scientific Specialty," *Social Studies of Science* 11 (1981): 275–303.

57. Camille Flammarion, "Le Congrès astronomique pour la photographie du ciel," *Astronomie* 6 (1887): 161–69, on p. 163. On mechanical objectivity, see Lorraine Daston and Peter Galison, *Objectivity* (New York: Zone Books, 2007), chapter 3.

58. For analogous views on the open-endedness of future research in medicine, see Andrew Mendelsohn, chapter 3 in this volume.

59. "Carte photographique du Ciel. Observatoires français. Conférence tenue à l'Observatoire de Paris, April 1898," Bibliothèque de l'Observatoire de Paris, MS 1060 IV-A-2.

60. "Projet de Budget pour 1895: Carte photographique du Ciel," Bibliothèque de l'Observatoire de Paris, MS. 1065.

61. Françoise Le Guet Tully et al., "Les traces matérielles de la Carte du Ciel: Le cas des observatoires d'Alger et Bordeaux," in Lamy, *La Carte du Ciel*, 213–35.

62. On the CIL see http://cil.bbaw.de/index.html, accessed 1 August 2014; on uses of the Carte du Ciel, see Alain Fresneau, "La détection de la matière interstellaire sur les plaques photographiques de la Carte du Ciel," in Lamy, *La Carte du Ciel*, 155–67.

63. On the necessity of stabilizing data, see the essays by Florence Hsia (chapter 1) and Vladimir Jancović (chapter 9) in this volume.

64. Peter Galison and Lorraine Daston, "Scientific Coordination as Ethos and Epistemology," in *Instruments in Art and Science: On the Architectonics of Cultural Boundaries in the 17th Century*, ed. Helmar Schramm, Ludger Schwarte, and Jan Lazardzig (Berlin: De Gruyter, 2008), 296–333.

65. Graeme L. White, "The Carte du Ciel—The Australian Connection," in *Mapping*, ed. Débarat et al., 45–51, on 48; cf. Lankford, "Impact," 32, on the converse advantages to American observatories that did not participate in the Carte du Ciel.

66. Jérôme Lamy, "La Carte du Ciel et l'ajustement des pratiques (fin XIXe–debut XXe siècle)," in Lamy, *La Carte du Ciel*, 45–67.

67. O'Hora, "Astrographic Catalogues," 137.

68. As a rough comparison, a woman working at this wage for fifty hours a week, fifty weeks a year, would receive an annual salary 1,875 francs; each of the measuring machines she used cost 3,300 francs. "Mesures des clichés [1891]," Bibliothèque de l'Observatoire de Paris, MS 1060 IV-A-2-3, Boite 25. There is a long history of using poorly paid (often female) labor for massive calculation and archival projects: see Ivor Grattan-Guiness, "Work for the Hairdressers: The Production of Prony's Logarithmic and Trigonometric Tables," *Annals of the History of Computing* 12 (1990): 177–85; Lorraine Daston, "Enlightenment Calculations," *Critical Inquiry* 21 (1994): 182–202; Simon Schaffer, "Babbage's Intelligence: Calculating Engines and the Factory System," *Critical Inquiry* 21 (1994): 203–27; David Alan Grier, *When Computers Were Human* (Princeton: Princeton University Press, 2006); and Daniel Rosenberg, chapter 11 in this volume.

69. Mme Isaac Roberts, née Dorothea Klumpke, Le Havre, 17 July 1921. Bibliothèque de l'Observatoire, MS 1021 (15). Klumpke married the British astronomer Roberts, whom she had first met through the Carte du Ciel, in 1901.

70. "Dîner du 24 Avril 1887," in Folder "Soirées. Dîners à l'occasion des réunions du Comité de la Carte du Ciel." Bibliothèque de l'Observatoire de Paris, MS 1060.IV-A-2-3, Carton 25.

71. Janet Browne, *Charles Darwin: Voyaging* (New York: Knopf, 1995), 357–65, 519–30; Adelheid Mommsen, *Theodor Mommsen im Kreise der Seinen*, 2nd ed. (Berlin: Verlag E. Ebering, 1937), 13–14.

72. Marchand, *Down from Olympus*, 50–51.

73. Chinnici, "La Carte du Ciel," 38.

74. Auguste Comte, *Cours de philosophie positive*, 6 vols. (Paris: Bachelier, 1830–42), vol. 1, 16.

75. Ernst Mach, *Die Geschichte und die Wurzel des Satzes von der Erhaltung der Arbeit* [1872], 2nd ed. (Leipzig: Johann Ambrosius Barth, 1909), 3.

76. Henri Poincaré, *La science et l'hypothèse* [1902] (Paris: Flammarion, 1968), 173.

77. Max Weber, "Wissenschaft als Beruf," in Max Weber, *Gesammelte Aufsätze zur Wissenschaftslehre*, ed. Johannes Winckelmann, 3rd ed. (Tübingen: J. C. B. Mohr, 1968), 612.

PART III

Problems and Politics: Controversies in the Global Archive

SEVEN

The "Data Deluge": Turning Private Data into Public Archives

Bruno J. Strasser

The "data deluge" is an interesting metaphor (fig. 7.1). Widely used since the 1990s, it attempts to capture the process resulting in the current "data flood," the immense amount of digital data about nature, people, and societies. But one of the most puzzling connotations of this metaphor is that data, like rain in the most famous deluge of all, seems to pour down naturally, pulled only by the laws of gravity, submerging the earth. All those who have paid close attention to archives and databases have however seen the data deluge in a very different light. Data did not fall naturally upon them. Data was something that had to be actively sought out. Creating a flow of data from source toward repositories was far more challenging than just waiting for rain to fall from the heavens. This chapter aims to understand how such data flows were created and sustained in the late twentieth-century experimental life sciences. It argues that the data deluge was not simply the product of technological revolutions in the modes of data production and a general increase in the amount of data being produced ("big data"). Rather, these developments resulted from two historically significant transformations: a redefinition of what counts as "data" and also of the obligations attached to possessing "data." These changes deeply affected how data came to be collected. But they did not upset the existing moral economy of the experimental sciences, based on individual authorship, credit, and rewards. Instead, individual and collective interests aligned in new ways. This chapter addresses a crucial aspect of the sciences of the archives: how the archives became filled with data available for public use.[1]

Fig. 7.1 *The Economist*, February 27, 2010, reported on the effects of the "data deluge" in a variety of fields, from the stock market to national security.

The literature about "big data" is growing almost as fast as "big data" itself. The popular *Big Data: A Revolution That Will Transform How We Live, Work, and Think* (2013), by Internet analyst Viktor Mayer-Schönberger and media commentator Kenneth Cukier, or the more scholarly *Reinventing Discovery: The New Era of Networked Science* (2012) by the physicist and writer Michael Nielson and *Too Big to Know* (2012) by the Harvard Internet scholar

David Weinberger contain many insights into how the availability of big data can change how knowledge is produced.[2] But all of these authors take for granted that the amount of data available publicly is solely determined by the rate of data production. Nielson, for example, summarized the situation for genetic data: "Each time [researchers] obtained a new chunk of genetic data in their laboratories, they uploaded that data to a centralized online service such as GenBank."[3] Had these authors been less impressed by the growing rate in the production of data, they might have paid closer attention to how data was actually collected and made publicly available. They might also have realized that researchers did not simply "upload that data to a centralized online services" once they had obtained it. Indeed, the single most important concern among all those who have developed the databases and other infrastructures of big data science was how to compel researchers to share their data.[4] In 1976, for example, the managers of the first international database in protein crystallography (the Protein Data Bank) admitted to the research community that "in spite of our recent rapid rate of growth . . . we are aware that the Bank lacks data." They went on to "urge investigators to deposit" their data as it became available.[5] Parting ways with recent concern over the overabundance of data, this chapter focuses on the *scarcity* of data, examining how the data deluge, far from being a natural phenomenon or the result of a technological revolution, was achieved only slowly and in the teeth of resistance by those who envisioned how publicly available big data could transform the production of scientific knowledge.

Under specific historical circumstances, various collections of scientific things and "data," to use today's term, have been turned into "archives," making possible the development of "sciences of the archives." Since the Middle Ages, the word "archives" has designated both a place and what it contains, namely documents about the history of a people or institution.[6] Archives were established as a link between the past and the future with imagined uses and users (see Lorraine Daston's introduction to this volume). State genealogical archives, for example, have permitted the authentication of family relations and patrimonial inheritance. In the sciences, when things and "data" began to be collected, not only as prized possessions for the present, but as a resource for the future, they became scientific "archives." In natural history, the term "archive" was commonly used to designate a collection of specimens (see David Sepkoski, chapter 2 in this volume) and these "scientific archives," such as herbaria and zoological collections, have played an essential rôle in the production of knowledge. Collecting, comparing, and classifying have been key epistemic practices applied to the archives of natural history. By contrast, in the experimental life sciences, scientific archives have been marginal, and the term "archive" rarely used, with the notable exception of experimental medicine, in which a number of journals were titled "Archiv," constituting a public collection of results intended for future use (see J. Andrew Mendelsohn, chapter 3 in this volume). However, there seems to have been no strict equivalent in the rest of the experimental life sciences, in which knowledge was produced through new experiments, not the systematic comparison of results of previous ones. Only in the late twentieth century did such comparisons of experimental results become a common practice in the experimental life sciences, transforming them into "sciences of the archives," as one might call them in view of how they function epistemically. Given that the experimental life sciences had rested on a very different epistemic tradition for most of their history, it is unsurprising that this recent transformation, culminating in the current data deluge, was far more laborious than current commentators have imagined.

COLLECTIVE COLLECTIONS

To understand the specific historical changes in data collection at the end of the twentieth century, it is useful to take a broader view and examine how things and data were collected in previous centuries and stored in scientific collections and archives—almost always a collective enterprise.[7] The sciences that relied on collections, from astronomy to zoology, have often involved a very wide range of participants, most of whom could be, since the nineteenth century, labeled as "amateurs," in that they did not make a living through their collecting practices. All of the great natural history collections of rocks, plants, and animals were gathered with the help of countless amateurs, often with exceptional levels of "lay expertise" in their field of specialty.[8] The same holds true for astronomy, in which amateurs were crucial, for example, to the collection of comet observations, or for meteorology, in which they played an essential role in the systematic recoding of rare and common phenomena across large spaces.[9]

The extensive involvement of amateurs in the sciences of the archives was no historical accident. Because these sciences required the collection of observations and things across vast geographic expanses or even the entire world, they needed observers who were physically present in diverse locales. The great scientific expeditions could collect large amounts of observations and things, but only at great expense and for a limited time period. They could hardly compete with the long-term presence of observers around the globe, provided these could be trained to supply standardized observations.[10] Thanks to intimate knowledge of their immediate surroundings, these "resident observers" could gather observations that the "traveling observer" would often overlook.[11]

The enrollment of large numbers of amateurs in a collective research project depended on the possibility of providing some kind of financial, symbolic, or personal reward. The commodification of natural objects stimulated a growing market where researchers could simply buy specimens from plant and animal dealers, recreational hunters, and private collectors.[12] More importantly, these sciences produced knowledge based on objects—ferns, crystals, clouds—that were visible to the (trained) human eye and had long been part of a vernacular culture. In nineteenth-century England, non-scientists could be passionate about ferns, discuss them in pubs, and hold them in their homes as prized cultural items, whereas for scientists these ferns were specimens to be named, classified, and theorized about.[13] Corals in eighteenth-century France tell a similar story: simultaneously beautiful ornaments of aristocratic salons and scientific objects for naturalists.[14]

With the rise of the experimental sciences, especially since the late nineteenth century, the relationship between professional and amateurs changed drastically. From its origins, the laboratory was a private space, located in the home of an experimentalist.[15] Strict control over who could access the laboratory was key to its epistemic function. The laboratory was accessible only to a select few gentlemen, not the broader public.[16] As the laboratory came to hold increasingly complex, expensive (and sometimes dangerous) equipment, it became almost exclusively located in research institutions, thus deepening the divide between professional scientists and the public. In the twentieth century, a few sciences, mainly those in the natural history tradition, still relied extensively on amateurs, but these sciences were becoming increasingly marginalized, in terms of both budgets and prestige, by the experimental sciences. The experimental sciences, on the other hand, became fully professionalized: there was no community of amateurs to rely on for the collection of experimental data. The production of experimental data was a matter for professionals. This fundamental difference between the experimental and the naturalist sciences had deep consequences for data collections.

WITHHOLDING DATA

In the molecular life sciences in the 1970s and 1980s, a number of new technologies resulted in improved methods for determining the structure of macromolecules. In 1977, for example, two new methods permitted determination of DNA sequences. Combined with the wide interest in the biological meaning of DNA sequences, these new methods resulted in an exponential growth in the production of sequence data.[17] In crystallography, more powerful methods, relying on digital computers, were also developed in the 1970s, speeding up the production of crystallographic data and the number of solved molecular structures.[18] This burgeoning store of data was often shared informally among researchers. It made possible a deep epistemic transformation in how data was used. As I have argued elsewhere, one of the most significant changes in the experimental sciences during the twentieth century was the increasing reliance on practices of producing knowledge based on collecting, computing, comparing, classifying, and curating large and diverse amounts of data.[19] These practices, so common in natural history and other observational sciences, became key to the experimental sciences as well. By adopting a different set of epistemic practices, whose potential became ever clearer starting in the 1970s, the experimental sciences were also confronted with new material and social challenges. But unlike the practitioners of natural history and other collecting sciences, who had long experience in resolving such problems, the experimentalists were at a loss to find solutions within the specific moral economies of their communities.[20] The single most important challenge was how to collect in a single place and make public the massive amount of data required to feed these epistemic practices.

Whereas most sciences of the archives had relied on and cultivated large networks of professional and amateurs, the experimental sciences had severed their ties with amateurs more than a century ago and consigned them to the role of distant spectators.[21] Thanks to the growth of the scientific workforce in the second half of the twentieth century, there was a much larger community of professionals able to contribute to a large collecting effort—on the condition that they could be persuaded to participate in this collective effort. Given a professional identity that valued individual achievement over collective participation, this was no simple task, especially if it implied sharing data difficult to produce and potentially rich in the epistemic rewards of new publications.

Since the Scientific Revolution of the sixteenth and seventeenth centuries, experimental results, "data" one could say, had been treated as the private property of the investigator who produced and carefully guarded it in laboratory notebooks. Data was disclosed publicly in exchange for scientific credit through oral communication in academies or publications in printed journals. This moral economy was still very much at work in the twentieth century. In 1968, American biologist James Watson revealed in his tell-all autobiography, *The Double Helix*, that he "was more aware of

[British crystallographer Rosalind Franklin's] data than she realized" and that "Rosy, of course, did not directly give us her data."[22] The data in question, communicated to Watson and his British collaborator Francis Crick though a confidential activity report, proved essential for the determination of the double helix structure of DNA. Scientists who reviewed Watson's book were almost unanimous in condemning his behavior and his shameless bragging about it.[23] In their view, Watson robbed Franklin of her data and thereby of her due credit. Data belonged to individuals, not to the scientific community as a whole.

COLLECTING DATA

Those who attempted to set up large collections of molecular data in the second half of the twentieth century experienced this problem firsthand, as the history of the Protein Data Bank and of GenBank, two of the major databases in the life sciences, make abundantly clear. The Protein Data Bank was set up in Brookhaven National Laboratory in 1973 to store all the existing data about the three-dimensional structure of proteins as determined by crystallographic methods. At the time of its creation, only a dozen protein structures had been determined by a small community of protein crystallographers who gathered at Cold Spring Harbor in 1971 for a symposium on the topic. Most of the researchers who had pioneered the determination of protein structures were present, including Max Perutz (hemoglobin), David Phillips (lysozyme), Frederic Richards (ribonuclease A), and William Lipscomb (carboxypeptidase A). Walter Hamilton, the young president of the American Crystallographic Association, aired the idea of a data bank for protein structures, a proposal made by two even younger colleagues, Helen Berman and Edgar Meyers. His colleagues responded very favorably to the idea, and the initial set of data was collected on the strength of friendships among Hamilton and some of the crystallographers present at the meeting.[24]

By May 1973, the data bank was "about ready to begin distribution" and a formal announcement was published in *Acta Crystallographica* and in the *Journal of Molecular Biology*.[25] The Protein Data Bank contained the coordinates of just nine proteins and anyone could obtain the entire data bank on a magnetic tape for a modest sum, covering shipping and the cost of a blank tape. The announcement repeated the call made two years earlier: the "usefulness of the system" would depend on "the response of the protein crystallographers supplying the data."[26] In the following months, researchers who determined structures began to acknowledge in their publications that they had deposited the coordinates in the Protein Data Bank. The number of available structures grew, but slowly. In January 1974, there were just twelve structures available, a year later fifteen, and the following year twenty-three.[27]

Most authors, who had been friends of Hamilton, agreed to release the data. However, not all of them must have been entirely comfortable with their decision. One crystallographer, for example, authorized the release of his data, but asked that he be informed about who would access it.[28] Others, such as Max F. Perutz, wished to hold back the data until it was further refined.[29] The Protein Data Bank managers tried to persuade him to release the data because "coordinates at any stage of refinement will be extremely interesting and very useful to many people,"[30] and Perutz eventually agreed, but only after a year (and after having submitted another paper based on the further refinement of the same data).[31]

In order to encourage researchers to deposit their data, while recognizing their interest in keeping them private so they could exploit them further, the managers of the Protein Data Bank, together with some journal editors, devised an original system in the early 1980s. Journal editors asked prospective authors that the data supporting the conclusions of their scientific paper be submitted to the Protein Data Bank. To overcome proprietary resistance, the data bank managers offered researchers the option to deposit their data but to restrict its access to the public for up to four years after the publication of a paper based on the data. In 1989, more than 75 percent of those depositing data chose to keep it private for the maximum period of four years.[32] Clearly, most of the community of crystallographers was not ready to make data communal property at the time of publication unless forced to do so.

To make matters worse, the very definition of what counted as data, especially "raw data," was not universally agreed upon.[33] Should only the atomic coordinates of the protein model be considered "raw data"? Or the "structure factors," a set of calculated data from measured diffraction intensities, from which the atomic coordinates were derived? Or the intensities of the diffraction spots measured on the diffraction images used to calculate the structure factors? Or the diffraction images themselves, produced directly by the x-rays going through a protein crystal? At first, the Protein Data Bank focused on the atomic coordinates that described with precision the position of each atom in a protein model. But these data were far from being "raw." They were the result of many steps of measurements, calculations, and interpretations. Without the structure factors, the proposed structure could not be challenged. Thus some crystallographers argued that coordinates were not data at all, but results derived from data.

As the American crystallographer Richard E. Dickerson put it in a letter to the president of the American Crystallographic Association, "Results without data are unproven, and interpretations without results are hearsay."[34] The Protein Data Bank managers therefore began to ask crystallographers to include structure factors along with atomic coordinates. But many researchers resisted, arguing that the structure factors should be considered research notes, not data, and thus felt no obligation to share them.

The situation was very similar among molecular geneticists who were facing their own data deluge in the same period. In 1983, the NIH funded the creation of a central database named GenBank for all DNA sequences.[35] Located at Los Alamos, its main architect, the physicist Walter Goad, had promised funding agencies that all the data published in the scientific literature would be included in GenBank "within a year" and all new data would be integrated as soon as they were published in journals.[36] But three years later, only 19 percent of the sequences published the previous year were publicly available in GenBank.[37] The gap between the data available in printed journals and in electronic databases was constantly growing. And there was an unknown amount of data that was neither published nor deposited in databases. Unlike the field of crystallography, much of the data associated with a published article was included in the scientific journals and thus publicly available. But printed DNA sequences (long strings of As, Ts, Gs, and Cs) were of little use for whoever wanted to analyze them with a computer. Worse, almost all the original published sequences were inaccurate (typographic errors were impossible to spot by a copy editor). Since researchers were eager to have correct sequences in electronic format available from a database like GenBank, managers of GenBank were faced with the daunting task of typing sequences manually from a journal into a computer, which was not only time consuming but also added its own set of errors.

As they were increasingly falling behind the growing amount of data available in the scientific literature, the managers of GenBank expected to enroll the community in the collecting effort. Like the managers of the Protein Data Bank, those of GenBank hoped that researchers would directly send their data in an electronic format to the database at the same time as they submitted a manuscript for publication in a journal—or at least that they would use the papers forms, sent out by journal editors, on which authors could carefully write down their DNA sequences and return it to GenBank. In numerous calls published in scientific journals, GenBank managers cajoled researchers to comply. They encouraged, threatened, persuaded, cheered, and appealed to moral obligations in order to change the behavior of experimentalists. Journal editors, who were swamped by sequence data and preferred to have it deposited in GenBank rather than printed in their pages, echoed these calls. In *Proceedings of the National Academy of Sciences*, for example, an editorial reminded the readers that "scientists who generate sequences . . . are also the users of sequences . . . self-interest should . . . dictate compliance."[38] All to little avail. As the editor of *Nucleic Acids Research* remarked pointedly, "Scientists would like access to everyone else's data through they do not necessarily wish to reciprocate."[39]

There were many reasons for the researchers' resistance to sharing data. Some simply felt that it was not worth the time needed to format and deposit the data in GenBank. Others were attempting to protect their data from potential competitors. Still others were concerned that it contained errors that could be spotted by others and tarnish their reputation. In the case of crystallographic data, researchers often wanted to "refine" the data further (a mathematical procedure), making it more precise, before they released it to the public. In practice, the basic principle that a publication requires all the data upon which the conclusions rest be made public (in print or upon request) was honored only in the breach. As Dickerson put it in no uncertain terms, "By the standards normally applied in other branches of science [the structures published without available data] are not really published at all, in the literal sense of making the information public."[40]

OPEN SCIENCE

The rise of open science in the last decade of the twentieth century was not the result of a spontaneous surge of altruism among researchers or a transformation in the moral economy of experimental science. Data collectors, such as those working for GenBank and the Protein Data Bank, began to pursue a new strategy more in line with the existing moral economy of the experimental life sciences. For many years, they had hoped that the communal ethos in science would allow them to collect data much as earlier natural history collectors had done. There, amateurs openly shared their "data," in the form of specimens or observations, with, for example, collectors affiliated with natural history museums or local naturalist societies. These amateurs received little, if any, credit for their contribution: at best, an acknowledgment in print or, exceptionally, a new species named after them. But in the experimental sciences, where professionals made a career (and a living) by turning data into credit, the naturalist system of data collection was bound to fail.

In this respect, the end of the 1980s marked a turning point. The National Institutes of Health (NIH) were particularly sensitive to the availability of data concerning genes or proteins related to diseases of great public concern, such as cancer or AIDS. In 1988, a group of researchers published a paper describing the structure of a protein from a cancer-causing gene, Ras. Two years later, the atomic coordinates describing the structure were still unavailable in the Protein Data Bank. When questioned about this omission, the article's lead author argued that he still needed to resolve some problems with the structure before depositing the data. An NIH official was annoyed by this prevailing attitude: the "data are good enough so that conclusions that are drawn from them can be published but not good enough to see the light of the day."[41]

To combat this problem, an NIH agency passed a resolution recommending that all grantees make their crystallographic data available within one year of publication, and that funding be withheld from those who did not comply.[42] Although this sanction sounded severe, it did not include any systematic enforcement measures. But it contributed, together with the growing pressure from professional societies, to the new system of data collection dependent on journal editors. By 1990, a number of them began to adopt policies mandating the sharing of crystallographic data with the Protein Data Bank. This proved far more effective.

This change in policy among journal editors resulted from the desperate efforts of database managers to solve their data collection problem and those of journals to preserve their epistemic authority while avoiding the costs of publishing growing amounts of data. Journal editors not only adopted but enforced mandatory submission policies, simply by deciding to publish only papers by authors who complied. These policy changes were almost always initiated by editorial board members with close ties to the data banks and who understood that data sharing was in the best interest of the research community. The molecular biologist Richard J. Roberts, for example, on the advisory board of GenBank and executive editor of *Nucleic Acids Research*, introduced the policy for that journal in 1988.[43] The principle was very simple. Journal editors would publish a paper only if the authors could provide an "accession number" demonstrating that the supporting data had been submitted to a public database, like GenBank or the EMBL data library, its European equivalent. Some journals, such as *Nature*, persisted in opposing any mandatory submission policy, and its editor-in-chief, John Maddox, encouraged other journals to resist "being turned into instruments of law-enforcement."[44] But *Nature* was becoming increasingly isolated. Authors also seemed to have some misgivings about having their data released too quickly, as almost 50 percent of those who submitted data to GenBank asked for confidentiality until their papers appeared in print.[45] Overall, the efforts of the EMBL and GenBank persuaded enough journals to adopt submission policies, essentially solving the problem of data collection for sequence databases. These policies had an immediate and dramatic effect: in 1990, 75 percent of all data submitted to GenBank came directly from authors. By attaching the rewards (priority, credit, and authorship) that go with publishing in a journal to data deposition, the experimental sciences solved in their own way an old challenge of the sciences of the archives.

DATA PUBLICATION

The open science revolution was thus no revolution at all, in the sense of a profound transformation in the political and cultural values governing individual and collective behavior. The rise of open science resulted from a new alignment between individual and collective interests within the existing moral economy of science. Researchers began to share data because it became a requirement in order to publish papers and thus reap the associated credit. More recent attempts to encourage data sharing illustrate further the conservative nature of the open access transformation. The mandatory data sharing enforced by journal editors only concerned data that was used as evidence for claims made in a scientific paper. The vast majority of data produced was unaffected by this policy and remained in private laboratory notebooks and computers. Thus scientists and science administrators imagined two different models to tie data sharing to the existing reward system in science: data authorship and data citation.

In the last decades of the twentieth century, as the amount of data produced by scientists increased dramatically, printed journals began to exclude the possibility of publishing data alone. Only when data was used as supporting evidence for a broader claim could it be included in a scientific paper or deposited in a public database. The production of data alone was no longer considered an intellectual achievement that could be rewarded by granting scientific authorship. A few decades earlier, the situation had been very different. In the 1950s, the biochemist Frederick Sanger published a series of papers describing, for the first time, the sequence of a protein, insulin—a publication rewarded by the Nobel Prize in Physiology or Medicine in 1958. But by the end of the century, protein sequences, and even more so DNA sequences, were determined in numerous laboratories, often through automated methods, without necessarily being published. In order to encourage researchers to submit these data and make them public, database managers attempted to rely on the same incentive as journal editors: the granting of authorship. The Protein Data Bank, for example, made it possible to cite "an entry without a published reference," including the name(s) of the author(s), a descriptive title, and a Protein Data Bank unique identifier (or a Digital Object Identifier, DOI).[46] A data entry could then be listed in an author's publication list, along with articles published in scientific journals. Journal editors have responded to databases' challenges to their exclusive rights to grant data authorship by launching new journals solely for the publication of data. Nature Publishing Group, for example, started *Scientific Data*, in 2014, for that purpose.[47] Databases and data journals both grant authorship, thus allowing researchers to claim the professional rewards attached to publications. Although this model for data sharing rests on the traditional reward system in the sciences based on publication records, its impact is limited by the fact that the value attributed to a publication depends on the reputation of the journal where it is published, a reputation that reflects how selective the peer-review process of the journal is perceived to be and various metrics of the journal's influence.[48] Although data deposition in a database or journal could be counted as a publication, its value in the scientific reward system thus remains low. So does the incentive to deposit data, limiting the impact of data authorship on data sharing.

An alternative model, based on data citations, was developed to overcome the limitations of data authorship. Along with the publication record, the scientific reward system has been based on citation records. Since 1964, the Science Citation Index, created by the American linguist Eugene Garfield, has tracked the number of times a given article is cited in the scientific literature.[49] This number is being used, in various combinations, as a way to measure quantitatively a scientist's impact on a scientific field. Since 2005, the number of citations is used to calculate the *h*-index, a measurement of a scientist's productivity and impact that has become a standard part of a scientist's resumé and that is often required by science funding agencies and academic search committees. As quantitative measurements of citations became increasingly influential in shaping scientific careers, proponents of data sharing—including database managers, journal editors, and funding agencies—encouraged authors to cite individual data entries, including the name of the researchers who had deposited the data, as they would for published papers. In the 1960s, when the first databases in the life sciences were established, researchers resented the fact that the database as a whole was cited, instead of the paper where they had first published the data, thus depriving them of the credit associated with a scientific citation. This citation practice also discouraged data sharing with a public database since it "anonymized" its origins.[50] Efforts by the Committee on Data for Science and Technology (CODATA), the US National Academy of Sciences, and various other groups of scientists led to the publication of a Joint Declaration of Data Citation Principles in 2013.[51] It emphasized the importance of minimal standards for data citations. Just a few months earlier, the media multinational Thompson Reuters, which maintains the bibliographic record and citation index Web of Science, launched its Data Citation Index.[52] This new tool made it possible to measure how often a particular data set is cited in the scientific literature, thus encouraging data sharing, even when data is not associated with the publication of research findings.

CONSERVATIVE REVOLUTION

It is too early to say whether these initiatives will have a significant impact on data sharing practices. But what seems historically significant is the degree to which they have retreated from the idealistic attempts of the 1960s and 1970s to transform the moral economy of experimental science. Instead, they all rely on the existing reward system based on the granting of authorship by community-based journals (or databases) and the citations of published work by members of the scientific community. Thus scientific journals, through their almost exclusive power to grant authorship, still hold the key to this reward system. Databases managers, after relying on journals to enforce mandatory data submission policies, began to challenge the exclusive rights of journals by arrogating to themselves the power to grant a form of authorship for data. The scholarly literature about the rise of open science has focused on the policies elaborated by governments and science funding agencies, overlooking the role of journal editors and database managers. In *Reinventing Discovery*, for example, Nielson claimed that "the granting agencies are the de facto governance mechanism in the republic of science, and have great power to compel change, more power even than superstar scientists such as Nobel prizewinners."[53] Others have described the "open science revolution" as essentially spontaneous, a revolution "from below," where individual researchers became committed to open science and shared data voluntarily in the best interest of the scientific community. This chapter argues that there was no revolution at all, or only a conservative one. The open science revolution might have changed how much data was made available publicly, a great collective benefit, but not the reason individual researchers shared data. By and large, researchers have shared data because it became in their own interest to do so, as defined by the existing reward system in the experimental sciences. Far from upsetting the current moral economy of science, the rise of open science illustrates how much it remained entrenched in current scientific practice. For this reason, researchers have even suggested that the term "data sharing," with its communitarian overtones, be abandoned and replaced by "data publication," a term perfectly in line with the individualistic ethos prevalent in the experimental sciences.[54]

In short, the data deluge was a product of two different transformations. First, it represents an expansion of what falls under the category of "data." Many research notes, preliminary measurements, and private observations did not count as data until the end of the twentieth century. The current data deluge is not simply the product of the increased amount of data being produced. It is also the result of the enlargement of what counts as data and thus merits preservation. The exact definition of "data," however, remains a moving target for all those involved in data policies, such as the National Science Foundation. Although since 2011 the NSF has required a "data management plan" (DMP) from its grantees, nowhere did it provide a definition of what counts as data, expecting that norms would be developed by the different research communities.[55] The NSF was treading lightly because labeling something as "data" created obligations to preserve it and make it publicly accessible. Second, the data deluge depended on the coupling of these obligations to the existing moral economy of the experimental life sciences. Data sharing became an obligation tied to gaining authorship and citations, the key components of the experimental sciences' individualistic reward system.[56] By making individual and collective interests coincide, the proponents of "open science" engineered a "data deluge," allowing the experimental science of the archives to flourish at the beginning of the twenty-first century.

NOTES

1. On the sciences of the archive, see Lorraine Daston, "The Sciences of the Archive," *Osiris* 27, no. 1 (January 1, 2012): 156–87; on collecting sciences, Bruno J. Strasser, "Collecting Nature: Practices, Styles, and Narratives," *Osiris* 27, no. 1 (January 2012): 303–40; on data flows, Stephen Hilgartner and Sherry I. Brand-Rauf, "Data Access, Ownership, and Control: Toward Empirical Studies of Access Practices," *Knowledge: Creation, Diffusion, Utilization* 15, no. 4 (1994): 355–72.

2. Michael A. Nielsen, *Reinventing Discovery: The New Era of Networked Science* (Princeton: Princeton University Press, 2012); Viktor Mayer-Schönberger and Kenneth Cukier, *Big Data: A Revolution That Will Transform How We Live, Work, and Think* (Boston: Houghton Mifflin Harcourt, 2013); David Weinberger, *Too Big to Know: Rethinking Knowledge Now That the Facts Aren't the Facts, Experts Are Everywhere, and the Smartest Person in the Room Is the Room* (New York: Basic Books, 2011).

3. Nielsen, *Reinventing Discovery*, 7.

4. See for example Rita R. Colwell, David G. Swartz, and Michael Terrell MacDonell, *Biomolecular Data: A Resource in Transition* (Oxford: Oxford University Press, 1989).

5. Protein Data Bank, *Newsletter* 3 (1976): 2.

6. Frédéric Eugène Godefroy, *Dictionnaire de l'ancienne langue française et de tous ses dialectes du 9e au 15e siècle* (Paris: F. Vieweg, 1881).

7. Nicholas Jardine, James A. Secord, and Emma C. Spary, eds., *Cultures of Natural History* (Cambridge: Cambridge University Press, 1996).

8. On amateurs in natural history, see David Elliston Allen, "Amateurs and Professionals," in *The Cambridge History of Science: The Modern Biological and Earth Sciences*, ed. Peter J. Bowler and John Pickstone (Cambridge: Cambridge University Press, 2009), 15–33; David Elliston Allen, *The Naturalist in Britain: A Social History* (London: A. Lane, 1976); Jim Endersby, *Imperial Nature: Joseph Hooker and the Practices of Victorian Science* (Chicago: University of Chicago Press, 2008); Kristin Johnson, *Ordering Life: Karl Jordan and the Naturalist Tradition* (Baltimore: Johns Hopkins University Press, 2012). On lay expertise, Steven Epstein, *Impure Science: AIDS, Activism, and the Politics of Knowledge* (Berkeley: University of California Press, 1966).

9. W. Patrick McCray, "Amateur Scientists, the International Geophysical Year, and the Ambitions of Fred Whipple," *Isis* 97, no. 4 (December 2006): 634–58; Dunlop Storm and M. Michèle Gerbaldi, eds., *Stargazers: The Contribution of Amateurs to Astronomy* (Berlin: Springer-Verlag, 1988); Jan Golinski, *British Weather and the Climate of Enlightenment* (Chicago: University of Chicago Press, 2007).

10. Peter Galison and Lorraine Daston, "Scientific Coordination as Ethos and Epistemology," in *Instruments in Art and Science: On the Architectonics of Cultural Boundaries in the 17th Century*, ed. Helmar Schramm, Ludger Schwarte, and Jan Lazardzig (Berlin: Walter de Gruyter, 2008), 296–333.

11. On residential science, see Robert E. Kohler, "Paul Errington, Aldo Leopold, and Wildlife Ecology: Residential Science," *Historical Studies in the Natural Sciences* 41, no. 2 (May 2011): 216–54.

12. Fa-ti Fan, *British Naturalists in Qing China: Science, Empire, and Cultural Encounter* (Cambridge, MA: Harvard University Press, 2003); Mark Barrow, "The Specimen Dealer: Entrepreneurial Natural History in America's Gilded Age," *Journal of the History of Biology* 33 (2000): 493–534.

13. Allen, *The Naturalist in Britain*.

14. Krzysztof Pomian, *Collectors and Curiosities: Paris and Venice, 1500–1800* (Cambridge, England: Polity Press, 1990).

15. Steven Shapin, "The House of Experiment in 17th-Century England," *Isis* 79, no. 298 (September 1988): 373–404.

16. Steven Shapin, *A Social History of Truth: Civility and Science in Seventeenth-Century England* (Chicago: University of Chicago Press, 1995).

17. On sequencing methods, see Miguel Garcia-Sancho, *Computing, and the History of Molecular Sequencing* (New York: Palgrave Macmillan, 2012).

18. Walter C. Hamilton, "The Revolution in Crystallography," *Science* 169, no. 941 (July 10, 1970): 133–41.

19. Strasser, "Collecting Nature"; Bruno J. Strasser, "The Experimenter's Museum:

GenBank, Natural History, and the Moral Economies of Biomedicine," *Isis* 102, no. 1 (March 2011): 60–96.

20. For a review of the notion of moral economy, see Didier Fassin, "Les économies morales revisitées," *Annales: Histoire, Sciences Sociales* 64, no. 6 (2009): 1237–66.

21. On the growing divide between the science and their publics around 1900, Bernadette Bensaude-Vincent, *L'Opinion publique et la science* (Paris: La Découverte, 2013).

22. James D. Watson, *The Double Helix: A Personal Account of the Discovery of the Structure of DNA: Text, Commentary, Reviews, Original Papers* (New York: Touchstone, [1968] 2001), 105

23. Watson, *The Double Helix*; Soraya de Chadarevian, *Designs for Life: Molecular Biology after World War II* (Cambridge: Cambridge University Press, 2002).

24. Interview with Helen Berman, July 17, 2009, New Brunswick, NJ.

25. Thomas Koetzle to Wayne Hendrickson, May 4, 1973, PDB Archives, New Brunswick, NJ (PDB Archives hereafter); anonymous, "Crystallography Protein Data Bank," *Journal of Molecular Biology* 78 (1971): 587.

26. Anonymous, "Crystallography Protein Data Bank," 587.

27. Protein Data Bank Annual Report to ACA, 1974, 1975, 1976, PDB Archives.

28. Wayne Hendrickson to Thomas Koetzle, April 6, 1973, PDB Archives.

29. Thomas Koetzle to Max F. Perutz, May 22, 1973, PDB Archives.

30. Thomas Koetzle to Max F. Perutz, May 22, 1973, PDB Archives.

31. Max F. Perutz, "Refinement of Hemoglobin and Myoglobin," *Acta Crystallographica Section A* 31 Supplement S (1975): 31.

32. Joel L. Sussman, "Protein Data Bank Deposits," *Science News Letter* 282, no. 5396 (December 11, 1998): 1993; Alexander Wlodawer, "Deposition of Macromolecular Coordinates Resulting from Crystallographic and NMR Studies," *Nature Structural Biology* 4, no. 3 (March 1997): 173–74.

33. Lisa Gitelman, ed., *Raw Data Is an Oxymoron* (Cambridge: MIT Press, 2013).

34. Richard E. Dickerson to Charles E. Bugg, July 27, 1987, PDB Archives.

35. Strasser, "The Experimenter's Museum."

36. GenBank Advisors Meeting, Minutes, November 6, 1987, EBI Archives.

37. Richard Lewin, "Proposal to Sequence the Human Genome Stirs Debate," *Science* 232, no. 4758 (June 27, 1986): 1598–1600.

38. Igor B. Dawid, "Editorial Submission of Sequences," *PNAS* 86 (1989): 407.

39. Richard T. Walker, "A Method for the Rapid and Accurate Deposition of Nucleic Acid Sequence Data in an Acceptably-Annotated Form," in *Biomolecular Data: A Resource in Transition*, ed. Rita Colwell (Oxford: Oxford University Press, 1989), 45–51.

40. Richard E. Dickerson to Charles E. Bugg, July 27, 1987, PDB Archives.

41. Abraham M. De Vos et al. "Three-dimensional Structure of an Oncogene Protein: Catalytic Domain of Human c-H-ras p21," *Science* 239 (1988): 888–93; Jim Cassatt to Helen M. Berman, January 4, 1990, PDB Archives.

42. John C. Norvell to principal investigators in NIGMS, April 23, 1990, Protein Data Bank Archive.

43. Patricia Kahn and David Hazledine, "NAR's New Requirement for Data Submission to the EMBL Data Library: Information for Authors," *Nucleic Acids Research* 16, no. 10 (May 25, 1988): I–IV.

44. John Maddox, "Making Authors Toe the Line," *Nature* 342 (1989): 855.

45. GenBank Advisors Meeting, Minutes, November 15–16, 1988, NCBI Archives.

46. Protein Data Bank, Policies & References, available at http://www.rcsb.org/pdb /static.do?p=general_information/about_pdb/policies_references.html, accessed July 10, 2014.

47. Anonymous, "More Bang for Your Byte," *Scientific Data* 1 (2014), accessed July 10, 2014, doi:10.1038/sdata.2014.10.

48. Björn Brembs, Katherine Button, and Marcus Munafò, "Deep Impact: Unintended Consequences of Journal Rank," *Frontiers in Human Neuroscience* 7 (2013). doi:10.3389/fnhum.2013.00291.

49. Eugene Garfield, "Citation Indexes for Science: A New Dimension in Documentation through Association of Ideas," *Science* 122, no. 3159 (July 15, 1955): 108–11.

50. Bruno J. Strasser, "Collecting, Comparing, and Computing Sequences: The Making of Margaret O. Dayhoff's Atlas of Protein Sequence and Structure, 1954–1965," *Journal of the History of Biology* 43, no. 4 (2010): 623–60.

51. CODATA-ICSTI Task Group on Data Citation Standards and Practices, "Out of Cite, Out of Mind: The Current State of Practice, Policy, and Technology for the Citation of Data," *Data Science Journal* 12, no. 0 (2013), CIDCR1–CIDCR75. See also Paul F. Uhlir, *Board on Research Data and Information, Policy and Global Affairs, National Research Council. For Attribution—Developing Data Attribution and Citation Practices and Standards* (Washington, DC: National Academy Press, 2012).

52. Thomson Reuters, "Thomson Reuters Launches Data Citation Index for Discovering Global Data Sets" (April 2, 2013) accessed November 17, 2014, http://thomsonreuters .com/content/press_room/science/730914

53. Nielsen, *Reinventing Discovery*, 191

54. Mark J. Costello, "Motivating Online Publication of Data," *Bioscience* 59, no. 5 (2009): 418–27.

55. Jacob Glenn, "NSF Data Management Plan" (2013), accessed July 11, 2014, http:// www.lib.umich.edu/research-data-services/nsf-data-management-plans

56. The sharing imperative became embedded within the system, as Chris Kelty has described for the creative commons license. See Christopher M. Kelty, *Two Bits* (Durham: Duke University Press, 2008).

EIGHT

Evolutionary Genetics and the Politics of the Human Archive

Cathy Gere

Evolutionary genetics is an archival science in two distinct senses. At the level of practice, it is based on the archival activities of collection and comparison, in the natural-historical as opposed to the experimental tradition.[1] On a more conceptual level, the genome itself is considered to be an archive, containing information about the deep evolutionary history of the species in question.[2] This paper will examine the political ramifications of the doubly archival character of this science when applied to the human species. In the second half of the twentieth century, it was determined that an evolutionary reading of the human genome depended upon the identification and sampling of human populations supposedly untouched by the mongrelizing and displacing effects of cosmopolitan modernity.[3] At different times, these groups of people were characterized as "remote tribes," "aboriginal populations," and "isolates of historical interest." This identification of the people of the periphery with the past of the metropolis conformed to the "spatialization of time" so cogently analyzed by Johannes Fabian in his seminal 1983 work of critical anthropology, *Time and the Other*.[4]

In a significant reversal of some of the ideological terms of nineteenth-century race science, however, post–World War II evolutionary genetics rested on the conviction that *indigenous* genomes partook of a unique genetic purity. This purity was precisely the quality that made the archival interpretation of their genetic markers plausible. Because these communities were genetically isolated, so the argument went, they represented the connection between genetic signal and geographical location required for a reconstruction of the so-called journey of human evolution—the spreading out of our species from East Africa to the rest of the globe.

This form of evolutionary genetics was developed in the 1930s in an explicitly anti-eugenic idiom, but it remains, at core, a science of racial classification. In the 1990s, amid concerns about the patenting of genes and the ownership of other forms of biocapital, the enormous epistemic value placed by evolutionary theorists on the racial purity of indigenous DNA seemed to recapitulate a long imperialist history of exploitation and abuse.[5] Indigenous activists mounted a successful boycott, and the whole endeavor had to reinvent itself in order to proceed. This paper examines some key episodes in the history of activist challenges to human evolutionary genetics, asking what they reveal about the politics of the human archive.

The paper is divided into four sections. Section one, "Antiracist Racial Science," supplies some historical background. From the 1930s to the 1950s, believing that good science was the best antidote to prejudice, population geneticists worked out the details of what they took to be a value-neutral definition of racial biology. The second section, "The Biocolonial Archive," reveals the limitations of this approach, telling the story of how population geneticists lost their rhetorical and political leadership on these questions as a result of the radicalization of antiracist politics and the postcolonial critique of scientific universalism. The section ends with the well-known case of the Human Genome Diversity Project (HGDP), an earnest attempt to address the unremarked whiteness of "the" human genome, which crashed and burned in the mid-1990s amid accusations of biopiracy and genetic vampirism. Section three, "The Neocolonial Archive" explores a less well known indigenous genetic sampling initiative, conducted by Arizona State University in the same period, which suggests that some of the fears attendant on the HGDP were well founded. In both these cases, the transformation of biological materials into legible archives propelled scientists into a semantic field replete with interpretative challenges that they were clearly unable to meet. Section four, "The Postcolonial Archive," examines the pragmatic solutions to these discursive problems arrived at by the Genographic Project, an initiative that succeeded where the others had failed. These episodes make visible the power asymmetries inherent in the practice of a human archival science. Who does the collecting, organizing, reading, and interpreting? And, conversely, who is collected, organized, read, and interpreted? What does it mean to be made legible as a human archive?

Evolutionary information was first deduced from the chromosomes of a living species by Theodosius Dobzhansky, a leading member of the group of population geneticists who mounted a scientific challenge to eugenics. Born in Ukraine in 1900, Dobzhansky trained as an entomologist in the (soon-to-be-suppressed) Russian Darwinist tradition. He came to the US in the last days of 1927 on a Rockefeller grant, to work in T. H. Morgan's laboratory at Columbia University, which had pioneered the technique of doing Mendelian breeding experiments to map the chromosomes of *Drosophila melanogaster*, the common fruit fly.[6]

Dobzhansky's Russian Darwinist training had given him an abiding interest in processes of speciation. In 1930, he embarked on a study of hybrid sterility in stocks of wild fruit flies, on the grounds that the emergence of varieties unable to interbreed with one another was an important stage in the splitting off of a subspecies. During the course of this work, however, he was diverted by the unexpected degree of chromosomal variation in the wild stocks. Used to working with highly standardized, purified, and manipulated strains of *D. melanogaster*, he was intrigued by the possibilities for studying genetic variability in their wild cousins, *D. pseudoobscura*.

D. pseudoobscura is found only in remote places, mainly in western North America. During the summer months this species is abundant in elevated forests of its host plant, the Ponderosa pine. As temperatures drop in the autumn, the populations move to lower elevations and can be found in the desert, breeding on agave and cactuses. Their shyness and the extreme specialization of their habitats—in contrast to the cosmopolitan and unfussy *D. melanogaster*—gives rise to rich genetic variability between subspecies, defined by chromosome breaks, loops, and inversions, which prevent interbreeding. Dobzhansky used these morphological differences as taxonomic markers in distinguishing different "races," as he called them. When the patterns overlapped—when they seemed to show that a second break had occurred within the arrangement of the first break—they could be used to create phylogenies or geographical descent trees of the subspecies.

In 1936 Dobzhansky cowrote a paper with Alfred Sturtevant (the prodigy who had first arrived at the chromosome mapping technique as an undergraduate in one of Thomas Morgan's classes), which laid out his argument about the phylogenetic relationships between different chromosomal patterns. The pattern of inversions and breaks that distinguished different races of *D. pseudoobscura* could be ordered historically. The first step was to analyze which of the chromosomal arrangements differed least from all the others. This one they designated the "ancestral" sequence, on the grounds that "it is the only type to which several others are directly related." With this, they established the basic reasoning for genetic ancestry testing that persists to this day.[7]

Although *Drosophila* was his model organism, Dobzhansky was always concerned with the bearing of the fruit fly work on large questions about human society. In 1936, he was invited by his friend the geneticist L. C. Dunn, to Columbia University to give a series of lectures, which later became the groundbreaking book *Genetics and the Origin of Species*. Columbia was a hotbed of antifascist sentiment, led by Dunn and the anthropologist Franz Boas.[8] In this context, Dobzhansky's discussion of the problem of evolutionary fitness in fruit flies was pointedly aimed at "those eugenical Jeremiahs" who overlooked the complexity of the function of deleterious mutations.[9] The antiracist dimension to Dobzhansky's work was made even more explicit in the second edition of *Genetics and the Origin of Species* (1951), in which he denounced the "misuse of Darwinism by propagandists and bigots."[10]

Dobzhansky always used the word "race" for subspecies. Races, he insisted, were open genetic systems, susceptible to changes from mechanisms such as random genetic drift, sexual selection, and changes in environment. Natural selection was environment-specific and so fitness was always relative and never absolute. This meant that a reservoir of recessive mutations was evolutionarily healthy, providing the raw material for natural selection to act on when the environment changed. These ways of characterizing evolution—in which the language of statistical rigor and mathematical sophistication was leveraged against the crudities of eugenic analysis—set the tone for the next three decades of human genetics. The reclusive, shy, wild *Drosophila pseudoobscura* was not only the model organism for a historical/evolutionary interpretation of chromosomes, but also for a new anti-eugenic politics of evolution, in which genetic variability was celebrated and natural selection underplayed.

Such was the anti-eugenic thrust of Dobzhansky's work that when, in December 1949, an international group of scholars gathered in Paris under the auspices of UNESCO to author an authoritative rebuttal to racism, they drew on his language in the opening paragraph of their "Statement on Race," ascribing "differences between different groups of mankind" to "the operation of evolutionary factors of differentiation such as isolation, the drift and random fixation of the material particles which control heredity (the genes), changes in the structure of these particles, hybridisation, and natural selection." Having opened with these technical points, however, the authors concluded that "'race' is not so much a biological phenomenon as a social myth."[11] This did not sit well with the geneticists, and UNESCO invited the dissenters, including Dobzhansky himself, to come up with a version more in line with scientific definitions of race. The first paragraph of version 2.0 accordingly gestured towards the artificiality of the race concept while acknowledging its scientific utility: "The concept of race is unanimously regarded by anthropologists as a classificatory device providing a zoological or biological frame within which the various groups of mankind may be arranged and by means of which studies of evolutionary processes can be facilitated."[12]

This definition of race was at the cutting edge of genetic science, and the "zoological or biological frame within which . . . mankind may be arranged" was already the subject of hot debate. In 1950, a symposium at Cold Spring Harbor had begun to discuss how to apply the insights of evolutionary genetics to the human species. The event was attended by 129 scientists, and the emphasis was on how to define the human group as a unit of analysis.[13] What the scientists needed to identify were communities of people who still lived on the same lands as their remote ancestors, and who were tightly interrelated and culturally homogenous. It was isolation that would enable scientists to establish the relationship between genetic markers and geographical location that they needed in order to establish ancient migration patterns. What they were looking for, in other words, was the human equivalent of *Drosophila pseudoobscura*.

The encroachments of urbanization, emigration, and cosmopolitanism, with all their attendant genetic intermixture, endowed the whole endeavor with a measure of urgency. As Dobzhansky noted: "Because of the rapid development of communications even in the most remote corners of the world, and the consequent mixing of previously isolated tribes, such investigation cannot be long postponed. In fact, it is possible that our generation is the last one which can still secure data of momentous significance for the solution of the problem of the origin of human races."[14] In one session, entitled "Race Concept and Human Races," Joseph Birdsell, a UCLA physical anthropologist suggested the "population or isolate" as the proper unit of study.[15]

In the early 1960s, following the decipherment of the genetic code, evolutionary genetics took on a more explicitly archival idiom. A 1965 paper by Linus Pauling and Emile Zuckerkandl, "Molecules as Documents of Evolutionary History," posed the question of "where in the now living systems the greatest amount of their past history has survived."[16] Importantly for the politics of human evolutionary studies, the authors suggested that the most concentrated source of historical information was to be found in DNA sequences that were phenotypically neutral, for example in so-called "junk DNA" that was not transcribed into amino acids. Because these genetic differences were not under selection pressure, they constituted a trace of random mutative events that could be assumed to occur with a fair degree of regularity, by analogy with radioactive decay. This emphasis on the epistemic value of nonphenotypic, nonfunctional traits further bolstered the claim of the population geneticists to be conducting ideologically neutral race science.[17]

THE BIOCOLONIAL ARCHIVE

The most prominent of the scientists to pursue this research program was the Italian geneticist Luca Cavalli-Sforza. Born in 1922, Cavalli-Sforza was an antiracist in the Dobzhansky tradition, countering arguments about the inheritability of intelligence with the rigorous statistical analyses of population biology.[18] His first foray into human genetics asked a classically Dobzhanskian question: was genetic drift or natural selection responsible for the genetic makeup of isolated populations in the region around Parma? Cavalli-Sforza later observed that genetic drift served as a better marker of human evolution than natural selection: "Deleterious or advantageous mutations tell us more about the history of the environmental challenges that organisms have had to meet over time. Thus, neutral mutations under the control of drift are more useful in tracing the history of organisms themselves."[19]

In the early 1960s Cavalli-Sforza made his first foray into the field, traveling to Zaire in the hope of making contact with the Pygmies who lived in the forest. For subsequent trips, he teamed up with an anthropologist who provided support, medical and demographic information, and contacts for collection of blood samples. In 1971 he published the results of his work in a book called *The Genetics of Human Populations*. While warning about the perils of racism, the work ended with a long chapter cautiously endorsing eugenics, on the grounds that it was now based on science rather than prejudice.[20] Overall, his work attempted to establish a new human genetics that repudiated politics altogether in the name of value-freedom and objectivity.

Cavalli Sforza was soon to find out, however, the discourse of race and science had become radicalized beyond the reach of the assumption that population genetics and racial prejudice were natural enemies. In 1990, as part of the first phase of planning of the Human Genome Project, he floated a proposal to sample human genetic diversity, as a counterweight to the idea that the chromosomes of a handful of scientists and laboratory technicians could stand as representative of the whole human species. As the Human Genome Diversity Project gained momentum, it attracted support from scientists active in organizations such as Amnesty International, Physicians for Human Rights, and the Argentine Grandmothers of Plaza de Mayo. A geneticist who had worked with the Kurds, a group in Turkey and northern Iraq who had been brutally repressed by Saddam Hussein, wrote of the urgency of sampling the DNA of endangered human populations in terms that indicated the link between his work against genocide and his preoccupation with such "populations in the Middle East, Asia and Africa [who] have been isolated by geography and limited technology, and represent rare and valuable resources to study human origins and patterns of population genetics."[21] Concerns that the HGDP would be a "sample and run" initiative were offset by promises that the DNA samples would be used to do medical analyses to benefit the populations.

The HGDP seemed to be the ultimate fulfilment of the project of an antiracist genetics, but by the last decade of the twentieth century the notion of a value-free racial science had become irretrievably problematic. A bioanthropologist from the traditionally African-American Howard University asked at one of the planning meetings for the HGDP why European "isolates of historical interest" were referred to as "ethnic groups," while their nonwhite counterparts were called "tribes." An editorial in the *New Scientist* in April 1993 denounced the HGDP as "21st century technology applied to 19th century biology"[22] Questions about commercialization had also become increasingly urgent since the landmark *Diamond v. Chakrabarty* case of 1980, which had set a precedent for patenting a living organism.[23]

In 1993, just as the HGDP was facing a wave of bitter rebellion from cultural anthropology, the Rural Advancement Foundation International (RAFI), an organization at the forefront of activism against the commercialization of traditional knowledge, discovered that the US government had filed patents for cell lines derived from four individuals: a man from Papua New Guinea, two Solomon Islanders, and a woman from the Guyami people of Panama.[24] Faced with this evidence of the monetization of indigenous DNA, RAFI accused the Human Genome Diversity Project of threatening the livelihood and autonomy of native groups. As a result, the claim that medical research would be conducted on the samples as a way of giving back to the donors merely stoked the fear that there were commercial imperatives behind the project. In December, a UN organization, the Indigenous Peoples Council on Biocolonialism, denounced the HGDP as the "Vampire Project."

Michael Dodson, Australia's first Aboriginal and Torres Strait Islander

Social Justice Commissioner, noted what he took to be indigenous people's central objections to the HGDP:

> Our core objection to the Human Genome Diversity Project is not its potential to undermine our cosmology: that would hardly be something new; it happens in every aspect of our interaction with nonindigenous culture. It is the denial of our humanity, it is that the project would have our bodies the sites of exploitation as our lands have long been.[25]

Some of this distrust can be put down to an enduring tone-deafness on the part of the scientists involved. The architects of the Human Genome Diversity Project stressed the urgency of sampling the "vanishing" DNA of indigenous people, whom they dubbed "isolates of historical interest." Anyone with even a passing knowledge of indigenous rights activism, in which "not vanishing" had become a rallying cry, could have identified this wording as a red flag.[26] The idea that scientists would be swooping down on communities of people whom they had identified as disappearing, in order to extract resources of scientific value, had undeniable resonance with the evolutionary triumphalism of Europeans who had regarded the march of so-called civilization with such infamous complacency in the imperial age. In a 1994 book Cavalli-Sforza had announced that "we restrict our study to aboriginal populations, which we define as those already living in the area of study in 1492."[27] But his continued obtuseness about the significance of this date in Native American history was apparent in an adulatory book about his work published in 2005. Upon being asked what his feelings were about the failure of the HDGP, Cavalli-Sforza flourished "a beautiful American expression that explains it all: 'It is easy to recognize a pioneer: his back is full of arrows.'"[28]

THE NEOCOLONIAL ARCHIVE

The UN boycott effectively derailed the HGDP. For Cavalli-Sforza and many of the other scientists involved, the resistance of native activists could only be founded on a gross misinterpretation of their motives. At the same time, however, a study was underway in Arizona that exemplified the way in which—individual motivations aside—the practices of human scientific archiving could exacerbate asymmetries of economic and political power. In 1990 researchers from Arizona State University (ASU) embarked on a research partnership called the Diabetes Project with the Havasupai Tribe, an American Indian community living in Supai Village on the floor of the Grand Canyon. The Havasupai suffer from extremely high rates of diabetes, with nearly half the adults affected by the disease. In 1989 one of the members approached John Martin, an ASU anthropologist who had worked with the tribe for many years, and appealed to him for help in dealing with this debilitating threat to Havasupai health. The anthropologist contacted Therese Markow, a geneticist colleague, who agreed to participate in a study of the genetic underpinnings of the condition. Markow expressed an interest in expanding the study to include research into the genetic basis of schizophrenia, but was told that this was unlikely to find favor with the Havasupai. Accordingly, the two researchers applied to ASU for a project that included health education, collecting and testing of blood samples, and DNA testing to search for links between genes and diabetes risk.

In June and July 1990, the blood draw began. The anthropologist had repeatedly stressed to the Tribal Council that the genetic research was strictly confined to diabetes, but the consent forms signed by participants were broad in scope. Entitled "Medical Genetics at Havasupai," the forms referred to "the causes of behavioral/medical disorders" and declared that "the possible benefits of my participation in the research are a better understanding and treatment of diseases in my family and tribe."[29] Noting that the participants were sometimes reluctant to sign forms, one member of the research team began to obtain only oral consent, once again stressing that the study was only about diabetes. Between 1990 and 1994 over 200 tribal members gave blood, as well as handprints, for research purposes that were characterized, at least on paper, in broad and vague terms.

Meanwhile Markow had obtained a grant from the National Alliance for Research on Schizophrenia and Depression to "study the role of genetic factors in schizophrenia . . . which occurs at a significantly higher rate among the Havasupai (7%) than in any other population (1%)." The grant application also contained the claim that "all [Havasupai] cases of schizophrenia occur in lineages tracing back to a single man (a shaman or medicine man) who lived in the 1880s." It is not clear what Markow's sources were for these claims, but a medical student on a psychiatry rotation in Phoenix in the summer of 1990 remembers that she asked him to help draw blood at the clinic at Supai Village, and to look through medical records after hours for evidence of a high incidence of schizophrenia. He recalls telling her that the files revealed no abnormal levels of mental illness.[30] The statement about the line of descent from the shaman seems to have derived from Markow's casual conversations with her anthropologist collaborator.[31]

The genetic work on diabetes did not produce any promising leads. The blood samples were stored in a freezer at ASU, and then periodically distributed and repurposed. More than a decade later, Carletta Tilousi, a member of the Havasupai tribe and a participant in the Diabetes Project, became a graduate student at ASU. In 2003, Tilousi attended a PhD defense that discussed the Havasupai samples in the context of research completely unrelated to diabetes. Disconcerted, she stood up and challenged the student to explain how he had got hold of the samples. He said he had authorization from Markow. It soon emerged that twenty-three papers had been published using the Havasupai samples, only eight of which concerned diabetes. The other fifteen were about migration, inbreeding and schizophrenia.

In 2004, the Havasupai Tribe filed two lawsuits against the Arizona State University Board of Regents and ASU researchers for misuse of their DNA samples. The lawsuits articulated concerns about lack of informed consent, violation of civil rights through mishandling of blood samples, unapproved use of data, violation of medical confidentiality, misrepresentation, infliction of emotional distress, violation of civil rights, and negligence. The first lawsuit was dismissed on the grounds that the research subjects had signed a broad consent form, and therefore given their permission for all the later studies. The second was dismissed due to a procedural error. The Arizona Court of Appeals later reinstated the second lawsuit, leading to a lengthy legal battle. In March 2010 the parties reached a settlement in which tribal members received $700,000 as compensation, funds for a clinic and a school, and return of the remaining frozen blood samples.

Archival practices were especially problematic in the Havasupai case. Storing the blood samples for future uses, impossible to define with any precision, broke the contractual ties that, at least in theory, bound the researchers and research subjects in a set of shared goals. In response to the legal judgment, the *New England Journal of Medicine* proposed that the solution might lie in a "tiered" consent protocol, in which participants would choose whether their samples could be used for further research, and if so, what.[32] This attempt to maintain the human archive within the contractual terms of informed consent fails to address the interpretative openness that is inherent in all long-term archiving activities. Storing biological materials for future research is predicated on a commitment to unpredictable applications, something that cannot be fully captured and tamed in a consent procedure, however many tiers it has.

In addition to these problems of storage and repurposing, the archival metaphor of evolutionary genetics generated further difficulties. One of the scientific papers interpreted the Havasupai genome as a document of evolution, arguing that it was a closer match with central Asian people than indigenous South Americans, thus slotting the tribe into its place in the Bering Straits migration narrative. The perennial indigenous criticism of such narratives is that they have the potential to undermine sovereignty claims based in traditions that Native peoples are created on and out of the lands they inhabit or have lost. The evolutionary reading of the Havasupai samples was also marked by the reintroduction of arguments about genetic fitness, assessed in both positive and negative terms. Arguing that "human colonization of the Americas involved a people adapted to Arctic conditions," the paper suggested that the Havasupai genome was suited to the conditions found in "much of North America, whereas survival in the biologically more diverse environments of Latin America was aided by the use of new recombinant HLA-B alleles."[33]

The discourse of evolutionary fitness became much more extreme in Markow's research into the genetics of schizophrenia. Markow had taken handprints from tribal members alongside blood samples in order to test a hypothesis about the developmental pathologies attendant on inbreeding. In 1986 she had published a speculative paper on the "genetics of liability to schizophrenia" that proposed a series of links between phenotypic asymmetry, mental illness, and reduced genetic variability.[34] By the time of the Havasupai project, she was using the term "behavioral phenodeviance" to signify a series of speculative links between inbreeding, asymmetry in traits such as fingerprints, impaired nervous system development, and behavioral disorders.[35] The same year she published a paper arguing that the handprints of the Havasupai showed the traits of behavioral phenodeviance: "Fluctuating asymmetry in two dermatoglyphic traits suggests that inbreeding significantly compromises developmental homeostasis in this population."[36] She did not mention schizophrenia directly in the paper, but the implication for her work as a whole was clear: here was a group of people who by virtue of their genetic isolation were at risk for elevated incidence of mental illness.

Schizophrenia is one of the most debilitating and disabling of psychiatric conditions, and Markow clearly thought she was onto a promising line of enquiry about its genetic cause, which she speculated lay not in a single gene, but in a cascade of developmental impairments linked to homozygosity and manifesting in detectable phenotypic asymmetries such as markedly different left and right handprints. When an anthropologist suggested that she work on a genetic study of a genetically isolated population, it must have seemed like a chance to make some real progress on an urgent question. But Markow's theory of phenodeviance among the Havasupai was highly speculative, and seems to have been undermined by her own analysis of the blood samples. A slightly later paper using the samples remarked on the "homozygote *deficiency*" (my italics) of the Havasupai samples at one particular locus, and speculated about how the tribe managed to maintain its genetic diversity in the face of its geographical isolation.[37]

One of the ways in which Dobzhansky sought to rid genetics of its unedifying political legacy was to deemphasize natural selection and adaptive fitness. Against the background of anxieties about overpopulation and environmental degradation, his successors moved into a more decisively positive register, expressing admiration for indigenous peoples' unique adaptation to a given environmental niche, understood as more a matter of cultural and technological adaptation than biological fitness.[38] In Markow's work, these midcentury idealized notions of cultural adaption and isolation gave way to a series of speculations about the pathologies of inbreeding, including, even if only glancingly, a retrospective diagnosis of shamanic practice as psychotic. The fact that she used hand prints to diagnose these latent pathologies raises the spectre of Francis Galton, who coined the word "eugenics," perfected forensic fingerprinting, and speculated about whether the patterns of whorls and ridges on the hands and feet could provide information about heredity and character. Galton was disappointed in this regard; he might have been excited by Markow's advances on his legacy.[39] This move from an idealized, genetically pure indigene to a pathologized interbred one surely reveals how easy it is for the human archive to reassume the hierarchical characteristics of the nineteenth-century evolutionary sciences in which it finds its conceptual origins.

THE POSTCOLONIAL ARCHIVE

The Havasupai samples have been returned to the tribe and the Human Genome Diversity Project is no more, but the pursuit of human evolution studies has been taken over by various initiatives, the most high-profile of which is the Genographic Project, privately funded by *National Geographic* magazine, IBM, and the Waitt Family Foundation. The Genographic Project set out to identify the genetic markers associated with human migratory patterns over the past 200,000 years. Just as with the HGDP, this work was urged onwards by the worry that genetic signal of indigenous bloodlines will soon be irretrievably muddied by mobility and urbanization. For the project to function at all, it was going to have to learn fast from the mistakes of its predecessors, and its partial success is down to the absorption of the lessons of history by its director, the American drosophilist-turned-human-geneticist Spencer Wells.

In 1996, just as the HGDP was falling apart, Wells scraped together some funds and set off to post-Soviet Central Asia to take samples from nomadic hunter-gatherers living in the frozen reaches of northeastern Siberia. This was an inspired move, both politically and scientifically. Not only had the

Iron Curtain kept these communities safely away from such organizations as the UN Indigenous Peoples' Council on Biocolonialism, but Central Asia also turned out to be an important crossroads of early human migration. Scenting a good story, a British film crew accompanied Wells on his next excursion, and the Genographic Project grew from there, piecing together a modus operandi as it went along. In 2005 it was officially launched as a five-year enterprise with the goal of sampling the DNA of 100,000 indigenous people from all corners of the globe and creating an "atlas of the human journey."

The Genographic Project has not escaped criticism from the same activists who condemned its predecessors,[40] but everything about the way it is organized and presented is a testament to hard-won lessons learned. The doomed Human Genome Diversity Project, for example, made the promise of reciprocity on the grounds of potential medical benefits springing from the genetic testing itself. This was condemned as profiteering by the activist critics of the HGDP, and although the motives of the individual scientists involved may have been misdiagnosed, Markow's work on schizophrenia among the Havasupai tends to support the paranoid interpretation. Markow thought that she was onto a great thing, and didn't scruple too much about how she pursued it. The neo-Galtonian result demonstrates what can happen when the human archive is deployed within the promissory speculations of entrepreneurial medicine. Geneticists now have to face the fact that DNA sampling is not necessarily its own reward, and the Genographic Project was reduced to *boasting* that it conducts "no medical research of any kind."[41] Instead, indigenous groups themselves now get to define what counts as a benefit to them, and the overwhelming verdict is that they could use some assistance with cultural survival against the encroachments of globalization. Accordingly, they can apply to the Genographic Legacy Fund—financed by selling $100 ancestry testing kits to members of the public—for initiatives such as reversing language loss, resurrecting craft techniques, and writing down oral traditions.

A narrative of African origins and human migration is the ultimate product of the Genographic Project. The repeated refrain of the work is that the diversity of the human species, to which we are so visually sensitive, is of recent origin, and that under the skin, we are all African. Its emphasis on human migration reshapes the old pools of racial types into the branching subway map of a global genealogy. This is both more individual—in that Oprah Winfrey can trace her ancestry back to a specific region of West Africa—*and* more inclusive—we can all trace our deepest human ancestry to a region of East Africa.

The problems with out-of-Africa migration theories for indigenous sovereignty claims remain, however, and perhaps the most interesting part of the Genographic Project Ethics Statement attempts to address this conflict. Drawing from the work of Mick Dodson, quoted above as one of the most vociferous critics of the HGDP, it tries to strike a delicate balancing act between the claims of genetic ancestry testing and those of indigenous tradition:

> Principal investigators are required to be (and are) sensitive to the fact that knowledge generated by the project may give rise to narrative accounts that function as an alternative to some traditional accounts of the origin of the cosmos (including people). All project participants understand that scientific narratives do not have priority over other types of narrative—and that indigenous communities will determine the extent (if any) to which such narratives might complement their existing world views.[42]

Of course, as various episodes in the promotional film about the Genographic Project make abundantly clear, Spencer Wells and his colleagues do not really believe that ancestry genetics is on an epistemological par with indigenous cosmology, so their intermittent presentation of their findings as "a" truth, rather than The Truth, is more a matter of courtesy than conviction.

It may well be impossible to advance a project like this at the same time as sincerely undermining its claim to describe the way things really are, but it is interesting to note that good twenty-first-century human-scientific etiquette demands a measure of enacted modesty about the scope and stability of scientific knowledge. There seems, in fact, to be a suggestive family resemblance between the Genographic Project's epistemic relativism and that cherished axiom of late-twentieth-century sociology of scientific knowledge, the symmetry principle, or equivalence postulate. The former says that "scientific narratives do not have priority over other types of narrative"; the latter that the "same [social] types of explanations must be used for both successful and unsuccessful scientific theories." Both of these postulates propose a form of epistemic egalitarianism between the incommensurable worlds of science and nonscience. Postcolonial human science seems to demand, if only as a pragmatic measure for maintaining the support of the research subjects, the basic insight that the laboratory is only one of the sites in which truth regimes are produced and perpetuated.

CONCLUSION

In 2006, Spencer Wells, the director of the Genographic Project, published a popular book in which he declared that evolutionary genetics was "the late 20th-century equivalent of studying ancient languages in order to decipher inscriptions uncovered at an archaeological site. With the right samples of DNA, and knowledge of how to read the text they carried, it would be possible to study the history of life on Earth."[43] I have argued elsewhere that these archaeological and antiquarian metaphors fulfill a particular rhetorical purpose: by comparing natural objects to inscribed ones, they legitimate acts of "reading" the material world. More specifically, in comparing biological objects to the inscribed monuments of antiquarian investigation—for which the inscription and the material form are equally important—these analogies hover between textuality and physicality, allowing language and nature to meet in new ways.[44]

The case of molecular evolutionary genetics is perhaps the most exemplary in this regard. Genome sequencing turns chromosomal material into strings of letters. These sequences are like texts, in that they can be shared and disseminated without being diminished, degraded, or used up. But their textual character is far upstream of any kind of complete interpretation. In this respect they still partake of much of the interpretive openness of biological samples. Semantic promiscuity is inherent in this combination of potentially unlimited circulation and hermeneutic openness. The evolutionary interpretation of genomic sequences then fits these data into a story of human origins, in which the historical specificity of the circumstances of their collection and analysis is necessarily stripped away in favor of a sweeping narrative told in geological time. Because of the emphasis on finding the pure genetic signal of people isolated from the imperatives of modernity, the biological samples in question are often collected from non-literate people, whereupon they enter a sort of Wild West of textual practice, open to assimilation into narratives in any genre, from the Rousseauesque (Cavalli-Sforza and others' wistful representations of the endangered ecological balance of native cultures), to the neo-Galtonian (Markow's theory of phenodeviant handprints) to the biblical (e.g., "Y-chromosomal Adam" and "mitochondrial Eve," outside the scope of this paper, but almost ubiquitous in popularizations of the topic).

In the immediate aftermath of the Second World War, population genetics seemed to offer the resources for the reform of the human sciences. By the end of the century, the (political if not economic) decolonization of most of the world, as well as the increasing visibility of indigenous peoples' movements, had radicalized the debate, and scientific universalism had emerged as more of a problem than a solution for the long and sorry history of racial theory. Despite the efforts of population geneticists to purge their science of its nineteenth-century ideological residue, the rhetoric of manifest destiny continued to haunt the human archive. The huge expansion of the market in biological derivatives and the controversy about patenting genes resonated with colonialism's more traditional forms of exploitation. The demise of the Human Genome Diversity Project and the settlement against ASU in the Havasupai case exposed the inadequacy of the discourse of science as a simple antidote to racial prejudice.

Postcolonial human evolutionary genetics has now learned to conduct itself in a different register. The values are explicit, and to the greatest extent possible aligned with those of the research subjects. The scrupulous maintenance of the contractual terms of the sampling activities has become necessary for the legal preservation of the human archive. As the medical anthropologist Emma Kowal has observed as a result of her fieldwork among geneticists in Australia, exchange value in this field of science is now bound up with certain kinds of affective ties:

> The postcolonial scientist must maintain the network. . . . Geneticists who neglect these affective networks (as some of my informants have) risk ethical suspicion from other geneticists, bureaucratic hurdles, and, most frightening, a media scandal. Just like DNA, ethical biovalue can degrade over time if not appropriately maintained.[45]

Another strategy for the maintenance of ethical biovalue has been to restrict the use of the samples, as the Genographic Project has undertaken to do: "All samples collected will be held under strict conditions maintaining confidentiality and may not be used for any purpose inconsistent with the strictly limited scientific objectives of the project."[46] The narrative of the Genographic "Atlas of the Human Journey" is saturated with the values of human universalism of the immediate postwar period: everything about the way that the findings are presented to the public hammers home the message of the unity of the human family and the interrelatedness of its branches. But the pragmatic postmodernism of the Genographic ethics statement is perhaps the most interesting strategy of all, acknowledging, sincerely or not, that such narratives are always partial, incomplete, and inconclusive.

NOTES

1. Bruno J. Strasser, "The Experimenter's Museum: GenBank, Natural History, and the Moral Economies of Biomedicine," *Isis* 102, no. 1 (2011): 60–96; Cathy Gere and

Bronwyn Parry, "The Flesh Made Word: Banking the Body in the Age of Information," *BioSocieties* 1, no. 1 (2006): 41–54; Bronwyn Parry, *Trading the Genome: Investigating the Commodification of Bio-Information* (New York: Columbia University Press, 2004).

2. Marianne Sommer, "History in the Gene: Negotiations between Molecular and Organismal Anthropology," *Journal of the History of Biology* 41, no. 3 (2008): 473–528; Staffan Müller-Wille, "Claude Lévi-Strauss on Race, History, and Genetics," *BioSocieties* 5, no. 3 (2010): 330–47; Keith Wailoo, Alondra Nelson, and Catherine Lee, eds., *Genetics and the Unsettled Past: The Collision of DNA, Race, and History* (New Brunswick: Rutgers University Press); Soraya De Chadarevian, "Genetic Evidence and Interpretation in History," *BioSocieties* 5, no. 3 (2010): 301–5.

3. Veronika Lipphardt, "The Jewish Community of Rome: An Isolated Population? Sampling Procedures and Bio-Historical Narratives in Genetic Analysis in the 1950s," *BioSocieties* 5, no. 3 (September 2010): 306–29.

4. Johannes Fabian, *Time and the Other: How Anthropology Makes Its Object* (New York: Columbia University Press, [1983] 2002).

5. Jenny Reardon, *Race to the Finish: Identity and Governance in an Age of Genomics* (Princeton: Princeton University Press, 2005).

6. Robert E. Kohler, *Lords of the Fly: Drosophila Genetics and the Experimental Life* (Chicago: University of Chicago Press, 1994); Mark B. Adams, ed., *The Evolution of Theodosius Dobzhansky: Essays on His Life and Thought in Russia and America* (Princeton: Princeton University Press, 1994).

7. A. H. Sturtevant and T. Dobzhansky, "Inversions in the Third Chromosome of Wild Races of Drosophila Pseudoobscura, and Their Use in the Study of the History of the Species," *Proceedings of the National Academy of Sciences of the United States of America* 22, no. 7 (July 1936): 448–50, 449.

8. John Beatty, "Dobzhansky's Worldview," in Adams, *The Evolution of Theodosius Dobzhansky*, 206.

9. Theodosius Dobzhansky, *Genetics and the Origin of Species* (New York: Columbia University Press, 1937), 126.

10. Theodosius Dobzhansky, *Genetics and the Origin of Species*, 2nd ed. (New York: Columbia University Press, 1951), 77.

11. UNESCO, *The Race Concept: Results of an Inquiry* (Paris: [UNESCO], 1952), 98, 101.

12. Ibid., 11.

13. Reardon, *Race to the Finish: Identity and Governance in an Age of Genomics*, 63.

14. Ibid., 68.

15. Ibid., 65. Also see Lipphardt, "The Jewish Community of Rome," for a fascinating account of the ways in which historical factors had to be radically simplified in order to yield a biologically salient, "evolutionarily coherent" definition of an isolated population.

16. Emile Zuckerkandl and Linus Pauling, "Molecules as Documents of Evolutionary History," *Journal of Theoretical Biology* 8, no. 2 (1965): 357–66, 357.

17. Sommer, "History in the Gene."

18. Linda Stone and Paul F. Lurquin, *A Genetic and Cultural Odyssey: The Life and Work of L. Luca Cavalli-Sforza* (New York: Columbia University Press, 2005), 98, 168.

19. Ibid., 65.

20. L. L. Cavalli-Sforza and W. F Bodmer, *The Genetics of Human Populations* (San Francisco: W. H. Freeman, 1971), chapter 12.

21. Reardon, *Race to the Finish*, 48.

22. Ibid., 192.

23. Cori Hayden, *When Nature Goes Public: The Making and Unmaking of Bioprospecting in Mexico* (Princeton: Princeton University Press, 2003).

24. Debra Harry, "Indigenous People and Gene Disputes," *Chicago-Kent Law Review* 84, no. 1 (2009): 147–96, 179–80.

25. Michael Dodson, "Indigenous Social and Ethical Issues: Control of Research and Sharing the Benefits," speech delivered at the Scientific, Social and Ethical Issues Symposium, 22 July 2007, quoted in Matthew Rimmer, "The Genographic Project: Traditional Knowledge and Population Genetics," *Australian Indigenous Law Review* 11, no. 2 (2007): 33–54.

26. David Hurst Thomas, *Skull Wars: Kennewick Man, Archaeology, and the Battle for Native American Identity* (New York: Basic Books, 2000).

27. L. L. Cavalli-Sforza, Paolo Menozzi, and Alberto Piazza, *The History and Geography of Human Genes* (Princeton: Princeton University Press, 1994), 4.

28. Stone and Lurquin, *A Genetic and Cultural Odyssey*, 176.

29. Katherine Drabiak-Syed, "Lessons from Havasupai Tribe versus Arizona State University Board of Regents: Recognizing Group, Cultural, and Dignitary Harms as Legitimate Risks Warranting Integration into Research Practice," *Journal of Health and Biomedical Law* 6 (2010): 175–225, 180.

30. Paul Rubin, "Indian Givers," *Phoenix New Times News,* May 27, 2004, 4.

31. Drabiak-Syed, "Lessons from Havasupai Tribe versus Arizona State University Board of Regents," 182.

32. Michelle M. Mello and Leslie E. Wolf, "The Havasupai Indian Tribe Case: Lessons for Research Involving Stored Biologic Samples," *New England Journal of Medicine* 363, no. 3 (2010): 204–7; Drabiak-Syed, "Lessons from Havasupai Tribe versus Arizona State University Board of Regents."

33. P. Parham et al., "Episodic Evolution and Turnover of HLA-B in the Indigenous Human Populations of the Americas," *Tissue Antigens* 50, no. 3 (1997): 219–32, 228.

34. Therese Ann Markow and Kevin Wandler, "Fluctuating Dermatoglyphic Asymmetry and the Genetics of Liability to Schizophrenia," *Psychiatry Research* 19, no. 4 (December 1986): 323–28.

35. Therese Ann Markow and Irving I. Gottesman, "Behavioral Phenodeviance: A Lerneresque Conjecture," *Genetica* 89, no. 1 (1993): 297–305.

36. T. A. Markow and J. F. Martin, "Inbreeding and Developmental Stability in a Small Human Population," *Annals of Human Biology* 20, no. 4 (1993): 389–94.

37. T. Markow et al., "HLA Polymorphism in the Havasupai: Evidence for Balancing Selection," *American Journal of Human Genetics* 53, no. 4 (October 1993): 943–52.

38. Joanna Radin, "Latent Life: Concepts and Practices of Human Tissue Preservation in the International Biological Program," *Social Studies of Science* 43, no. 4 (August 1, 2013): 484–508.

39. Jane Caplan and John Torpey, *Documenting Individual Identity: The Development of State Practices in the Modern World* (Princeton: Princeton University Press, 2001), 275.

40. Kim Tallbear, "Narratives of Race and Indigeneity," *Journal of Law, Medicine, and Ethics* 35, no. 3 (2007): 412–24.

41. https://genographic.nationalgeographic.com/faq/about-project/#profit, accessed April 23, 2014.

42. https://genographic.nationalgeographic.com/wp-content/uploads/2012/07/Geno2.0_Ethical-Framework.pdf, accessed April 28, 2014.

43. Spencer Wells, *Deep Ancestry: Inside the Genographic Project* (Washington, DC: National Geographic, 2006), 11–12.

44. Cathy Gere, "Inscribing Nature: Archaeological Metaphors and the Formation of New Sciences," *Public Archaeology* 2, no. 4 (2002): 195–208. See also Sepkoski, chapter 2 in this volume.

45. Emma Kowal, "Orphan DNA: Indigenous Samples, Ethical Biovalue, and Postcolonial Science," *Social Studies of Science* 43, no. 4 (August 1, 2013): 577–97, 590.

46. Spencer Wells, *Genographic Project Ethical Framework* (n.p., n.d.).

NINE

Montage and Metamorphosis: Climatological Data Archiving and the US National Climate Program

Vladimir Janković

The climatological record contains data and information on what nature has done and, in the actuarial sense, is likely to do in the future.
—A Strategy for the National Climate Program: Report of the Workshop to Review the Preliminary National Climate Program Plan

Data compilation for its own sake should of course be avoided.
—Social Science Climate Impact Research

Within a decade following the decision of the US Congress to enact the National Climate Program Act in 1978, the National Research Council published a series of reports on the state of climate research and institutional infrastructure intended for acquisition and management of climate data, products, and services. These reviews took place during a period characterized by a major international effort to generate a global climatological record, an effort that capitalized on the developments in atmospheric modeling, new information and monitoring technologies, and the increasing importance of climatological archives in reconstructing recent and past climatic variability in the context of human-induced atmospheric processes and their economic implications. The perceived sense of urgency and the political significance that these developments had for climate research led to a series of high-level meetings organized by the US National Research Council and other institutions designed to consider the adequacy of existing practices of data management and the challenges concerning the methods through which new data might be archived and used.[1]

This chapter explores the methodological, institutional, and economic dimensions of climatological data archiving argued in these meetings. I discuss the elements of the cognitive politics associated with the notion of the climatological archive in relation to the US National Climate Program's objective to provide robust climate products for industrial use and the society at large. As the program's architects gave climatological archiving a vital role in public policy and economic planning, they stimulated a major investment of intellectual capital in addressing the problems of collection, processing, and dissemination of climate data products. Deliberations on these issues have affected the conceptualization of the climate archive as its own working world, combining methodological protocols, institutional politics, environmental research, and the complexities arising from the materiality of the archive itself. How did these processes affect the *purpose*, *practice*, and *status* of modern climatological archiving? What were the energy flows that shaped the emergence of the climatological archive as one of the epicenters of the contemporary environmental sciences and, with the growing concern over the anthropogenic climate change, a key site of environmental politics? What inspired sociologist Andrew Ross to suggest that "climatology, hitherto considered a second-class adjunct to the more exciting field of meteorology, or at best a branch of physics that had more in common with geography, has seen its object—knowledge about a stable archive of climate statistics—transformed into a volatile, political commodity of the first-rate importance."[2]

Ross's observation underscored the politicization of recent climatological research—although climatology and politics have intersected on many occasions in the past[3]—but also the fact that "a stable archive" had traditionally been considered a central object of climate research, the sole empirical "basis" of climate dynamics. More recently, developments in climate simulation and "scenario science" have revealed the political importance of historic climate data and their role as the empirical yardstick in the sometimes contentious discussions about Earth's climatic futures, as witnessed during the recent Climategate episode.[4] Yet both the conventional and the recent conceptualizations of climate data have produced the stereotype of climate archive as stable and definitive, a place of "absolute commencement" that vouchsafes the "principles of credibility" and provides a ground zero for all claim-making activities.[5] The climate archive is often imagined as an information *warehouse* in which raw data, having been harvested from the natural environment, remain deposited in a secure space until they are retrieved for "higher" theoretical or modeling purposes.

In this paper, however, I challenge the warehouse model on several counts. I explain why the climate archive should primarily be seen as a *process* and a *practice* rather than a "repository" and how such a process relates to specific objectives of climate research in the society at large.[6] Within the context of the National Climate Program, climatological data archiving comprised a series of scientific and physical tasks of acquiring, indexing, digitizing, cataloguing, describing, interpreting, analyzing, editing, and referencing archival materials. Bypassing these practices as merely incidental to the archive's function as an information warehouse obscures the fact that the archive's authority derived precisely from these practices and the decisions about the mode of their execution, rather than a presumed factual reality inscribed in the archival content. Those charged with the job of considering the place of climate information in the program argued that the climate archive ought to be designed as a space where people, machines, and institutions engaged with materials that have dimensionality, weight, and texture and where "the physical embodiment of the documents we possess" affected archival work experience and the interpretation of sources[7] (see also Rebecca Lemov, chapter 10 in this volume). The climate archive would be an infrastructure capable of meeting the challenge of increasing information loads while answering the needs of the research community and the society at large.

Furthermore, I argue that the warehouse stereotype of climate archives fails to account for the fact that climatological "archiving" is never *posterior* to the goals of climatological science itself. This is for two reasons. First, the climate archive is not a result of some practical negotiations on how to keep and employ "facts" in representing the history of climate variability and geographic spread of climatic zones. The climate as a scientific concept cannot exist prior to data and data prior to archives. What one finds in the archives are not simple mimetic inscriptions referring back to an observable reality. Consider, for example, the fact that the World Meteorological Organization defines climate as the mean values of environmental parameters over a minimum of thirty years so that current climatological "normals" derive from datasets prepared for the period from 1961 to 1990 but which will be replaced, in 2020, by the new normals based on the period from 1991–2020. Current climate is a statistical metric calculated from data collected more than twenty-four years ago: it is a conventional proxy to the past environmental trends.

Secondly, there isn't much sense in assuming that climate data come first and climate archiving second—not even in a temporal sense, as we will see below. The acquisition, storage, and use of climate information are coconstitutive rather than serialized: it is wasteful to collect data not destined for an archive. Just as there would be little logic in recording information that would become irretrievable, there would be equally little sense in thinking of a repository of information that could not be collected. The archive in this sense both justifies the labor expended on data collecting and supplies the criteria of information gathering. It acts as a precondition of climatological research rather than a methodological tool. The very definition of climate thus *requires* data-management tasks that in turn *require* a pool of accredited evidence that could be put to end use. There has never been a climatological datum that had not already been destined to be archived, regardless of the fact that many climate data have not been, nor need to be saved—but that is a different matter altogether.

These preliminaries would have been familiar to practicing climatologists in the archival services of the National Oceanic and Atmospheric Administration during the late twentieth century. They have been keenly aware of the inseparability of data from climate concepts and the amalgamation of research with seemingly exogenous factors such as information technologies, policy concerns, storage facilities, and labor costs. These interdependencies became of special concern during the 1970s, when the advances of satellite and telecommunication systems employed in a global environmental surveillance presented researchers with an unprecedented volume of data as well as new data formats.[8] To complicate matters, the incoming terabytes coincided with the century's most volatile weather anomalies, which inflicted major damages to world agriculture and utilities. As we will see, it is not by accident that the passing of the US National Climate Act came at the tail end of these events, which dramatized the need to deal with the climate/economy complex, to be explored in the next section. A critical role in delivering these objectives was given to data management.

PULLING THE CLIMATE TRIGGER

The decade of the 1970s was characterized by a growing climate consciousness, both public and scientific. The interest was triggered by a series of high-impact weather events and by scientific discussions about the causes of increased climate variability and climate change. Three sets of factors mobilized this interest. The first, in 1972, was the effect of bad weather on a range of social and economic matters, including the quadrupling of commodity prices, food shortages in the Sahel of West Africa and in South Asia, a failure of coffee production in Africa and South America, and a major fall in the anchovy fishery of the Pacific. The net effect was the first drop in the world's food production since 1945. The second factor was a growing consensus that man-made changes in the chemistry of the atmosphere—due to transportation, industrial emissions, and nuclear weapons—could cause regional, even global, increases in acid rain, ultraviolet radiation, precipitation, and temperatures.[9] And the third factor was a call for an institutionalized role of *service* climatology, a field of applied study mandated to deal with the relationships between climate conditions and weather-sensitive industries in the context of changing climates.[10] While in the past climatologists had provided advice to economists on the assumption of relatively stable climates, the unprecedented climate anomalies of the 1970s challenged that assumption and called for a more robust understanding of climatic variability based on a reexamination and further assimilation of historic and contemporary data.

It is hard to overestimate the level of uncertainty then associated with the causes and the likely return of climate shocks from the 1970s. Revising perceptions and changing attitudes presumed knowledge about what had actually happened and whether anything could be learned from short-term deviations from normals, and that knowledge simply wasn't there. When leading atmospheric scientists convened in a 1974 meeting to review the progress of the Global Atmospheric Research Program their recommendation on the problem was marked by a considerable humility: "In making its recommendations, the Panel is aware of what has been called the problem of '(don't know),' i.e., those who are called on to implement the program may not know that we don't know the answers to the central questions. The presentation of this report at least makes it clear that we don't know, and thereby reduces the exponent to unity."[11]

The priorities emerging from these chastened discussions crystallized in the launching of several national climatic programs—as well as the World Climate Program in 1979—whose objectives were to enhance climate research, improve data management, and, in particular, streamline the provision of weather and climate products for sectors such as agriculture, transport, hydrological services, construction, and so on.[12] One aim of the new climatology—apart from removing the remaining "don't know"—was to act as a macroeconomic tool designed to hedge against market uncertainties arising from future climate anomalies. Climate science was to become key to socioeconomic planning. In the United States, the National Research Council's Climate Research Board, the preeminent advisory body made up of leading climate scientists and representatives of industry, developed studies to understand how people act in uncertain conditions and whether the methods used in nonenvironmental fields could apply to climate-society interactions. For example, the center of Advanced Engineering at MIT and the NRC's Climate Research Board examined applications of planning around climate risk in areas of agriculture, air transportation, snow management, offshore drilling, heating fuel consumption, and electric transmission.[13] The American Meteorological Society sponsored the Conference on Climatic Impacts and Societal Response in 1980, and a Social-Science Climate Impact Research Committee was proposed to assist the National Climate Program Office in developing a socioeconomic climate impact program.[14]

Data management issues were recurrent agenda items. The early consultations among climate scientists highlighted weaknesses in the existing uses of climate information and called for "deliberate speed" in improving the utility of climate data services. In mid-July of 1978, a group of climate scientists and policy makers met at the National Academy of Summer Studies Center at Woods Hole Marine Laboratory to discuss the roadmap towards the US Climate Program Plan, coordinated by the US Climate Research Board and in accord with the provision of the recently passed National Climate Program Act (1978).[15] The documentation produced in the wake of this and the follow-up meetings is extensive, covering a host of climate/policy factors raised by select working groups between 1978 and 1986. Among the key findings was the recognition of the tight correlation between climatic variations and economic security and the realization that the existing climate information had not been used to its optimal capacity nor disseminated to stakeholders in weather-sensitive sectors. It was noted that the United States currently "lacked a well-defined and coordinated program in climate-related research, monitoring, assessment of effects, and information utilization."[16]

To address these weaknesses, three aspects of the program received special consideration: data and services, climate research, and impact assessment.[17] Leaving aside the latter two areas, I focus on how scientists worked with policy advisors to define the societal role of data management, especially with respect to the form and function of climate archives. The guiding principle in defining this role was the decision that the program's major outcomes would be climatic data and services intended for private and public users, including farmers, water managers, manufacturers, and consultancies. Architects of the program believed it would be judged on the basis of its economic impact and the extent to which the dissemination of climate data could increase the industrial productivity and GDP: "A major objective of the USCP [United States Climate Program] is to improve productivity in all sectors of US economy through effective application of knowledge of climate."[18]

The applied science focus of the program strongly influenced data management activities such as collection and selection of data, archiving criteria, and custom design of data derivatives. Yet the Climate Research Board raised questions as to what data should be extracted from the existing record and what data should be considered a priority, given their likely destinations. In what form could such data be made most useful and how should the anticipated uses of the record influence the selection and preservation of climate data?[19] To handle this cluster of questions, a dedicated advisory committee was charged with the role of obtaining feedback from client communities on how to *design* and *tailor* data products for postarchival use—the process to be further examined below. The client communities ranged widely, including individual farmers, plant and animal breeders, agricultural scientists, irrigation managers, commodity traders, industrial suppliers, food distributors, the Economic Research Service, the Foreign Agricultural Service, the Soil Conservation Service, and the Bureau of Reclamations and Land Management of the Department of the Interior.[20]

In the Woods Hole report, the Climate Research Board prioritized the formation of a single, *unified* data archival system but acknowledged difficulties arising from the complexities of past assimilation and storage methods. One problem was with the extraction and storage of information from the already published reports that included the state-organized *Monthly Climatic Data*, *The Climatological Data National Summary*, *Monthly Climatic Data for the World*, *World Weather Records*, and so on. Disparate formats found in these repositories required management of decentralized databases and "remote interfacing" to address problems of incompatible units of measurement, gaps in record series, and space-time resolution discrepancies. Even before the most appropriate architecture of the archival space could be considered, a large-scale assessment and validation were required to ensure compatibility of records.[21] Furthermore, the need for a centralized archive was connected with an unprecedented increase in data streaming from both land and space instrumentation. If there were no adequate storage and retrieval facilities to absorb the flow, there would be little sense in improving monitoring technologies or expanding surveillance to remote areas.

Indeed, in 1979, the archived data within the repositories of National Oceanic and Atmospheric Administration exceeded twenty terabytes (the size of the Library of Congress), while the number of annual requests from outside users reached 95,400. To cater for this demand, it was necessary to coordinate between at least forty-four sites, including the National Climatic Data Center, National Environmental Satellite, Data, and Information Service, the US Air Force Environmental Technical Application Center, and many smaller institutions, federal agencies, laboratories, and universities.[22] The coordination and processing would remain a headache for every future data administrator. In 1984, for example, a dedicated Panel on Climate Related Data, consisting of nine members including Helmut Landsberg, the nation's leading climatologist, struggled to come to terms with the processing needs for the undigested multiformat information streaming from satellites, radars, radio-sondes, remote sensors, buoys, and local automated observing systems: "the explosive growth in the quantity and diversity of weather and climate data, growth that could overwhelm the capabilities of current data systems in the US."[23] The increase was so great that even keeping the track of where the data went defined a genre of its own, comprising successive updates of holdings in the form of "selective guides."[24] Similar concerns over the magnitude of incoming data flows were endemic across environmental sciences of the period.[25]

The gargantuan task of managing the data flux became a matter of national priority. A specialized interagency Climate Data Management Workshop was convened in May 1979 in Harpers Ferry, West Virginia, to help outline the roadmap ahead. Sponsored by the National Climate Program Office and NOAA, the workshop participants, divided into seven working groups, found themselves embroiled in two days of deliberations on data management priorities. The published report featured summary recommendations that, even in their terse form, outlined a menacing silhouette of a hugely complex organizational problem of handling a range of technical activities and theoretical deliberations. Some of these included—in the report's own technical language—a permanent mechanism for coordination, evaluation, publication, and update of the inventory; an objective, program-wide coordinated system for compilation of data sets; a real-time review process of operational data products; and an ongoing documentation of changes in operational data analysis procedures, among others.[26] Most importantly, data management and archiving procedures had to deal with both the endogenous issues of existing data sets and the exogenous requirements to provide customized, user-oriented climate data *products* that often had little connection with the motivations behind the original data collection. Examples of climate products include seasonal climate forecasts, drought predictions, ENSO forecasts, crop-moisture index, degree days, temperature and radiations sums, and so on.[27]

In this context, Working Group 1 (Basic Atmospheric Data) laid out four requirements for the data to be archive-worthy: relevance, credibility, utility, and accessibility. It is symptomatic of the group's priorities that three out of the four criteria prioritized the needs of user communities. Only "credibility" referred to epistemic considerations. The reason for this bias was found in the realization that without the knowledge of the users' needs, data managers could not set priorities nor be expected to respond in a timely and efficient manner when requests were put to the archivist.[28] Such user-oriented decisions, made early in the process of establishing a data management system for the National Climate Program, suggested a landmark revaluation of the semiotics of atmospheric information from one that (mostly) served the research community to one increasingly oriented towards the needs of the market and public welfare. The resulting changes in the ways the US science administration handled the complexities of environmental trends had a lasting effect on the status of climate data and, in particular, on the *transformative* nature of climate archiving, to which we now turn.

THE CRUCIBLE OF DATA

The organizational challenge was not the only problem facing the program officers. The archival issues went beyond the concerns with the infrastructures of compatibility. There was the issue of data *selection* and data *manipulation*.[29] Scientific archives of necessity remain incomplete, for at least two reasons: first, scientists care only about a slice of observable reality; and second, even that slice may not be archive-worthy. Not all that is collected is archivable; and not all that is archivable is fit for use. The necessary shedding of "raw" data is governed by preferences that have to do with the historically situated criteria of data value. Equally important are considerations on the type and extent of interventions that need to be applied to data destined to cater for postarchival purposes. In this regard, the members of the Climate Research Board argued that "the framework for developing a comprehensive [Climate Information System] would provide an effective management of climate data, *the transformation of these data into useful climate information*, and the rapid delivery of data."[30]

What would this transformation entail? What is being transformed and for what reasons? And how do these internal workings bear on the conceptions of the archive as a definitive place of information? Does not the archive's status as the *locus terminus* give it canonical status? The authors of the 1979 National Research Council report made such questions central to the development of the National Climate Program (fig. 9.1). The authors argued that "*intermediate* between the observations and applications is data management, quality control, processing, retrieval, and dissemination to users. Indeed, planning for data management should *precede* the development of the observing system."[31] The authors saw the organization of data management as one of the key planks of the program for two main reasons. First, there was the concern over the accuracy and authenticity of

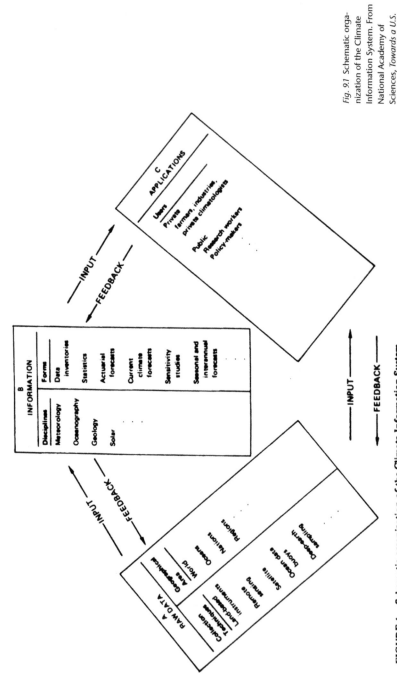

Fig. 9.1 Schematic organization of the Climate Information System. From National Academy of Sciences, *Towards a U.S. Climate Program*, 17.

FIGURE 1   Schematic organization of the Climate Information System.

data intended for operational use such as weather forecasting. Second, data management had a monetary value: the resources involved in minimizing data loss, eliminating errors, reducing ambiguities, and correcting inconsistencies in time series had a much lower cost than observations themselves, while the return on investment was high. It was made clear that the investments in the "preparation" of data obeyed the rule that the cost of (entering data into an archive) + (extracting the data from it) = constant. Cheap entry entailed expensive retrieval and vice versa. Given the user-oriented, economy-driven approach taken by the Climate Research Program managers, it was logical that their preference was the combination of expensive archiving and cheap/simple use.[32]

Their recommendations confirmed that the "stabilization" of data into archivable form entailed a series of actions aimed to ensure the quality of the content and create a range of synthetic products (for example, Drought Index, Soil Moisture Index, Degree Days) prepared for external access. During these stabilization procedures, unprocessed data could be discarded, corrected, synthesized, or left in their original form. In these procedures, no epistemic privilege was granted to original, firsthand observations: if anything, such privileges were reserved for the secondhand data derivatives. Raw data, argued participants of the Harpers Ferry workshop, "are often by-products, acquired for other purposes, and they may not be the best set or in the best form to solve the problems of a specific user."[33] In addition, producers of raw data rarely extracted their full research value because of their focus on the original purpose, and because of limited resources and time.[34] Needless to say, any attempt to make use of raw data called for an expenditure of labor. NOAA described this as the "data management life cycle," in which data managers and archive curators had freedom to improve (or reject) the incoming data that did not meet the set list of requirements.[35]

The interlinked workloads involved in these activities resulted in a particular *montage* of archival content characterized by diversity, modularity, and user-orientedness. As a technique used in film production to describe the selective arrangement of raw shots into a sequence, montage captures the climatological archive's content as a collection of elements subjected to a series of editorial, often automated, interventions. In such a montage, "virgin data" could be discarded, corrected, combined, adjusted, or left untouched, depending on how reliably they are judged to represent elements of a meaningful climate series. Outliers are most commonly corrected, but all data are routinely put through a quality control sieve before entering the archival stage. The National Oceanic and Atmospheric Administration refers to these processes as "data management life cycle." In the prearchival life form, the data and metadata enter the archive and access system;

in their intra-archival life form, they undergo further analysis. And if in the process the users wish to "improve" (or discard) the data, they open a new cycle or archiving, evaluation, and improvement. Resulting collections of data contain virtually nothing that is "primary, originary and untouched."[36] To the contrary, only in their state of montage is it possible to speak of archived data as "authentic" documents of and products for entities outside the archive. If for Lorraine Daston statistical derivatives "hover between the realms of the invented and the discovered,"[37] then climate statisticians placed their collections in both of these realms. And it has already been noted that "[the archive] reveals the rules of a practice that enables statements both to survive and to undergo regular modification; it is the general system of the formation and transformation of statements."[38]

As an archival act, the montage-like transformation of data stands in contrast to the privileged status that raw documents enjoy in the conventional imaginaries of archives as canonic warehouses. Archivists perform a number of tasks with their materials to make their classification meaningful, their storage optimal, and their retrieval convenient, but these tasks do not—usually or intentionally—infringe on the content of that which is to be classified, stored, and retrieved. It is different with climate data. Their content can and does change. The climate archive has an inner metabolism, churning received observations into verified records that may or may not resemble their earlier incarnations, but that often morph into something different. Sometimes it is improved data, and sometimes it is their synthetic statistical proxies that await further montage depending on the decisions from quality controllers, users, and climate policy advisers. As a result, the editorial montage becomes an ongoing process that involves manipulation of at least three kinds of entities: "raw observation data" (patchy and subject to error and physical biases); "artifact-free data" (revised and adjusted to render the "raw data" free from biases and errors); and "algorithmically compressed data" (products of complex modeling and simulation).[39]

The level of intervention required to manage and improve data quality could be extraordinary. In *Meeting the Challenge of Climate* (1982)—a report authorized by the National Research Council and the four highest ranking US climate boards—authors made clear the central role given to these interventions and identified several data needs, listed in the table (fig. 9.2).

The authors argued that an archive for such a range of information required a modular design that changed in size and content as the function of incoming data streams, their selection and reanalysis, and the production of *calculated* entities such as composite and nonstandard data products. During the early 1980s, these activities in turn required synchronization between the massively expanding upstream supply chain of information

Type of data	Description
Tailored information	Gives the probability of certain climate events and trends (droughts, hurricane strengths, high winds, flooding) for legal purposes and contingency planning.
Metadata	How data are collected, processed, stored and validated. Changes in methods of observation, location of stations, urbanization and replacement of instruments compromise the quality of time series. Metadata require more storage but warrant confidence in data products by providing an information about the history (lineage) of data
Composite indices	Ready-to-use indices such as wind-chill, growing degree days, heating and cooling degree days, etc.
Nonstandard data	Snowpack, iceberg frequency and spread, windiness, water runoff, etc. Nonstandard data reflect policy priorities: climate change, solar and wind technologies, deep-sea mining, urban design. etc.
Compatible data	Those used with nonclimate data
Coverage data	Provides full information about *current climate* and a suite of *climate forecasts* for planning purposes.

Fig. 9.2 Types of climate data identified in *Meeting the Challenge of Climate* (Washington, DC: National Academies Press, 1982).

and the computational downstream processing work. The responsibility for the former—the collection of all weather observations and a limited climate analysis—fell to the National Weather Service and its Climatic Analysis Center, which performed an initial validation of select data sets and stored them for three years in preliminary databases. But the task of longer-term storage of fully prepared data—the downstream processing works involving quality assurance, indexing, archiving, and retrieval—lay with the National Climatic Data Center (NCDC) and the National Environmental Satellite, Data, and Information Service (NESDIS).[40]

DATA HYGIENE, ARCHIVES IN MOTION

Any proposal related to the steering of the massive data flows required taking into account not only a theory of data management but also, perhaps primarily, the materiality of the media and the work measured in terms of resources necessary to overcome what has been termed "data friction."[41] Data friction refers to every physical or computational resistance that data offer in transit from collection to retention, measured by the expenditure of labor and resources needed to smooth out the transit. Errors (in the climatological jargon known as artifacts) could proliferate not only in the gross form of instrumental misreading or a faulty instrumentation but also in the recording and reception of data, their misplacement or misreading, and as a result of gaps in records. Errors tend to cascade, sprouting during statistical analyses or as a result of mis-inscription, calculation glitches, mis-indexing, and cataloguing or physical handling of data. Needless to say, every kind of data—raw or otherwise—is under suspicion of being an artifact unless proven sound.

In the Climate Research Board documents from the late 1970s, the hygiene of primary data and a suite of quality checks worked to complement the format adjustment protocols and revalidation procedures. "We believe that it is most cost-effective for the data banks to spend a good deal of time in cleaning the various errors out of the data and in making certain that the formats are as stated," wrote Roy L. Jenne, resident data analyst at the National Center for Atmospheric Research at Boulder, in 1975. "Many of the data sets that NCAR receives are processed to reduce the error content, and are put into formats that often reduce the data volume and the time necessary to unpack the data. Such cleanup work is always in progress at other data banks also."[42] With hundreds of millions of observations and the constraints of manual intervention, the cleanup was usually partial at best. Once transcribed, the records were subjected to a semiautomated review designed to flag suspicious items for manual inspection by the validators. Using manual techniques and quality control software, validators were meant to eliminate, correct, or interpolate missing data. The work sequence from the check-in of data, through their sorting, digitizing, running of the error-detection software, and reviewing the output was to be completed within a working month.[43] Afterwards, portions of artifact-free data would be sent for further control and applications to either the National Climatic Data Center, private climate consultants, or state climatologists, all of whom were expected to create value-added products by repackaging information in user-friendly forms or making customized composites to meet clients' specifications.[44]

We will return to the purpose-oriented criteria of archiving in the final section. For now, I note the iterative and open-ended character of climate archiving. The scientists and decision makers involved in the shaping of the National Climate Program considered this openendedness to be essential in order to meet the demands of data management, the changing and future research interests, and an expanding clientele for products and services. They also put a premium on the reduction of data friction and errors. Yet the decisions on how to proceed with the planning of archival process and architecture had to be based on the *anticipated* outside use and framed in consultation with the representatives of user communities. The program's archivists were thus responsible not only for the platforms for users' interactions with the content but for the archive content itself. As library scholars have recently argued, the archivist's "finding aid is a type of expert annotation, and the processes of appraisal, deaccessioning, and arrangement impact the decisions about selection and style that establish the archive."[45] Put differently, "the archivization produces as much as it records the event."[46]

It is evident that, in applying these procedures, the US climate administration conceived of the climate archive as a "content in the making"—very much in the sense of the montage procedures I introduced earlier. The authors of official documents conceived of the archive as regenerative rather than static, self-correcting rather than final, and user-oriented rather than repositorial. The archive was a *procreational* data space in which the successive stages of content existed only temporarily and in transition, and that laid claim to its canonical value in between successive iterations of reanalysis and cleanups. The program's documents from the early 1980s are replete with claims explaining the need for this *interim* value of archived materials, recognizing that even the fully accredited information required a retrospective adjustment, especially in light of new measurements or new clients' needs. New satellite measurements, for example, could challenge the quality of the already archived surface measurements. The otherwise inviolable record may then be dug up, corrected, and reentered. Or the archived content could change in response to changing research agendas, such as reexamining empirical signals of anthropogenic climate change through ocean circulation shifts, sea level rise, deep-sea sedimentary stratigraphy, hurricane paths, paleotempestological record, glacier meltdown, tropical cloud albedo, etc. Moreover, since the 1980s, climate scientists have assembled from historical sources a number of special-purpose archives of clouds, radar images, global energy balance measurements, radiosonde measurements, aircraft data, arctic records, snow cover, historical ocean temperatures, and so on. But whereas these records might be crucial for the science in one decade, they might grow obsolete and be ignored in another—which does not invalidate their canonical status in the interim.

The concerns over the incoming *and* outgoing data streams positioned the archive in between two cultures of work. On the one hand, the archive's design influenced observing systems and selection mechanisms, its operability and interface design. On the other hand, this design depended on what the user communities expected to find in the archives. It would be incorrect to suggest that the archive worked as a conduit through which end users *controlled* the choices made in climate monitoring and data assimilation, but it should be kept in mind that the political consequences of the 1970s climatic shocks—and the threat of long-term greenhouse gas buildup—had made the socioeconomic utility of climatic information a priority in economic strategy and science policy. The Climate Research Program officers developed an archiving philosophy that expressly endorsed this political framework and worked to find ways of better liaising with user communities. This, to a certain extent, did gradually come to determine the emphasis on the data perceived to have "epistemic potential" for emerging scientific research.[47]

Within this framework, some of the requirements addressed epistemic features of the archival content. This content was to have depth, completeness, homogeneity, relevance, and quality; its data were to be artifact-free and supported by metadata. But other features were downstream-oriented: the content was to be easy to access, tailored for diversity of needs, and compatible with external user systems. Archivists were advised to consult with clients on the meaning, processing, and interpretation of data. They were asked to be alert to what users wished to find in data. The archive itself was serviced by an independent facility for external requests where staff could explain methods and answer complaints and questions regarding documentation.[48] Climatological archiving, in summary, demanded vigilance in keeping the data in constant motion between their sources and their users (and back), resembling what Antonia Walford described as "data moves," i.e., a flow of information between people and institutions, a flow that itself constitutes data in the first place.[49] Meeting these requirements for mobility and transformation within the administrative cultures of NCDC, NESDIS, and NOAA has been a complex affair that deserves a separate discussion. What merits emphasis here, however, is the "orientedness" of archiving practices and the anticipatory mechanisms of data arbitration and customization, which the program's directors deemed a central policy goal. Whereas some scholars have, perhaps rightly, linked the archive to the image of "the temple and the prison"[50]—and so played on its combined canonicity and terminality—it is significant that the archive could also be a space of significant intervention and knowledge procreation.

BUILDING DATA POTENTIALS

I have suggested that the ethos of climate data archiving expressed in the government documents surrounding the 1978 National Climate Program

Act responded to environmental and political anxieties triggered by the climate anomalies of the 1970s and the resulting scientific impasse emblematized as the "don't know" problem. Forging an infrastructure of research *and* a security safeguard has shifted the meaning of climatological archiving away from preservation and research toward responsiveness to the needs of the public sector and national economy. This usually implied that data be organized and documented for a larger community of stakeholders, for which curators needed an empathetic sense of users' needs and likely developments in applied research. Thus in addition to recognizing the continuing importance of improving the repositories' depth, extension, and completeness, climate policy makers during the 1980s made a plea for enhancing access, relevance, and reuse.

This was a major new development. Information and library science scholars have suggested that the archival data management has traditionally been centered on collecting and appraisal of archival content, privileging data retention over release for public consumption.[51] The data "sitting behind" published research have usually commanded greater respect than those relevant for nonacademic stakeholders such as retail consultancies, feasibility analysts, futures markets brokers, land planners, and construction companies (all of whom require reliable data products based on archival materials). But while the archives of climate (and other scientific) data have been and will continue to be the "fundamental infrastructural component of the modern research system," they have, as we seen above, also become increasingly a tool of policymaking and socioeconomic analysis and advice.[52]

Without going into whether the same applies to other environmental archives, it should be pointed out that the climate archive has become both a normative pillar of research architecture and an axiological (value-based) instrument of social policy. In this sense the archived content has acquired both an "analytic potential"—the likelihood that a data set will be used in future analysis—and an "axiological potential"—the likelihood that a data set will, in some derivative/montaged form, be consulted by policy makers. More generally, the likelihood that *any* part of archival content may hide a latent value for future use is acknowledged by the current managers of climate data, a legacy of the National Climate Program engagement with the climate crisis of forty years ago. Thus the World Meteorological Organization stresses that "many of the ultimate uses for climate data *cannot be foreseen* when the data acquisition programs are being planned. [. . .] The global climate change issue, for example, is stretching the requirements for climate data and data management systems far beyond those originally conceived when the original networks were established."[53]

The fact that we design archives as sites holding *historical* records of atmospheric events reflects not only our wish to preserve the past, but our expectation that such a record may serve a *future*—and not in the sense of random serendipity, but by virtue of the archive's diachronic growth. Accumulation reveals patterns, and patterns lead to the recognition of emergent properties (on the future use of archives see Lorraine Daston, chapter 6 in this volume). The archive is this sense is *jussive* (commanding) in that it carries the past into the future. It carries a content that reflects the curatorial anticipations of the future, right now. It is characterized by a "prospective retrospective—that is, imagining now what we will want to remember in the future."[54] And it thus relates to "a promise, and of a responsibility for tomorrow."[55] Furthermore, instead of indiscriminate inclusion, the archives' jussive potential works to make archives exclusionary,[56] or "reconstituted collections, in which presences as well as absences mold a space within which it is possible to imagine" the past.[57] In the context of policy discussions about the goals of national climate archiving surrounding the passing of the National Climate Act in 1978, this selectivity can be seen to represent a transition from "archive-as-source to archive-as-subject,"[58] a move from a statistical inventory of unknown purpose into a meticulously montaged constellation of data products for the use in climate-sensitive sectors of public and commercial life.

In light of the archival infrastructures of the National Climate Program, it is possible to see the outlines of the climate archive as a multidimensional object shaped by the input from a number of stakeholders: climate scientists, data managers, government policy advisors, economists, administrators, and the private sector. Analytically speaking, the archive's multidimensionality may be parsed into several interrelated aspects. The climate archive is *architectural* in the sense of enabling the flow of and communication between data, their ordering, montage, curation, and ease of access. The archive is also *theoretical* in that it embodies assumptions about the methodologies of environmental research that impact data selection criteria. The archive is *computational* as incoming information undergoes quantitative control and transformation into composites, indices, and modeled data. The archive is also *teleological* in that its contents and structure reflect anticipated use. And the archive is of course a *material object,* a structure made up of concrete, glass, plastic, paper, and computing infrastructures.

But perhaps most importantly, the climate archive of the early NCP can be best described as *metamorphic*: a space where computing-intensive labor sifted, remade, and stabilized the empirical content in accordance with user-oriented policies designed to promote the public good and economic security. Items were archive-worthy only when they ceased to act as stable markers of a climate "reality" and when, through calibrations, moves, and fixes, became creations fit for future and outside use. Neither static nor passive, the climate archive was to be the site of suspension and metamorphosis, resembling a chrysalis in which the organism remains temporarily silent to allow for growth and differentiation that leads from one form of life to another.

NOTES

1. On the early US National Climate Program, see Gabriel Henderson, "Governing the Hazard of Climate: The Development of the National Climate Program Act, 1977–1981," *Historical Studies in the Natural Sciences* 46 (forthcoming). Spencer Weart, *The Discovery of Global Warming* (Cambridge, MA: Harvard University Press, 2008); Nicolas Nierenberg, Walter R. Tschinkel, and Victoria J. Tschinkel, "Early Climate Change Consensus at the National Academy: The Origins and Making of *Changing Climate*," *Historical Studies in the Natural Sciences* 40 (2010): 318–49.

2. Andrew Ross, "Is Global Culture Warming Up?" *Social Text* 28 (1991): 3–30, 7.

3. Vladimir Janković, "Climates as Commodities: Jean-Pierre Purry and the Modelling of the Best Climate on Earth," *Studies in History and Philosophy of Modern Physics* 41 (2010): 201–7.

4. Marianne Ryghaug and Tomas Moe Skjølsvold, "The Global Warming of Climate Science: Climategate and the Construction of Scientific Facts," *International Studies in the Philosophy of Science* 24, no. 30 (2010): 287–307.

5. Thomas Osborne, "The Ordinariness of the Archive," *History of the Human Sciences* 12 (1999): 51–64; Jacques Derrida, *Archive Fever: A Freudian Impression*, trans. Eric Prenowitz (Chicago: University of Chicago Press, 1995), 17.

6. On scientific data, Paul Edwards, Steven J. Jackson, Geoffrey Bowker, and Cory P. Knobel, "Understanding Infrastructure: Dynamics, Tensions and Design," NSF Report of a Workshop on "History and Theory of Infrastructure: Lessons for New Scientific Cyberinfrastructures," January 2007, http://www.si.umich.edu/cyber-infrastructure /UnderstandingInfrastructure_FinalReport25jan07.pdf, accessed 14 August 2014; Sabina Leonelli, "Packaging Small Facts for Re-use: Databases in Model Organism Biology," in *How Well Do Facts Travel?*, ed. Peter Howlett and Mary S. Morgan (Cambridge: Cambridge University Press, 2011), 325–48.

7. Maryanne Dever, "Provocations on the Pleasures of Archived Paper," *Archives and Manuscripts* 41 (2013): 173–82, 182; Gabrielle Dean, "Disciplinarity and Disorder," *Archive Journal* 1 (2011), http://www.archivejournal.net/issue/1/archives-remixed/the -archeology-of-archival-practice/, accessed 9 June 2013; Marlene Manoff, "Theories of the Archive from across the Disciplines," *Libraries and the Academy* 4 (2004): 9–25.

8. Etienne Benson, "One Infrastructure, Many Global Visions: The Commercialization and Diversification of Argos, a Satellite-based Environmental Surveillance System," *Social Studies of Science* 42 (2012): 843–68; Eric Conway, "Drowning in Data: Satellite

Oceanography and Information Overload in the Earth Sciences," *Historical Studies in the Physical and Biological Sciences* 37 (2006): 127–51.

9. Robert W. Kates, "The Interaction of Climate and Society," in *Climate Impact Assessment: Studies of the Interaction of Climate and Society*, ed. R. W. Kates, J. H. Ausubel, and M. Berberian, ICSU/SCOPE No. 27 (New York: John Wiley, 1985); Committee on Atmospheric Sciences, "The Atmospheric Sciences: National Objectives for the 1980s," *Bulletin of the American Meteorological Society* 62 (1981): 226–31; R. W. Katz, "Assessing Impact of Climatic Change on Food Production," *Climatic Change* 1 (1977): 85–96; Stephen H. Schneider, "Climate Change and the World Predicament: A Case Study for Interdisciplinary Research," *Climatic Change* 1 (1977): 21–43.

10. Stanley A. Changnon, "The Past and Future of Climate Related Services in the United States," *Journal of Service Climatology* 1 (2007): 1–7. See also Alan Hecht, "Meeting the Challenge of Climate Services in the 1980s," *Bulletin of the American Meteorological Society* 65 (1984): 365–66.

11. National Academy of Sciences, *Understanding Climatic Change: A Program for Action* (Washington, DC: National Academy of Sciences, 1975), xi.

12. Vladimir Janković, "Working with Weather: Atmospheric Resources, Climate Variability, and the Ascent of Industrial Meteorology," *History of Meteorology* 7 (2015): 98–111.

13. *Climatological Data: Pennsylvania, January 1984* (Arlington, VA: NOAA, 1984).

14. V. K. Smith, "Economic Impact Analysis and Climate Change: An Overview and Proposed Research Agenda," *Final Report to the National Climate Program Office*, NOAA (Washington, DC: Department of Commerce, 1980).

15. In 1978, the members of the National Climate Research Board included Robert M. White (chairman), Francis Bretherton (NCAR director 1974–80), Dayton H. Clewell (Mobil Oil Corporation), Herbert Friedman (Naval Research Laboratory), John Imbrie (Brown University), Joseph Smagorinsky (GFDL, NOAA), John E. Kutzbach (University of Wisconsin), Thomas Malone, William Nierenberg (Scripps), Verner Suomi (University of Wisconsin), Charles W. Howe (Resource Economics, Colorado), and Karl K. Turekian (geochemistry, Yale).

16. National Climate Program Act, 95-367 (September 17, 1978), www.epw.senate.gov /ncpa.pdf, accessed 1 July 2014.

17. National Academy of Sciences, *Toward a U.S. Climate Program Plan: Report of the Workshop to Review the U.S. Climate Program Plans* (Washington, DC: National Academy of Sciences, 1978), ix.

18. National Academy of Sciences, *Toward a U.S. Climate Program Plan*, 3.

19. National Academy of Sciences, *A Strategy for the National Climate Program: Report of the Workshop to Review the Preliminary National Climate Program Plan, Woods Hole, MA, July 16–21, 1979* (Washington, DC: National Academy of Sciences, 1980), 11. See also Peter J. Robinson, "Use of the National Environmental Data Referral Service," *Bulletin of the American Meteorological Society* 65 (1984): 1310–15.

20. National Academy of Sciences, *Toward a U.S. Climate Program Plan*, 19.

21. National Academy of Sciences, *Report of the Climate Board Ad Hoc Panel on Climate Impacts to the National Climate Program Office Regarding Social Science Climate Impact Research* (Washington, DC: National Academy of Sciences, 1981).

22. *Government Data Center: Meeting Increasing Demands* (Washington, DC: National

Research Council, 2003), 11; National Academy of Sciences, *Toward a U.S. Climate Program Plan*, 22–23; on data sources in 1979, see C. F. Ropelewski, M. C. Predoehl, and M. Platto, eds., *The Interim Climate Data Inventory: A Quick Reference to Selected Climate Data* (Washington, DC: US Department of Commerce, 1980).

23. *Atmospheric Climate Data: Problems and Promises* (Washington, DC: National Academy Press, 1986), 1. For the satellite data archiving see J. A. Leese, A. L. Booth, and F. A. Godshall, *Archiving and Climatological Applications of Meteorological Satellite Data*, ESSA Tech. Report NESC 53 (Suitland, MD: NESS, 1970).

24. *Selective Guide to Climatic Data Sources: Superintendent of Documents, U.S. Government Printing Office, Washington, DC* (Asheville, NC: National Climatic Center, 1969); Keith D. Butson and Warren L. Hatch, *Selective Guide to Climatic Data Sources* (Washington, DC: US Department of Commerce, 1979); Warren L. Hatch, *Selective Guide to Climatic Data Sources* (Washington, DC: US Department of Commerce, 1983).

25. Conway, "Drowning in Data"; Ron Doel, "Constituting the Postwar Earth Sciences: The Military Influence on the Environmental Sciences in the USA after 1945," *Social Studies of Science* 33 (2003): 635–66; Elena Aronova, Karen S. Baker, and Naomi Oreskes, "Big Science and Big Data in Biology: From the International Geophysical Year through the International Biological Program to the Long Term Ecological Research (LTER) Network 1957–Present," *Historical Studies in the Natural Sciences* 40 (2010): 183–224.

26. National Academy of Sciences, *Report of the Climate Data Management Workshop* (Washington, DC: US Department of Commerce, 1980).

27. *Selective Guide to Climatic Data Sources,* passim.

28. National Academy of Sciences, *Report of the Climate Data Management Workshop*, 2.

29. For detailed discussion see Roy L. Jenne and Dennis H. Joseph, *Techniques for the Processing, Storage, and Exchange of Data*, Technical Note NCAR-TN/STR-93 (Boulder, CO: National Center for Atmospheric Research, 1974).

30. National Academy of Sciences, *Toward a U.S. Climate Program Plan*, 24.

31. National Academy of Sciences, *A Strategy for the National Climate Program*, 11.

32. Edward Linacre, *Climate Data and Resources: A Reference and Guide* (London: Routledge, 1992), 61–62.

33. National Academy of Sciences, *Report of the Climate Data Management Workshop*, 3.

34. Carolyn L. Geda, "Social Science Data Archives," *American Archivist* 42 (1979): 158–66, 159.

35. *Environmental Data Management at NOAA: Archiving, Stewardship, and Access* (Washington, DC: National Academy Press, 2007), figure 4.1.

36. Ann Laura Stoler, *Along the Archival Grain: Epistemic Anxieties and Colonial Common Sense* (Princeton: Princeton University Press, 2009), 44–45.

37. Lorraine Daston, "Why Statistics Tend Not Only to Describe the World but to Change It," review of *The Politics of Large Numbers: A History of Statistical Reasoning* by Alain Desrosières, *London Review of Books* 22, no. 8 (2000): 35–36.

38. Michel Foucault, "The Historical *a priori* and the Archive," in Charles Merewether, ed., *The Archive* (London: Whitechapel, 2006).

39. James McAllister, "Climate Controversies and the Demand for Access to Empirical Data," *Philosophy of Science* 79 (2012): 871–80.

40. *The National Climate Program: Early Achievements and Future Directions*, Report of

the Woods Hole Workshop 15–19 July 1985 (Washington, DC: National Academies Press, 1986), 44.

41. Paul N. Edwards, *A Vast Machine: Computer Models, Climate Data, and the Politics of Global Warming* (Cambridge, MA: MIT Press), 2010, 84.

42. Roy L. Jenne, *Data Sets for Meteorological Research*, NCAR Technical Note IA-111 (Boulder, CO: National Center for Atmospheric Research, 1975), 2.

43. N. Guttman, C. Karl, T. Reek, and V. Shuler, "Measuring the Performance of Data Validators," *Bulletin of the American Meteorological Society* 69 (1988): 1448–52.

44. *Atmospheric Climate Data*, 12.

45. Tanya Clement, Wendy Hagenmaier, and Jennie Levine Knies, "Toward a Notion of the Archive of the Future: Impressions of Practice by Librarians, Archivists, and Digital Humanities Scholars," *Library Quarterly* 83 (2013): 112–30, 114.

46. Derrida, *Archive Fever*, 17.

47. On epistemic potential, see Birger Hjørland, *Information Seeking and Subject Representation: An Activity-theoretical Approach to Information Science* (Westport, CT: Greenwood, 1997).

48. *Atmospheric Climate Data: Problems and Promises*, passim.

49. Antonia Walford, "Data Moves: Taking Amazonian Climate Science Seriously," *Cambridge Anthropology* 30, no. 2 (2012): 101–17.

50. Randall C. Jimerson, "Embracing the Power of Archives," *American Archivist* 69 (2006): 19–32.

51. Carole L. Palmer, Nicholas M. Weber, and Melissa H. Cragin, "The Analytic Potential of Scientific Data: Understanding Re-Use Value," ASIST 2011, October 9–13, New Orleans, LA, http://www.asis.org/asist2011/proceedings/submissions/174_FINAL_SUBMISSION.pdf, accessed 14 September 2014.

52. P. F. Uhlir, "Information Gulags, Intellectual Straightjackets, and Memory Holes: Three Principles to Guide the Preservation of Scientific Data," *Data Science Journal* 10 (2010): 1–5.

53. http://www.wmo.int/pages/themes/climate/climate_data_and_products.php, accessed 3 July 2014.

54. Elizabeth Churchill and Jeff Ubois, "Designing for Digital Archives," *Interactions* (2008), accessed 10 July 2014, doi: 10.1145/1340961.1340964.

55. Derrida, *Archive Fever*, 36.

56. Geoffrey C. Bowker, *Memory Practices in the Sciences* (Cambridge, MA: MIT Press, 2005), 12.

57. Marija Dalbello, "Digitality, Epistolarity and Reconstituted Letter Archives," *Information Research* 18 (2013), http://www.informationr.net/ir/18-3/colis/paperC26.html #Williams77, accessed 1 July 2014.

58. Stoler, *Along the Archival Grain*, 44–45.

PART IV

The Future of Data: Archives of the New Millennium

Archives-of-Self: The Vicissitudes of Time and Self in a Technologically Determinist Future

Rebecca Lemov

Even when one is no longer attached to things, it's still something to have been attached to them; because it was always for reasons which other people didn't grasp . . . Well, now that I'm a little too weary to live with other people, these old feelings, so personal and individual, that I had in the past, seem to me—it's the mania of all collectors—very precious. I open my heart to myself like a sort of vitrine, and examine one by one all those love affairs of which the world can know nothing. And of this collection to which I'm now much more attached than to my others, I say to myself, rather as Mazarin said of his books, but in fact without the least distress, that it will be very tiresome to have to leave it all.

—Charles Swann in Marcel Proust, *Remembrance of Things Past*

But the global panoptic mind-thing is made of . . . us! Literally. However weirdly.

—Twitter feed of William Gibson @GreatDismal 11:40 PM, June 1, 2013

Microsoft senior researcher Gordon Bell, a pioneer in the once and future art of self-archiving, offers in his book *Your Life, Uploaded* a presumably beguiling vision. One day not far from now, a child asking her grandfather how he met her grandmother will be able to receive not just a fond-if-time-varnished reminiscence but also to watch documentary footage of the actual event. Her grandfather will have recorded it, along with snapshots of every other moment in his life (at a standard interval of ten seconds between shots), using a range of lifelogging and sensory data-gathering technologies. The self-archivist will then be able quickly and efficiently to search his personal digital archive—a "complete e-memory of [his] time on earth"—to retrieve the relevant information. MyLifeBits, the name of Bell's ur-archive, has been growing apace since 1998 and now amounts to 261 gigabytes stowed in a cloud-based system. And in fact, around the same time Bell purveyed this vision in print, one of his associates, the Scottish researcher Cathal Gurrin, who was wearing the technological aid Sense-Cam for an experimental sixteen months, did capture the moment in his "lifestream" when he met the woman who would soon become his girl-friend. If things work out between them, fifty years from now he will be able to inform his grandchildren about that meeting with moment-by-moment visual accompaniment.[1] And then, even after he is dead, his progeny will be able to ask a simulacrum invested with his stored data the same question, according to Bell's vision:

> Your digital memories, along with the patterns of fossilized personality they contain, may be invested into an avatar (a synthesized persona) that future gen-erations can speak with and get to know. Imagine asking your great-grandfather about what he really loved about your great-grandmother. Your digital self will reach out to touch lives in the future, allowing you to make an impact for genera-tions to come.[2]

As Gurrin remarked, the goal is to produce "a non-forgetting, non-decaying archive of everything you have done."[3] Or as self-archiving re-searcher Deb Roy of MIT's Laboratory for Social Machines observes, you can "sift through and discover traces of patterns in your own life that previously had gone undiscovered."[4] Your data—in the form of an archive of your day-to-day experiences—will outlast you. *You,* in some possibly new sense, will outlast you. Your time on earth will no longer be bound by the existence of your solid flesh. What remains to be carried over will be a self-amassed, interacting archive. Impact will continue to be made.

Bell's reassuring vision is also *unheimlich,* uncanny. Something that might have remained hidden has nevertheless come to light. This, one of Freud's definitions of the uncanny (which he takes from Schelling), applies in a mechanical sense to Bell's vision of future encounters of the living with the dead.[5] Erstwhile private, fungible, nonretainable moments—"these old feelings, so personal and individual" as Proust's Swann had them—such memories are capable of coming to light with the merest tap of the proper search-engine technology. Something seemingly *echt* human, made of the stuff of emotional memories and the texture of everyday experience, is rendered alien. Lifeloggers and lifebloggers (two very different occupations, it turns out), technologists and futurists (occupations often not so different) are drawing on a rich history of Western knowledge's technologies of self including the diary, the essay, the lifechart, and the various forms of modern self-fashioning to craft an emerging sense of what an archive can do.[6] This technologizing of the self, as it might be called, is both a reflection of and a further goad to ongoing transformations in scientific and social relationships.

The archiving of lives, through which each life amounts to an ongoing archival accounting and ends with a lump sum of data: this is more and more a modus operandi via a plenitude of instruments arriving or about to arrive in the marketplace, from the wrist-borne Apple Watch and FitBit to the neck-borne Microsoft SenseCam and adaptably velcro-able GoPro to phone-based apps such as Shadow that collect one's nightly dreams (see below). The human and social sciences are rife with resonant precursors in the field of intensive personal-data gathering, for example in the "human document" movement, the British Mass Observation study, the Kinsey Report, the American Soldier project, the Projective Test movement, the Harvard Refugee Interview Project, and other large-scale enterprises of the past century.[7] Before these and other twentieth-century "big social science" efforts, however, the longue durée suggests that a pivotal moment occurred during the eighteenth century when empirical practices of knowing oneself shifted and the new psychology emerged as queen of the sciences in Europe. At its heart was the conviction that "psychological ways of understanding the human being and of grounding knowledge, from logic to legislation and from aesthetics to pedagogy" constituted routes "to propel humanity into enlightenment and enable human perfectibility to be realized."[8] To know oneself directly through observation was at the heart of this variety of empiricism.

Two centuries of the "sciences of self" foreground the new data-driven self-archiving projects currently under way. Although it is not possible here to trace the full range of those scientific practices, it is useful to bear in mind an often-ignored feature of their history: self-initiation, with the emphasis on the two senses of self implied in the term; first, self-oriented (treating the self as object of research) and, second, self-driven (the self as subject initiating collection activities).[9] What happens when the imperative to engage in empirical data-collecting on the self and the intimate precincts of subjectivity takes shape not *outside* of the subject—via a scientist choosing to study an S—but *inside*, within the family circle, or by the autobiographical self, or in the hands of the experimentalist focusing on himself? A self-scrutinizing lineage developed via empirical studies of one's own dreams in the eighteenth century, for example, in the numerous self-observations of dreams and altered states published by Prussian self-investigators in the *Magazin zur Erfahrungsseelendunde* in the 1770s;[10] one's own fluctuating daily emotions as reflected in rectal temperature in the nineteenth century, as in Ugolino Mosso's study;[11] and one's own children as experimental subjects in the early twentieth century, as in the intensive baby documentation studies of California investigator Millicent Shinn and her network of home-based observers.[12] When self-initiated, the sciences of self played out according to a different trajectory from that of other mass collections. They raised the possibility of alternative forms of surveillance (not Big Brother but perhaps Big Mother)[13] and tracking regimes that were objectifying yet curiously intimate. They further offer a contrast to some modes of "archival thinking," as historians have pursued them. For example, Guatemala expert Kristen Weld examines in her recent book, *Paper Cadavers,* an untoward state archive containing seventy-five million once-secret police documents that describe the disappearances of protesters, farmers, students, and others who opposed the CIA-installed regime that held sway over the country during a thirty-five-year civil war. Weld characterizes archival thinking as a frame that allows the simultaneous examination of past conditions in relation to present possibilities for political action. Collections such as the National Police Historical Archive, considered at a historical remove and painstakingly resurrected from ruin and decay, are ways to "understand the larger systems of power, control, and legibility that record keeping necessarily enables."[14] Such pivotal archives are nearly opposite in impact to the ever waxing, seemingly trivial, self-based, and seemingly narcissistic archives described in what follows.

Self-initiated nonstate archives tend to embody a different set of power and control nodes, a difference perhaps most easily embodied in the contrast between the relations Michel Foucault described in *Discipline and Punish* (in which the pervasive "eye of power" spread disciplinary and dressage-like techniques that are absorbed through a network of power relations) and the processes he examined in *The History of Sexuality* volumes 2 and 3 (in which a set of techniques since Antiquity addressed the care and illumination of the self). In the self-archive, a powerful paradox is at work. The imperative to optimize the self through archiving it is accompanied by a concomitant desire to "outsource" responsibility for choices, as Natasha Schüll's recent ethnographic work on designs of wearable tracking technologies shows. This contradiction is an unexpected and largely unexamined engine driving growth: the consumer is taking responsibility for documenting her own physical and mental health, yet also delegating the mechanics, rationality, and even the resulting data caches to external corporate entities and the promptings of their algorithms.[15]

This essay investigates the question of how and in what sense the "self" is becoming more and more an archive made up of all the moments of a human life through which it constitutes itself. My thesis is that this is increasingly the case, aided and accelerated by technologies of moment-by-moment data capture, and that historical precedents can be useful guides to such seemingly unprecedented procedures. Looking historically at the sciences of self and their archiving practices also reveals that the simple designator "digital divide" fails equally to mark a before-and-after moment of self-archiving and to explain concomitant changes in the quality and quantity of subjectivity.

A good place to start in tracking the current vision of futurist self-archiving is with what can be called the fantasy of total information: the dream of condensing or shrinking immensities of text into small spaces, often advanced, as by a British commentator on microphotography in 1859, in the form of a striking vision of compression: "The whole archives of a nation might be packed away in a snuffbox," he announced.[16] With the aid of micro-technologies, you could haul away all the world's knowledge in a van, Vannevar Bush would claim almost a hundred years later in his manifesto for the Memex imaginary machine, "As We May Think."[17] "A future five-foot shelf may be no bulkier than a pack of playing cards," a writer for *Time* magazine observed excitedly in 1944 of a new analog data-storage technology, the Microcard.[18] Expressions like these—in which the all is packed away in the small—pepper technological writings from the mid-nineteenth century on, featuring snuffboxes, matchboxes, and small rooms. Physicist Richard Feynman debuted the concept of nanotechnology in 1959 by speculating about how nanoscience could shrink all the world's knowledge into a manageable size—an allocated thirty-five pages of the *Encyclopaedia Britannica,* as he put it.[19] This is not the place to explore the early modern emergence of this fantasy at length, but it would be interesting to look for its roots in an idea such as the "chain of being," which offered, since the Neoplatonists, a concrete conception of a linked and knowable world in which each denizen was arrayed by the principles plenitude and gradation. Yet the chain itself was a compact mechanism for holding—at least conceptually—the totality of creaturely, divine, and near-invisible existents in the world. Likewise, encyclopedic devices such as the Enlightenment *Encyclopédie* entailed a scaling-down effect. As Anke te Heesen demonstrates in her innovative study of an Enlightenment-era German pedagogical cabinet that required children actually to assemble and order the world themselves through cutting and pasting images and arranging them in a box, the fantasy, often grandiose and homely in equal parts, has metamorphosed alongside the devices available to hold the immensity of what it known. Its storage is a technical and miniaturizing matter.[20]

How does this fantasy of total information via technological tininess apply to the self? As with personal devices such as cell phones that store repositories of data—each device in effect a data bank—such collections simultaneously mirror and produce the self.

One way to look at it is that the self is both amenable and resistant to capture by managerial technologies. Accounts of the self have always been held in miniature, it can be argued. Devices that qualify as "technologies of self" include letter writing, photograph albums, diaries, commonplace books, Benjamin Franklin's "virtues chart," and the essay form as pioneered by Michel de Montaigne. In addition, Michel Foucault investigated the Stoics' use of the *hypomnemata*, reading diaries that should "not simply be placed in a memory cabinet but deeply lodged in the soul, 'planted in it,' according to Seneca"—for "they must form part of ourselves." Echoing the recent history of the onset of the information age, in which the latest gadgets are praised for their "at your fingertips" availability, the Stoics also valued hypomnemata for this quality: to have them "near at hand," *prokheiron, ad manum, in promptu,* Foucault explains, was to be able to use them in action, whenever the need was felt. "It is a matter of constituting a logos bioethikos for oneself, an equipment of helpful discourses, capable—as Plutarch says—of elevating the voice and silencing the passions like a master who with one word hushes the growling of dogs."[21] They were incorporated into the process of self-fashioning. Such writing books concerned with reading and rereading key texts formed, thus, an "important relay in th[e] subjectivation of discourse."[22]

In common parlance, the self is considered somehow greater than that which can be reduced to form, contained in a room, tracked in a test, or kept on a gadget (this is a leftover of the discourse of the soul extending to early modern sciences of the soul).[23] Yet as the density of recording devices observing human behavior of all kinds grows—and especially those tuned to the self—this will also impact the casement of the self and the very sense of the self as an object. As Gordon Bell points out, soon we will all be festooned with tiny data-recording devices that will feed large data sets: "this will be nearly effortless, because you'll have access to an assortment of tiny, unobtrusive cameras, microphones, location trackers, and other sensing devices that can be worn in shirt buttons, pendants, tie clips, lapel pins, brooches, watchbands, bracelet beads, hat brims, eyeglass frames, and earrings. Even more radical sensors will be available to implant inside your body, quantifying your health." In what follows, the interplay between format and contents in archives-of-self will be explored in three cases ranging from the early twentieth century to the twenty-first.

BUCKMINSTER FULLER'S DYMAXION CHRONOFILE

Buckminster Fuller lies in Mt. Auburn Cemetery in Cambridge, Massachusetts, under a stone urging visitors to "Call me Trimtab," although while living he labeled himself with the moniker "Guinea Pig B." With both appellations he declared his life an experiment. The trimtab is the tiny part of a boat that creates a small vacuum allowing the rudder to operate properly and steer without getting locked in position. It creates the preconditions for steering, exactly Fuller's conception of his own visionary role in the twentieth century. Likewise, Guinea Pig B was a self-nominated experimental creature who set forth by offering his own experience as the basis for the modification of others'. He was both overseer and subject of experiments. The fact that Fuller called his overarching lifetime job description "Comprehensive Anticipatory Design Scientist" is thus not surprising.

All of these threads—Trimtab, Guinea Pig B, anticipatory scientist—came together in what was perhaps his lowest-tech but farthest-seeing work, the instantiation of his own life as a chronological and exhaustive file, the Dymaxion Chronofile. From 1920 to 1983 he documented his life every fifteen minutes (via a scrapbook), and also added to the file his letters, notes, designs, doodles, memoranda, dry cleaning bills, receipts, and leases. "I decided to make myself a good case history of such a human being and it meant that I could not be judge of what was valid to put in or not. I must put everything in, so I started a very rigorous record."[24] With over 140,000 pieces of paper, media archives including 64,000 feet of film, 1,500 hours of audio tape, and 300 hours of video recording, and 13,500 5×8-inch cards cross-referencing the Chronofile alphabetically from 1970 to 1980, along with retrospectively added childhood photos from the age of four, Fuller's large file has caused some to label his the "Most Documented Human Life in History" (*Wired* magazine).[25] It was deliberately based on the unfolding of his life, a "come-as-it-may chronological—rather than an alphabetical or a categorical—record of my activities," as he wrote in "The Self-Disciplines of Buckminster Fuller."[26]

In his early twenties, his Chronofile still in infancy, Fuller married Anne

Hewlett, with whom he had a daughter named Alexandra, who contracted infantile paralysis and spinal meningitis, and died on her fourth birthday. During the five years following this tragedy, he developed a design and building business but failed to make it profitable and "became discredited and penniless." Around the same time his business went under, in 1927, his second daughter Allegra, was born in good health. He was then thirty-two and entered a crisis to which suicide seemed the only reasonable solution for a time. He consulted his Chronofile. It "clearly demonstrated that in my first thirty-two years of life I had been positively effective in producing life-advantage wealth—which realistically protected, nurtured, and accommodated X numbers of human lives for Y numbers of forward days—only when I was doing so entirely for others and not for myself." The Chronofile further indicated that the larger the numbers for whom he worked, the better the effect. Thus working "always and only for all humanity," he could achieve optimal effectiveness. Instead of accommodating "everyone else's opinions, credos, educational theories, romances, and mores," as he had earlier done, he sought to "do my own thinking, confining it to only experientially gained information."

This spurred further experimentation. "Finding myself a 'throwaway' in the business world, I sought to use myself as my scientific 'guinea pig,' my most objectively considered research 'subject' in a lifelong experiment designed to discover what—if anything—a healthy young male human of average size, experience and capability with an economically dependent wife and newborn child, starting without capital or any kind of wealth, cash savings, account monies, credit, or university degree"—he had been kicked out of Harvard—"could effectively do that could not be done by great nations or great private enterprise to lastingly improve the physical protection and support of all human lives, at the same time removing undesirable restraints and improving individual initiatives of any and all human aboard our planet Earth."[27] Further inventions followed, including the Dymaxion house and car, self-governing suburbs that glowed in the dark, strange boats, geodesic architecture, and the World Game.

The Chronofile, unlike Fuller, rests at Stanford University Library, which acquired it in 1999. In physical size, it is large and cumbersome, its 270 linear feet of material, from brochures of Bermuda to his daughter Alexandra's health records, occupying full-sized shelves. Last year, the renowned environmental architect William McDonough followed self-consciously in Fuller's footsteps by declaring himself a "Living Archive" via the same library, his minutiae and worldly doings to be collected in real time, but presumably much less space.

Gordon Bell's 2009 book *Total Recall,* republished in 2010 as *Your Life, Uploaded* to avoid being mistaken for an Arnold Schwarzenegger vehicle, chronicles the building of the MyLifeBits archive, his "project to record everything we do in life." Bell, often called a pioneer of the information age, was an employee of DEC (Digital Electronic Computers) from from 1960 to 1966, where he was responsible for designing several iterations of their PDPs (programmed data processors). Until a heart attack in 1983, he continued as a vice president at DEC and one of the first engineers to arrange networks among computers. Later, at the National Science Foundation, he worked with supercomputers to further their networking capabilities in an early version of the internet. Subsequently named a senior researcher at Microsoft, he still works there with something of a free hand. In starting his self-archiving project, Bell's initial impulse was to reduce paperwork. A friend, Raj Reddy of Carnegie Mellon, was attempting to scan a million books to his hard drive and asked if Bell would mind contributing his own book on software business ventures. Bell said yes, and began to think of what it would mean to digitize his personal output.

In 1998, Bell began scanning stacks of papers and letters, which process he enjoyed for its decluttering effects. If Fuller's Chronofile is clutter end to end, Bell's is a machine to eliminate clutter. Filing cabinets and bankers boxes full of bills disappeared, and Bell's passages describing this process are some of the most emotional in the book. He decided to expand his remit to include "everything" in his life, including:

1. digital copies of everything from Bell's past
2. recording and storing everything he saw, heard and did from that point forward
3. figuring out how to organize the information in my digital corpus.[28]

In quest of a record of everything, he hired a personal assistant in 1999 to begin scanning en masse. They watched paper "dwindle away like dirty old winter snow in the spring thaw."[29] They included his collection of mugs festooned with eagles—memorialized in the form of digital photos, with the objects themselves conveyed to the Computer History Museum, which Bell founded around this time. Posters and commemorative t-shirts too disappeared into digital format. They also transferred home movies from 8mm to VHS to CDs. By 2007, his archive contained 122 thousand e-mails; 58 thousand photographs; many recordings of phone calls and instant-messaging exchanges since 2003; his desktop activity (windows he opened); health records; labels of bottles of wine he consumed; his pulse; and his books, both those he had written and those he had read.[30]

His phase-two inspiration, after gathering up all this and miniaturizing it, was to make it searchable, manageable, and, really, livable. How could each person live with such a collection, and what would be the point? Bell and his colleagues were by this time sure that everyone fairly soon would have her own capacious archive. It would be a matter of "opting-out" if one didn't want to have some event or stream of life recorded. "You become the librarian, archivist, cartographer, and curator of your life." You have no choice, once you refuse to opt out. (Not remembering something, say a traumatic life event, is, however, an option: one can simply "lock" the file away.) Exploring what all this would mean and lead to, Bell came across Vannevar Bush's now much heralded piece of premonitory writing, his 1945 piece for the Atlantic, "As We May Think," which galvanized him. Here Bell found his chosen ancestor, the visionary who invented an imaginary machine for holding massive amounts of microfilmed material within a single oak desk, to be called up at will to desktop screens of translucent glass. The operator, wearing a "tiny camera the size of a walnut" on his forehead, would be able to hop from screen to screen or then again to conduct experiments in front of himself, then photograph and store them. He could create "trails" through the data, and at will call up his particular train of association for the day. "Wholly new encyclopedias . . . will emerge," Bush predicted, and Bell self-consciously sees his own encyclopedic collection as inheritor of this lineage.[31]

Bell's collection is a product of the ability to amass data by technological advances (he cites the Moore's law increase of data-storage capacity as well as the inevitable growth of tracking gadgetry). Undoubtely MyLifeBits is borne along by a technological determinist vision.[32] Even as technology increases its penetration of quotidian spaces of activity, the imbrication of self within these processes personalizes them, makes them increasingly intimate. In Bell's language, it is no longer your desktop but *you* that is the technological hub of your system: "you, not your desktop's hard drive, are the hub of your digital belongings." It is a kind of somaticization of data.

Meanwhile, the Quantified Self movement grows apace and new boosters have come along to share Bell's mantle of most enthusiastic self-archivist. Youthful Russian entrepreneur and "immortalist" Dmitry Itskov is the organizer of the Global Future 2045 International Congress, which took place June 15–16, 2013, in Lincoln Center, New York. It featured future-lookers from Marvin Minsky to Ray Kurweil, as well as the unveiling of Hanson Robotics' new Philip K. Dick Android, "a state-of-the-art robot with a large vocabulary, complex facial expressions, a sense of humor and something of an ego." Itskov, characterized in a fulsome *New York Times* profile as a man with "a colossal dream," envisions "the mass production of lifelike, low-cost avatars that can be uploaded with the contents of a human brain, complete with all the particulars of consciousness and personality."[33]

The conference's promotional Youtube video, titled "2045: A New Era for Humanity," is a riveting mix of high-futurist techno-optimism and dire predictions (with storm-tossed seas) of coming cataclysms—and it is also a meditation on time, or perhaps an exploration of the fantasy of trying to control time. The narrator intones, citing natural and human on-going and upcoming disasters, "The time we have to act grows shorter and shorter." And yet, if the technology is adopted, the time we *will have*, once invested in consciousness-storing databases, grows longer and longer. That the manipulation and storage of digital memories is equivalent to work-ing with time in a material form is an article of faith in such projects. This vision resonates with the ambitious scope of the University of Cambridge's Computer Lab, where the Memories for Life project has launched.[34] Fit-tingly enough, here one finds a reprise of the information-reduction trope, recalibrated as the self in a grain of sand: "If we imagine someone with a camera strapped to his or her head for 70 years (2.2×10^9s), that is something of the order of 27.5 terabytes of storage required, or about four hundred and fifty 60GB iPods. And if Moore's Law continues to hold over those 70 years (admittedly a large assumption!), it would be possible to store a continuous record of a life on a grain of sand."[35] Such ratios are contagious. Boosters of self-archiving feel that through the proper focus on new arrangements, not only human consciousness will be transformed but also the very meaning of what it is to be human.

CHRIS DANCY'S INNER NET AND QUANTIFIED SELFLESSNESS

Forty-seven-year-old Denver-based engineer Chris Dancy nominates him-self as the most quantified human in existence. He and those who report about him use other coinages, too, including "Data Exhaust Cartographer," "The World's Most Connected Human," "The Versace of Silicon Valley," and "Cyborg."[36] He employs gadgetry of a wide variety (ten devices on his person, thirteen at home, more in his car) and rich intensity (some 300 or perhaps 700 systems or online services to crunch, save, and collate the data). Metrics he tracks include pulse, REM sleep, skin temperature, and mood, among many others. As an "extreme life hacker," he takes screen-shots of his work, and everything else he encounters, to create an ongoing stream including "every meeting, every document he creates, every Tweet he sends, every file he shares, every screenshot he takes . . . providing him with a timeline [of] his entire work life."[37] He tracks his online presence with Memolane, keeps all articles he reads in Evernote, stores his stream feed in Google Plus, and uses his Memento narrative camera to video and snap shoot his life every thirty seconds. He fantasizes about ever smaller gadgets inserted in his clothes, buried in his shoes, or implanted in his skin, and a widely circulating photographic portrait shows him with wires protruding from the veins of his wrist.

In March 2014, after using his self-archiving system to lose one hundred pounds, Dancy quit his job as an information technology director at a software firm to pursue self-archiving full time. Like Fuller's and Bell's tales of data pursuit, Dancy's is one of transformation, but transformation in a different register. Whereas Fuller and Bell experienced life crises (bankruptcy and a heart attack, respectively) before they turned to self-documentation—so that self-documentation served as the redemptive path to satisfactory transformation—Dancy experienced a crisis *in and through immersion* in his self-generated data.

When Dancy was a child, he recounts, his father would measure his three sons' height against the doorjamb once a year. Unlike his brothers, however, Dancy didn't want to wait a year, and began measuring his own height regularly. In his twenties, he tracked his finances and purchases on paper. He became a predigital tracker who adjusted easily to computer devices. Around the age of forty, as he told an interviewer about his turn to self-archiving, it became important to him to collecting everything he produced so as to "have it." "It's embarrassing, but I didn't want to lose anything I did online. I said, 'How can I save this stuff?'"[38] Starting with Yahoo Pipes, he began saving every online interaction in Evernote. (It was not possible at the time to save one's Facebook feed, or other feeds.) One day, rummaging around in his digital collection, he found his Facebook post from six years before and experienced a dizzying sensation, a sort of epiphany: "It literally was a wormhole back to that moment in time. All of a sudden, all the business of my life collapsed, and I was at one with a piece of data that I left some time back." Oneness with his own past was what he sought.[39]

He started experimenting with services beyond Pipes: IFS, Zapier, and others. Each year, he tried to bring three more systems online, adding apps, devices, senses, and services. (An example of a service is Dropbox.) After four years, he had three to four hundred systems simultaneously collecting data at any moment. Some he considered "me" and some he considered environmental. Surprisingly, he found himself most drawn to the "reflections" of other people interacting with the digital version of himself. (A simple example of this "data on data" is a Facebook record of people interacting with you via reshare or "like.") Initially, becoming aware of other people using or interacting with his data was equivalent to "creat[ing] Skinner boxes for the mind in themselves." His anxiety intensified.

As Dancy became "aware of my own existence" via constant input about what he was doing—taking pictures, sending emails—he became aware of how systems of relationships worked. Through Map 10 he projected his data onto Abraham Maslow's "hierarchy of needs" in order to measure his relative level of self-actualization.[40] Finally, "I became really afraid." Before this intensive tracking of data-on-data, he had had no idea "how much impact I had on my environment" and on others. "It's one thing to be contemplative. It's another to stumble into it, to be almost waterboarded with awareness. It's one thing to Google yourself. It's another to Google . . . your life. I could see too much. . . . I was coming slightly unhinged with the amount of information I had about myself. It started to make me feel slightly detached from reality."

This fear of what might be called data ripples or the revelation of intense interconnectedness in a web of living and nonliving things caused him to consult a spiritual advisor, who told him (as he paraphrases the diagnosis), "You've become aware, but you're not aware of what you're aware of." Formal meditation helped him to realize that what he was seeing was a manifestation of what Buddhists call interdependent coarising, the continuous codetermination and mutual constitution of human, nonhuman, and indeed all experience. As Dancy recalled the turning point, "I'm almost sort of emotional saying this. . . . There was such a beautiful symbiosis between my head and my heart, and all the systems I could see starting to create each other. . . . If I was kinder, gentler to myself, literally I could exponentially see it in my work and in my personal life." What he did, in short, affected other people, and now he could see this to be so.

Contemplative endeavor became the goal of his self-archiving. What began as (potentially) an infinite loop of self-absorption turned into an experiment in dissolving the distinction between self and world. The challenge is ongoing. "I call it the Hubristic Butterfly Effect. It is very easy to fetishize my actions once I see them. Oh look, what a good person I am." This danger is at the core of taking the self as a data stream. "It is very dangerous. . . . It is at the point where I would not recommend this to anyone. Maybe as a practicing Buddhist, [it is possible] but . . . self-awareness is not easy." He noted that comments on a recent Youtube video labeled him "Thought Police," and a "Digital Hitler." Online feedback to a *Guardian* profile of Dancy comprised an insulting litany of Britishisms: from "nit" to "knob" to "nitwit" to "tool" to various adjectives for self-absorbed. "I can get where people would see that," Dancy said. "From my experience, I might be unique in how I'm gathering what I'm seeing; but I'm not unique in what I'm seeing." He finds himself to be a pioneer, in a sense, of extreme embeddedness, someone who has confronted possibilities for a degree of self-absorption that is undoubtedly happening and apparently intensifying. "People are even aware of it. We are aware of our relationships to others, but not our intimacy with others. [We need to] respect our intimacy with ourselves and others." It is necessary to work through this, and not simply to blame or eliminate tracking technology: "We're not going to have *less* technology in a year."

As if in confirmation of Dancy's bon mot, a recent "massive . . . experiment on Facebook" manipulated the feed streams of 689,003 users with positive or negative emotional valences, tipping them in one direction or the other. (That is, it seems Facebook already selects what the user sees in her feed, and can algorithmically alter the emotional tone of that feed on command.)[41] Among other things, the study showed that "advanced emotional engineering" is likely to grow in subtlety and prevalence. Yet controversy has erupted not so much over the results of the study—which demonstrated that a slightly more upbeat news feed causes users to post slightly more upbeat self-reports—but rather the (in retrospect unsurprising) fact that Facebook would so cavalierly experiment with its users. Double knowledge is hard to avoid: first that influence of each upon each is constant, and second that social media participation renders one a de facto or sometime de jure experimental subject. To return, then, to Dancy's point, which in turn recalls his precursors': "We need a better relationship between ourselves and the data, one that doesn't allow you to fetishize it or pathologize it" (fig. 10.1). Certainly these tools can function as forces of unfreedom but they can also act "like your mom pulling out the family album . . . that bad glazey stuff from Sears [covering the photos] . . . There are wormholes that allow us to explore ourselves."[42]

A consideration of the trio of Fuller, Bell, and Dancy suggests that there is possibly more to the technological-determinist forward edge than the simple eradication of shadows and the boldness of constant forward motion. The question becomes: how might different means of collecting and storing data through devices mirror different aspects of the self and thereby produce different kinds of selves? For example, Apple Watch, Shadow, and SenseCam each have a quite different logic of what the self is, how it's mirrored, how it's produced. The present moment of digital self-archiving is not one seamless mode but a site of proliferation. How does this compare

Fig. 10.1 Slide from a Chris Dancy powerpoint presentation: "Real You" by Aaron Jasinski. Courtesy Chris Dancy: chrisdancy.net

with earlier moments in the sciences of self, when technologies and modes of experimentation multiplied?

CONCLUSION

A literary critic recently identified a thriving subcategory of autobiography, "the genre of the un-remembering memoir," in which the narrator, usually the victim of a brain fever, organic disorder, or early trauma, must reconstruct his own inner life out of its absence by accessing myriad documents, tracking down eyewitness accounts, and researching her own self as if it were another's.[43] These are books of forgetting and compensatory documentation. An accident has imperiled memory, and the plot of the story is to restore it through the collection of the documents of one's life. In the case of self-archiving, by contrast, we find a constantly observing self so diligent that the excess of data overflows and produces, out of these ongoing emotional exhaust trails, a vast and ever-ramifying file. Similar to the subgenre of the self-remembering memoir, however, there is a species of uncertainty and a potential accident at the core of this relentless self-archiving process.

Constant observation, a kind of self-surveillance and self-tracking that is ever more the norm, can be seen as the result of uncertainty entering the domain of self. Recent work by Limor Samimian-Darash and Paul Rabinow identifies a new conception of uncertainty as inherent to systems, particularly to knowledge systems. More knowledge or evidence does not reduce uncertainty, but instead enhances it in new ways.[44] If we see the self within an archive of self-produced, ever-more-automatically-gathered evidence as a thing produced through tracking and traces—always populating itself with further evidence though never reaching completion—then the element of uncertainty will be built in and grow with and through the accumulation of self-knowledge. Every attempt to shore up the provinces of the self against time and decay adds to its uncertainty. If Gordon Bell imagines a future in which his assistant's grandchildren will call up the memory of their grandparents' meeting, this possibility is attended by uncertainties both technical and existential. The "sciences of subjectivity" already are reckoning with these kinds of uncertainties even as they offer programs that promise new regions of knowledge command. Indeed, ever-growing streams of data issuing from self-tracking and self-archiving projects constitute the "data rich future of the social sciences."[45]

Finally, there are at least two specific ways in which psychic databases reflect and affect the constitution of the self: (1) in their collective quality; and (2) in their peculiar temporality.

Collectivity. The data-archiving future is found not just in the individual maintaining ties to an individualist self,[46] augmenting enlightenment promises of self-determination, and renewing a great instauration of privacy, but also in a collective psyche emerging from penetrating data-gathering devices. Even Gordon Bell predicts (but perhaps does not fully explore the ramifications of) sensors embedded in gadgets and "peppered throughout your environment" so that you might record as much or as little as you want of what happened to you and around you. Still, it may not be up to you. Recently the Harvard-based bioengineers George Church and Sriram Kosuri used standalone DNA itself as their data-storage vehicle and found they could lodge 700 terabytes on a single gram. The binary pairs of DNA nucleotides function as "bits," 0's and 1's, so that they can treat "DNA as just another digital storage device." Merging medium and message in an unprecedented way, they published their own book about DNA data banks *on* DNA, including images, text, formatting and all. Imagine, they say, holding entire libraries in vats. To publish you can spray it on walls.[47] These DNA-based storehouses could also act as ubiquitous recording devices. In other words: it is not only *you* who will be archiving yourself. The self will be constructed from a variety of data streams. And it is not really "yourself" in the traditional European or North American (non–American Indian) sense of *having a self* that is the ultimate target, but a phenomenon much more fleeting and dispersed.

The ephemerality and insecurity of the archived self—in the face of its grounding in attempts to preserve that which is fleeting—is counterbalanced by several aftereffects that materialize and solidify the self through its enumeration and collection, a process the result of which has been labeled the "algorithmic self."[48] As critic Rob Horning describes, "Self-commodification does not diminish the user's self-conception but rather makes the self conceivable, legible. The self as product is inherently not guilty of some of the deauthenticating aspects of agency which threaten the integrity of other versions of the self: being calculating, unspontaneous, manipulative, phony, etc. The self as product can be seen as something that simply is, a given thing articulated in a definite form."[49] Authenticity is, as yet, a possible reward for the extensive use of devices for self-documentation and the reinforcement of self that they provide, in part through data collection and archiving of one's likes, preferences, dislikes, associations, free associations, friends' associations, comments, commentary, etc. Yet the solidifying of the self, paradoxically, renders it at risk of being tired, old, disposable. It must be perpetually renewed, as Horning points out: "So the self, as a product, loses its enchantment for us and needs to be revitalized to the extent that it becomes familiar, known, understood. We love ourselves only as a novelty, a mystery, not as a staple product." At the core is a desire to catch a glimpse of ourselves—the self—objectively, as others see us. In this way, the dynamics of self-archiving reanimate traditional "view-from-nowhere" forms of social science that attempt to objectify the subjective, yet always fall short of that empyrean realm of completion. The archive (in this case, of the self-documented self) cannot complete itself.[50]

Temporality. Here we turn to the question of how time operates in relation to these new compilations of self. From the Carte du Ciel's compendious archives to museums of thunderbolts to unpotentiated climate data to elegiac collections of stop words,[51] are there common answers to the questions, What does an archive do? What makes an archive an archive? Can an archive contain time's flow?

In light of the shifting practices of the archive as well as the shifting constitution of human "innerness" or subjective space, it seems clear that the self-archive is defined most immediately by the relatively-less-plastic limits of the human life in its corporeal form. In Bell's or Itzkof's system, you will no longer be recording experiences once your data is invested in an avatar. The lifespan itself is thus reified and circumscribed in the process

Fig. 10.2 Shadow, an app for a global dream archive—fully funded via Kickstarter as of 2013, though currently still in beta testing—promises to capture the 95 percent of dreams forgotten when people neglect to record them after waking up. Image courtesy Hunter Lee Soik.

of being purportedly extended. It is rendered a site for intensive data collection in the name of enhanced memory, transhumanism, and the "Good of All Mankind."[52] MyLifeBits is at its core a new carapace.

This erasure of time's finality also suggests an erasure of other things in the name of memory itself. Despite the aim to collect everything, there are some things that would evade the data set. Like Rutger Hauer's replicant in the film *Blade Runner* who has seen worlds, "things you people wouldn't believe," all of which, "lost in time," will die with him, what will happen to the memories of the nonhuman or the nonrich, those who can't afford avatars or don't want them? As in anthropologist Claude Lévi-Strauss's address to a "collection of oddments left over from human endeavors," or director Andrei Tarkovsky's "filmic archeology of the discarded," what can be made of the awkwardly fitting and the left out, the things that remain outside searchable categories?[53] Rallying cries for endless mnemonic enhancement make no room for the left over, dreamlike, boring, or unbelievable. The self-archive starts to be about erasure as much as enclosure. Art critic Hal Foster recently wrote, considering the endless activity that characterizes new forms of ceaseless curation and their concomitant collections, "It prompts one to wonder for what present, let alone what future, such archives are compiled."[54]

The more it can contain, the more apparent are the things it cannot contain. There are even algorithms to select moments of the life-stream that "stand out" so that only those will be collected. Dancy's "Inner Net," on the other hand, attempts to pick up the detritus: "There's an abundant amount of information we're losing moment to moment. How can we harvest this information?" A new app to archive dreams, Shadow, likewise promises to gather up the immense worlds of data lost every night in sleep and store it in a collective clearinghouse (fig. 10.2).[55] What is missing from futurist archives such as MyLifeBits, but what may be present in new as-yet-unmade archives or in just-resurrected files full of details of the credenzas of forgotten research subjects, is finally a sense of time detached from self-concern and technologically fueled nonreflection. A ghostly, often unexpressed tide of melancholy tugs at the positivist attitudes of self-archives and robust enthusiasms of self-archivers. Paradoxes, as detailed above, abound: the human life span and its limits are reinforced even as they are confronted; the ephemerality of personal experience is further highlighted within attempts to solidify that which passes away. As yet, there is little explicitly remarked on sense of loss, of what William Gibson calls the Gone World: "Somewhere, surely, there is a site that contains . . . everything we have lost," he suggests, and it is not clear whether he believes one could actually find it.[56]

NOTES

1. "Look," he says, "here's a picture of the first moment I met my girlfriend—not that I knew she'd become my girlfriend at the time." Quoted in Gordon Bell and Jim Gemmell, *Your Life, Uploaded: The Digital Way to Better Memory, Health, and Productivity* (New York: Plume, 2010), 48 (orig. *Total Recall* [New York: Plume, 2009]).

2. Bell and Gemmell, *Your Life, Uploaded*, 6.

3. Cathal Gurrin, "Quantified Self—Second Nature," vimeo video, http://vimeo.com /32054542. Accessed August 2016.

4. Deb Roy, "The Birth of a Word," TED talk http://www.ted.com/talks/deb_roy_the _birth_of_a_word?language=en. Accessed July 2016. This is the account of Roy's three-year panoptic study of his infant son, which resulted in an unprecedented (approximately) 200 terabytes of data.

5. Sigmund Freud, "The Uncanny," *The Standard Edition of the Complete Psychological Works*, ed. and trans. James Strachey et al. (London: Hogarth, 1955), 7:217–56, on 241. The translation provided here is "something which ought to have been kept concealed but which has nevertheless come to light" ["das Unheimliche sei etwas, was im Verborgenen hätte bleiben sollen und hervorgetreten ist"].

6. On technologies of self, see the late writings and discussions of Foucault: Michel Foucault, *History of Sexuality*, vol. 3, The Care of the Self (New York: Vintage, 1988), and *Technologies of the Self: A Seminar with Michel Foucault* (Amherst: University of Massa-

chusetts Press, 1988), a lecture delivered in 1981 at a seminar at University of Vermont. The difference between lifelogging and lifeblogging is simple but important: lifebloggers wish to publicize their data and reflections, whereas lifeloggers collect massive amounts of self-generated data but maintain it privately unless they decide otherwise. Bell is adamantly a lifelogger but not a lifeblogger, as he maintains that the former allows strict controls over what is remembered and accessible by others, and even by oneself. I.e., you may wish to lifelog an event but "lock" it from further scrutiny.

7. For an overview of efforts to "file the human" see Rebecca Lemov, "Filing the Total Human Experience: Anthropological Archives at Mid-Twentieth Century," in *Social Knowledge in the Making*, ed. Charles Camic, Neil Gross, and Michèle Lamont (Chicago: University of Chicago Press, 2011), 119–50. On the Harvard Refugee Interview Project see David Engerman, "The Rise and Fall of Wartime Social Science: Harvard's Refugee Interview Project, 1950–54," in *Cold War Social Science*, ed. Hamilton Cravens and Mark Solovey (New York: Palgrave, 2013), 21–43.

8. Fernando Vidal, *The Sciences of the Soul: The Early Modern Origins of Psychology* (Chicago: University of Chicago Press, 2011), 2. Vidal traces a turn to directly observed knowledge; as the authors of the *Encyclopédie*'s "Preliminary Discourse" remarked approvingly of John Locke, "In order to know our soul, its ideas, and its affections, he did not study books, because they would have instructed him badly; he was content with probing deeply into himself, and after having contemplated himself, so to speak, for a long time he did nothing more . . . than to present mankind with the mirror in which he had looked at himself" (quoted in Vidal, 16).

9. For the point about self-initiation as key, thanks go to Bruno Strasser. Self-collected data is "data . . . gathered in the first place by individuals for their own purposes. Problems can result when the data is then later not available to the self who generated it (is absorbed by corporations or health institutions) or when it is not updated properly." Gary Wolf and Ernesto Ramirez, *Quantified Self-Public Health Symposium*, 3, http://quantifiedself.com/symposium/Symposium-2014/QSPublicHealth2014_Report.pdf.

10. On dream collectors see Doris Kaufmann, "Dreams and Self-Consciousness: Mapping the Mind in the Late Eighteenth and Early Nineteenth Centuries," in *Biographies of Scientific Objects*, ed. Lorraine Daston (Chicago: University of Chicago Press, 2000), 67–85, pp. 74–75.

11. Ugolino Mosso's study described in Otniel Dror, "Seeing the Blush: Feeling Emotions," in *Histories of Scientific Observation*, ed. Lorraine Daston and Elizabeth Lunbeck (Chicago: University of Chicago Press, 2011), 326–48, p. 326.

12. Millicent Shinn's network described in Christine von Oertzen, "Science in the Cradle: Milicent Shinn and Her Home-Based Network of Baby Observers, 1890–1910," *Centaurus* 55, no. 2 (2013): 175–95.

13. On NSA-style surveillance in relation to archives, see Matthew Jones, chapter 12, this volume.

14. Kristin Weld, *Paper Cadavers: The Archives of Dictatorship in Guatemala* (Durham: Duke University Press, 2014), 13.

15. Natasha Dow Schüll, "Data for Life: Wearable Technology and the Design of Self-care," *BioSocieties* special issue on Big Data, edited by R. Rapp and L. Hogle (7 March 2016): 1–17.

16. Quoted in Frederic Luther, *Microfilm: A History, 1839-1900* (Annapolis: National Microfilm Association, 1959), 23.

17. Vannevar Bush, "As We May Think," *Atlantic*, July 1945, 101–8.

18. "Education: Book on a Card?" *Time,* September 4, 1944, 60.

19. "Tiny technologies" is a phrase borrowed from Hallam Stevens, which he used on his syllabus for a course on nanotechnology. See Davis Baird, Alfred Nordmann, and Joachim Schummer, eds., *Discovering the Nanoscale* (Amsterdam: IOS Press, 2004), including Joseph C. Pitt, "The Epistemology of the Very Small," 157–63.

20. As mentioned, it is impossible to describe fully the large literature that comprehends universal knowledge devices, which themselves may be said to participate (although in vastly changing ways) in the fantasy of total information. Several useful sources include: Anke te Heesen, *The World in a Box: The Story of an Eighteenth Century Picture Encyclopedia* (Chicago: University of Chicago Press, 2002); Robert Darnton, *The Business of Enlightenment: A Publishing History of the Encyclopédie, 1775–1800* (Cambridge: Harvard University Press, [1979] 2009), Such a history would also include the literature on curiosity cabinets and *Wunderkammern*, including Paula Findlen, *Possessing Nature: Museums, Collecting, and Scientific Culture in Early Modern Italy* (Berkeley: University of California Press, 1994); Lorraine Daston and Katharine Park, *Wonders and the Order of Nature 1150–1750* (New York: Zone Books, 2001). A. O. Lovejoy's William James lectures from 1936, published as *The Great Chain of Being* (Baltimore: Johns Hopkins University Press, 1948), are irreplaceable.

21. Michel Foucault, *The Care of the Self*, 210.

22. Ibid., 210.

23. Cf. Vidal, *Sciences of the Soul*.

24. Buckminster Fuller, Oregon Lecture #9, 324, 12 July 1962. Buckminster Fuller Collection Stanford University, Series 8, Manuscripts, Mixed Materials 28, Folders 1–2.

25. http://www.wired.com/table_of_malcontents/2006/12/the_dymaxion_ch/. Accessed August 2016. See Victoria Vesna, "Seeing the World in a Grain of Sand: The Database Aesthetics of Everything," in *Database Aesthetics: Art in the Age of Information Overflow*, ed. Victoria Vesna (Minneapolis: University of Minnesota Press, 2007), 1–39, 37n20.

26. Fuller describes beginning the file in 1907 at the age of twelve, inspired by Robert Burns's "Oh wad some power the giftie gie us to see oursels as others see us." Ten years later, he officially named the project the Dymaxion Chronofile. Buckminster Fuller, *Critical Path* (New York: Macmillan, 1981), 124. Using Chronofile documents as evidence, Loretta Lorace contests this version of Fuller's story as self-mythologizing in *Becoming Bucky Fuller* (Cambridge: MIT Press, 2009).

27. Fuller, "Self-Disciplines of Buckminster Fuller," *Critical Path,* 138, 25.

28. Alec Wilkinson, "Remember This?" *New Yorker*, May 28, 2007, 38–44.

29. Bell and Gemmell, *Your Life, Uploaded,* 30 (incidentally, this is perhaps the most poetic line in the book).

30. For further details of what is contained in MyLifeBits and its evolution, see Wilkinson, "Remember This?"

31. Bell and Gemmell, *Your Life, Uploaded,* 5, 10. However, there are significant differences. To mention only one, Bush envisioned Memex as an aid to scientific discovery; experiences the scientist experienced, or experiments he ran, were saved in order to

lead to new knowledge, not the curating of one's own personality and future family life. Bush's image of self was quite different from Bell's.

32. Technological determinism, a critical concept developed within the history of science, has of late been growing unabated without, in a celebratory sense. For an introduction, see Merritt Roe Smith and Leo Marx, eds., *Does Technology Drive History?: The Dilemma of Technological Determinism* (Cambridge: MIT Press, 1994). However, see the recent defense from an STS scholar of seven varieties of technological determinism: Sally Wyatt, "Technological Determinism Is Dead; Long Live Technological Determinism," in Edward J. Hackett, Olga Amsterdamska, Michael Lynch, and Judy Wajcman, eds., *Handbook of STS Studies* (Cambridge: MIT Press, 2008), 165–80; 196. Also relevant to the question of technological determinism are several bodies of theoretical writings, including the new materialism. Anthropologist Jason Pine likewise argues that tech determinism has new life if one accepts a broadened sense of agency in technological systems. Jason Pine, "Meth Labs, Alchemical Ontology, and Homespun Worlds," Harvard University Anthropology Department Colloquium, March 2013.

33. David Segal, "This Man Is Not a Cyborg. Yet," *New York Times,* June 1, 2013 http://www.nytimes.com/2013/06/02/business/dmitry-itskov-and-the-avatar-quest.html?_r=0. Accessed July 2016.

34. A. Fitzgibbon and E. Reiter, "'Memories for Life': Managing Information over a Human Lifetime," 22 May 2003 (via www.memoriesforlife.org), the starting paper for the UK Memories for Life Grand Challenge activity. See also Karen Sparck Jones, "Four Notes on 'Memories for Life,'" and Karen Sparck Jones, "Comments on Grand Challenge Document (GCD): 'Memories for Life': Managing Information over a Human Lifetime."

35. Alan Dix, quoted in Kieran O'Hara et al, "Memories for Life: A Review of the Science and Technology," *Journal of the Royal Society Interface* 3, no. 8(2006): 351–65, on 352.

36. See, e.g., Ira Boudway, "Is Chris Dancy the Most Quantified Self in America? One Man's Project to Keep Track of Everything He Sees, Does, Thinks, Eats," *Business Week,* June 5, 2014, http://www.bloomberg.com/news/articles/2014-06-05/is-chris-dancy -the-most-quantified-self-in-america. Accessed August 2016. Sometimes he claims he uses 300–700 sensors, devices, applications, and services: http://www.chrisdancy.com/. Accessed April 2016.

37. Klint Finley, "The Quantified Man: How an Obsolete Tech Guy Rebuilt Himself for the Future," *Wired,* February 22, 2013, http://www.wired.com/2013/02/quantified -work/all/. Accessed July 2016.

38. Dancy's narrative of existential and spiritual crisis is from an extended interview recorded in 2013 with Vince Horne of Buddhist Geeks, a podcast, available in audio and transcription at http://www.buddhistgeeks.com/2013/10/bg-298-quantified -selflessness/. Accessed May 2016.

39. Natasha Schüll points out that this desired mode of self-encounter is very different from that cultivated by Quantified-Self-ers, who often don't look back at their data at all, viewing it as a means for mindfulness in the moment of tracking. Looking back at one's data is a way of finding patterns useful to alter possible future outcomes, as in a self-tracker who told Schüll he thought of himself as a "time series self" [Schüll pers. comm. 2014].

40. Maslow's hierarchy of needs is a pyramidal scheme ranging from "physiological" motivations at the base to "self-actualization" at the tip. This was first expressed in Abraham Maslow, "A Theory of Human Motivation," *Psychological Review* 50, no. 4 (1943): 370–96.

41. Adam D. I. Kramer, Jamie E. Guillory, and Jeffrey T. Hancock, "Experimental Evidence of Massive-scale Emotional Contagion through Social Networks," *Proceedings of the National Academy of Sciences* 111, no. 24 (June 17, 2014): 8788–90.

42. Interview with Chris Dancy, Buddhist Geeks Episode 298 http://www.buddhistgeeks.com/2013/10/bg-298-quantified-selflessness/.

43. Cara Parks, "Books of Forgetting: Why We Can't Stop Writing about What We Can't Remember," *New Republic,* May 11, 2014.

44. See Limor Samimian-Darash and Paul Rabinow, eds., *Modes of Uncertainty* (Chicago: University of Chicago Press, 2015), and the cases published within, e.g., Adriana Petryna's on horizons of knowledge in climate science prognostication.

45. Gary King, "Ensuring the Data Rich Future of the Social Sciences," *Science* 331 (2011): 719–21.

46. Recently Lina Dib has suggested that Bell's archive project, in promising to lead to "greater self-management," embodies "that imperative of neoliberal times." Lina Dib, "The Forgetting Dis-Ease: Making Time Matter," in *Differences: A Journal of Feminist Cultural Studies* 25, no. 3 (2012): 42–73 on 43.

47. Bell and Gemmell, *Your Life, Uploaded*, 4. At a scale a 5¼-inch floppy disc once held, each synthesized strand of DNA contains ninety-six bits of storage; at the head of each strand is a nineteen-bit "address" used to order the strands properly. A whole trove could be held in a vat or sprayed on a wall, and subsequently re-sorted into usable data. George M. Church, Yuan Gao, and Sriram Kosuri, "Next Generation Digital Information Storage in DNA," *Science* 337, no. 6102 (September 28, 2012): 1628, http://www.sciencemag.org/content/337/6102/1628.abstract; http://www.extremetech.com/extreme/134672-harvard-cracks-dna-storage-crams-700-terabytes-of-data-into-a-single-gram. Accessed August 2016.

48. Frank Pasquale, "The Algorithmic Self," *Hedgehog Review* 17, no. 1 (Spring 2015), http://www.iasc-culture.org/THR/THR_article_2015_Spring_Pasquale.php. Accessed May 2016.

49. Rob Horning, "Know Your Product," *New Inquiry,* July 29, 2015, http://thenewinquiry.com/blogs/marginal-utility/know-your-product/. Accessed May 2016.

50. Cf. discussion at "The Total Archive," a conference held at CRASSH, University of Cambridge, March 2015. On the paradoxical pursuit of objectivity as the view-from-nowhere, see Lorraine Daston, "Objectivity and the Escape from Perspective," *Social Studies of Science* 22 (1992): 597–618.

51. Daston (chapter 6), Jankovic (chapter 9), and Rosenberg (chapter 11) respectively, this volume.

52. Giddens identifies as a distinctive feature of the modern self a mode of activity in which "self-actualisation implies the control of time." Insisting on the primacy of personal time becomes a way of controlling the available time of the lifespan. Anthony Giddens, *Modernity and Self-Identity: Self and Society in the Late Modern Age* (Stanford: Stanford University Press, 1991), 70.

53. Michel Foucault, *The Order of Things: An Archeology of the Human Sciences* (New York: Pantheon, 1970); Ridley Scott, director, *Blade Runner* (1982); Claude Lévi-Strauss, *The Savage Mind* (Chicago: University of Chicago Press, 1966), 17. Cf. Tarkovsky's "filmic archeology of the discarded" capturing stillness, described in Geoff Dyer, *Zona* (New York: Vintage, 2012), 117. Bell writes that one advantage among many of self-archiving is one's "ability to have the Proustian details of one's own life" at your fingertips," but are they still Proustian if they are at your fingertips? Bell and Gemmell, *Your Life, Uploaded*, 8. Cf. Viktor Mayer-Schönberger's work on "the virtue of forgetting in the digital age": *Delete* (Princeton: Princeton University Press, 2011), and the "right to forget" movement.

54. Hal Foster, "Exhibitionists," *London Review of Books*, 4 June 2015.

55. Dancy quotation from Youtube promotional video, *Christory 2014*, http://www.youtube.com/watch?v=jZOQkp80_BU; Accessed July 2016.

56. William Gibson, "The Net is a Waste of Time," *New York Times*, July 14, 1996, http://www.nytimes.com/1996/07/14/magazine/the-net-is-a-waste-of-time.html. Accessed August 2016.

ELEVEN

An Archive of Words

Daniel Rosenberg

> Instead of seeing, on the great mythical book of history, lines of words that translate in visible characters thoughts that were formed in some other time and place, we have in the density of discursive practices, systems that establish statements as events (with their own conditions and domain of appearance) and things (with their own possibility and field of use). They are all these systems of statements (whether events or things) that I propose to call archive.
>
> —Michel Foucault, *Archaeology of Knowledge* (1969)

Institutions called *archives* still matter greatly, preserving records of public and private interest. But in contemporary culture, an archival sensibility— even an archival fatalism—has become more general. The line between writing and archiving has become blurry. Writing in networked space passes effortlessly into unseen repositories and systems. We no longer *delete*, we *archive*. And, whether we intend it or not, our communications are constantly being archived by data-gathering entities. The distinction that Michel Foucault drew between visible characters "formed in some other time and place" (the traditional materials of the archive) and "systems that establish statements as events" (discourse and its conditions of possibility) collapses in the electronic ecosystem.[1] In important ways, writing and archiving become one and the same.

But what is the electronic archive? The contemporary discourse of big-ness seems to suggest that it is everything. It is far from that, but certainly is *a lot.* And self-conscious archiving efforts such as those of the NSA, Google, and Facebook are only the tip of an iceberg, just as self-conscious acts of electronic writing are only the surface ripple of a deep current of signals automatically generated, sent, and received by devices connected in an im-mense Internet of things. The difficulty, of course, is what to do with all those signals, and in the case of language produced by human beings, what to do with all those words.

The challenge isn't entirely new. In its oldest form, it pertains to a long history of documents and traces back importantly to a discourse on information overload more than four centuries old.[2] But the introduction of electronic information systems in the late twentieth century did alter key dynamics. In electronic space, objects once traditionally thought of as documents mingle, disintegrate, and recombine according to protean systems and rules. For understanding recent history, these systems and rules are themselves objects of great archival importance, though their traces are not often intentionally conserved. Figuring out how to archive this archive is no small matter. It will be the foundation for the history of the epistemol-ogy of our contemporary era.

THE AGE OF THE WORDLE PICTURE

As a thought experiment, consider the *stop list,* one of the notable, little-noticed mechanisms of late-twentieth century information culture, an ephemeral artifact if there ever was one. A stop list is a list of words that a computer is instructed to ignore when searching or processing text. Stop lists are designed to reduce storage and processing burdens on computers, and, while they are still used widely today, in the earliest days of computerized text analysis, systems could not have functioned without robust stop lists. Though such lists have been of great importance in computer processing for the past half-century, stop lists themselves are typically neither saved nor collected, and writing their history presents some thorny archival problems.

There are many kinds of stop lists, long and short, general and topical, statistically generated and editorially designed. Stop lists are devices of convenience. A word appears on a stop list because it is considered "nonin-forming" for a given inquiry, not because it lacks value in a broader sense. In a typical keyword search, for example, so-called "function words" includ-ing articles, pronouns, prepositions, and conjunctions are excluded. It is still the case, for example, that many computer programs will not permit a search for the English definite article, *the*. "Function words" such as *the* are, of course, perfectly good words, essential to the production of meaning in ordinary language and interesting in and of themselves in another setting, but as individuated keywords, not so much. They include the most common words in the language.

To understand the impact a simple stop list can have, consider two word clouds generated with the same computer program, Wordle, from the same source text, the Project Gutenberg edition of Charles Dickens's novel *Great Expectations* (1861).[3] Wordle creates simple word clouds in which the graphic size of individual words represents their relative frequency in the source text. The most frequently occurring word in the source text is the largest in the graphic; the least frequent is smallest.

The first word cloud in figure 11.1a was produced without a stop list; the second, with the standard stop list included in the program. (Users may also define their own.) The first cloud, unstopped, is dominated by the words *the, i, and, to, of, a, that, was*. The second, stopped cloud in figure 11.1b, highlights character names, *pip, joe, herbert, biddy, havisham, pumblechook, estella,* and so forth, and some other assorted nouns, verbs, and adjectives including *time, old, young, hand, eyes, pockets,* which suggest themes that are likely of more interest to most users than common English words such as *the, I,* and *and*. You might say that the second word cloud more obviously *characterizes* the novel.

Yet there are good reasons not to employ stop words. From the unstopped word cloud above, for example, we can make an accurate guess that *Great Expectations* is a first-person narrative. The word *I* appears with great frequency, as does the past tense. The value of stop words, then, depends on the aim of the analysis. If the goal is to identify characters or themes, for example, a typical stop list full of function words may be of substantial heuristic value. If the aim is to discover genre or some other feature of writing with a characteristic grammatical signature, those same words may be better left in play. The same considerations apply to sophisticated applications in topic modeling, latent semantic analysis, and a wide variety of computational approaches to language.

What is more, there are semantic and stylistic aspects of texts for which attention to function words is of special value. The Jesuit scholar Roberto Busa, a pioneer in humanities computing, argued that "all functional or grammatical words (which in my mind are not 'empty' at all but philosophically rich) manifest the deepest logic of being which generates the basic structures of human discourse. . . . In the works of every philosopher there are two philosophies: the one which he consciously intends to express and the one he actually uses to express it."[4]

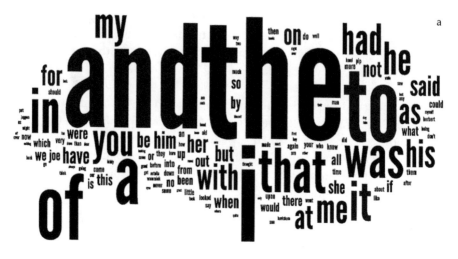

Fig. 11.1 Wordle of Great Expectations: No Stop List vs. Stop List.

Busa's 1946 dissertation on the theme of immanence in the works of St. Thomas Aquinas makes a case in point. When designing the project, Busa had initially intended to approach the problem by studying Aquinas's use of thematic terms such as *presence* (*praesens, praesentia*) as identified in extant subject indexes. But early research suggested that the problem of presence was too ubiquitous in Aquinas to be approached through an examination of only his explicit, guided discussions. What is more, as Busa says in the quotation above, it played itself out not only in *what* Aquinas wrote but also in *the way* he wrote. Busa became convinced that his questions could be answered satisfactorily only through study of the mechanisms of Aquinas's writing alongside his arguments. The resulting work was still about immanence and about words, but the key word that Busa tracked in his dissertation was not the conceptually rich *praesens*, as he had originally planned, but rather the humble Latin preposition *in*.

It may also be that linguistic structure is what interests us most. In 1971, the linguist Harald Weinrich demonstrated how much one could learn about a literary text simply by mapping out the relative frequency of indefinite to definite articles throughout. Weinrich showed that ordinary articles—*the, a, an*—have a deictic function that extends beyond the single sentences in which they appear: among other functions, definite and indefinite articles help the reader understand where she or he stands in the flow of a narrative or argument. Weinrich writes, "It is usual for the indefinite article to occur at decisive points in texts at which the recital of information takes a new and unprecedented direction."[5] Perhaps a new subject is introduced, a new idea, a new character. In the story "Little Red Riding Hood," for example, Weinrich observes that the title character is introduced first using the indefinite article, "*a* little girl of the village." Thereafter, she is "*the* little girl." Of course, exceptions may be found. Nonetheless, Weinrich argues, the distribution of function words such as articles, pronouns, and prepositions in a text provides a kind of formal sketch of the narrative. The texts with with Weinrich demonstrated his insight, the fables of Charles Perrault, the short stories of Albert Camus, and the plays of Molière, were short and easily anatomized by hand. And Weinrich makes his case through close reading. But textual maps such as these may be created without actually reading a text in any traditional sense. This would be an example of what Franco Moretti has referred to as "distant reading."[6] And, with the propagation of out-of-the-box programs for quantitative text analysis ranging from command-line programs such as the topic modeling software, Mallet, to GUI text-analysis dashboards such as Voyant, this kind of reading is becoming both easier and more common.

William James spotted regularities in usage patterns of function words and derived from them a decisively humanist conclusion. For James, the function words express agency in language even more than words such as nouns and adjectives, which are richer from a semantic standpoint. "There is not a conjunction or preposition, and hardly an adverbial phrase, syntactic form, or inflection of voice, in human speech, that does not express some shading or other of relation which we at some moment actually feel to exist between the larger objects of our thought. If we speak objectively, it is the real relations that appear revealed; if we speak subjectively, it is the stream of consciousness that matches each of them by an inward colouring of its own. . . . We ought to say a feeling of *and*, a feeling of *if*, a feeling of *but*, and a feeling of *by*, quite as readily as we say a feeling of *blue* or a feeling of *cold.*"[7] Depending on the kind of inquiry in which we are engaged, those very terms most likely to appear on a stop list may be most telling for analysis. The use of stop lists, then, both implies and imposes a certain agenda in reading.

NEGATIVE DICTIONARIES

Stop lists are tools of data removal, lists of words to be left off of lists of words. Thus, they are also sometimes referred to as *negative dictionaries*. In the historical thesaurus of the *Oxford English Dictionary*, a wonderful list in itself, *stop word* falls under the following hierarchical classification, in the subheading, *omitted data.*

> the external world > relative properties > number > computing or information technology > data > database > data entry > [noun] > processing > omitted data

For *stop word* in programming, the *OED* gives a first citation from 1969 and for *stop list*, 1970, though both terms were in use a decade earlier (fig. 11.4).[8] The *OED* entry for *stop word* reads:

> **stop word** n. a word (usu. one of a set of the words most frequently occurring in a language or text) that is automatically omitted from or treated less fully in a computer-generated concordance or index. **1969** *Computers & Humanities* 3/135. If stop words are desired, the user can either specify his own or request a standard list which is encoded within BIBCON. **1982** *N. & Q.* Oct. 385/1. I understand that a microfiche concordance of the stop words will soon be available.

By 1969, basic stop lists were included with indexing and concordance software, such as the BIBCON program used at the University of Wisconsin referred to in the *OED* citation above.[9] By 1982, the status of stop words had already changed.[10] The citation from *Notes & Queries* announces the impending publication of a concordance of *stop words* only, the *Microfiche Concordance to Old English: The High-Frequency Words.*[11] This marks a notable development in the history of computer-generated concordances, as computing speed and memory capacity began to achieve levels sufficient to justify processing those high-frequency words—prepositions, articles, pronouns, conjunctions, and so forth—that commonly appeared on stop lists.

Compared with *stop word*, the term *stop list* has a longer and more checkered history. Early in the twentieth century, according to the *OED*, a *stop list* was most commonly "a list of persons, etc., deprived of particular rights."

Etc. is an intriguing category here. As early as 1930, people who were *stop listed* could be arrested, denied entry to a public place, refused a loan, and so forth because they were considered dangerous or undesirable. But stop lists referred not only to *people* to be stopped; they could refer to anything. *Stop lists* have also been used for a century to ban books and other media, for example.

These directly political senses of *stop list* are still operative today, and in common usage they mingle importantly with those of "omitted data" because of the centrality of computer technology to contemporary governance. A public controversy took place several years ago when it was discovered that a demographic search service funded by USAID had used a stop list to prevent the return of articles that contained the word *abortion* in the title.[12] In China, Internet services employ stop lists to block online searches of a number of topics, including pornography and subjects of political controversy.[13] Terms related to the subject of censorship themselves appear on Chinese stop lists.

Because *search* has become a lucrative business, stop lists have become a subject of much study and rumor in the search engine optimization (SEO) quarter. SEO services propose to help clients achieve high visibility in Internet searches. This involves, among other things, clustering highly sought-after keywords on a website while avoiding stop words so far as practical. An Internet search for the terms *stop list* + *name of any major search engine* will produce dozens of sample stop lists generated by SEO companies using algorithms, trial and error, and sometimes just wishful guesswork.

In January 2008, tech bloggers observed the disappearance of notifications of stop word removal from Google search results and a new sensitivity in Google and some other major search engines to words that had previously been stopped.[14] Earlier, certain words, notably function words and words repeated in a phrase, had been automatically scrubbed from text strings searched in Google, a fact indicated in a brief message at the head of the Google result page showing which words in a search had been excluded by graphically crossing them out. Now, Google had stopped crossing these out, and apparently, started processing them instead. Without fanfare, a sea change had taken place: Google had noticed *the*. On the front end, this meant that users could increasingly write to search as if they were writing to a person. On the back end, it created a wealth of new analytic objects. These days, even the notoriously search-resistant British band, The The, leaps right to the head of a simple search.

The blogs noted a number of background developments at Google that might help explain the change including Google's acquisition of Applied Semantics, a company specializing in linguistic analysis, as well as its pur-

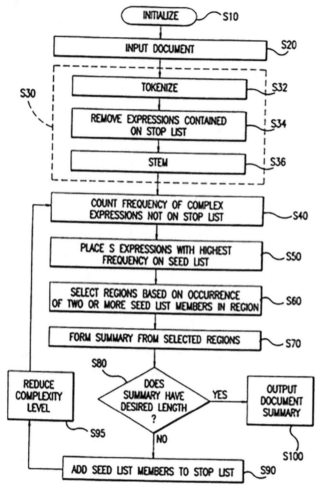

Fig. 11.2 Method and apparatus for summarizing documents according to theme, specifying removal and augmentation of stop list. From U.S. Patent 5384703. Xerox Corporation, 24 January 1995.

suit of several patents including one that specifically eliminated stop lists (fig. 11.2). There, Google proposed accelerating searches on phrases containing high-frequency words by accessing them through relationships of adjacency with lower-frequency terms in the same search phrase.[15] In general, across the industry, the same period saw increasing application of term frequency-inverse document frequency (tf-idf) approaches, which control for terms that occur with high frequency across a corpus without eliminating them. Related approaches tended to lessen the demand for stop lists overall. The great era of the stop list, roughly 1958–2008, began to draw to a close.[16]

THE AUTOMATIC INFRAORDINARY

How does a stop list work? In its typical application, the stop list belongs to a step in the information workflow known as *preprocessing*. In effect, words on stop lists are removed from texts before the main program ever gets to

```
                                ESPACE
                                ESPACE LIBRE
                                ESPACE CLOS
                                ESPACE FORCLOS
                  MANQUE D'ESPACE
                                ESPACE COMPTÉ
                                ESPACE VERT
                                ESPACE VITAL
                                ESPACE CRITIQUE
              POSITION DANS L'ESPACE
                                ESPACE DÉCOUVERT
            DÉCOUVERTE DE L'ESPACE
                                ESPACE OBLIQUE
                                ESPACE VIERGE
                                ESPACE EUCLIDIEN
                                ESPACE AÉRIEN
                                ESPACE GRIS
                                ESPACE TORDU
                                ESPACE DU RÊVE
                    BARRE D'ESPACE
          PROMENADES DANS L'ESPACE
          GÉOMÉTRIE DANS L'ESPACE
          REGARD BALAYANT L'ESPACE
                                ESPACE TEMPS
                                ESPACE MESURÉ
            LA CONQUÊTE DE L'ESPACE
                                ESPACE MORT
                                ESPACE D'UN INSTANT
                                ESPACE CÉLESTE
                                ESPACE IMAGINAIRE
                                ESPACE NUISIBLE
                                ESPACE BLANC
                                ESPACE DU DEDANS
              LE PIÉTON DE L'ESPACE
                                ESPACE BRISÉ
                                ESPACE ORDONNÉ
                                ESPACE VÉCU
                                ESPACE MOU
                                ESPACE DISPONIBLE
                                ESPACE PARCOURU
                                ESPACE PLAN
                                ESPACE TYPE
                                ESPACE ALENTOUR
                  TOUR DE L'ESPACE
          AUX BORDS DE L'ESPACE
                                ESPACE D'UN MATIN
          REGARD PERDU DANS L'ESPACE
                LES GRANDS ESPACES
            L'ÉVOLUTION DES ESPACES
                                ESPACE SONORE
                                ESPACE LITTÉRAIRE
              L'ODYSSÉE DE L'ESPACE
```

Fig. 11.3 Composition of phrases containing the word space from Georges Perec, *Espèces d'espace* (1974), capturing the iterative visual and conceptual logic of Keywords-in-Context.

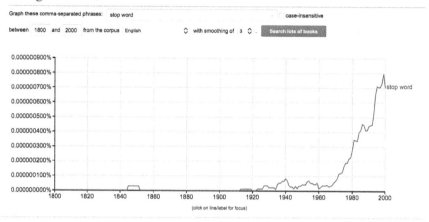

Google books Ngram Viewer

Graph these comma-separated phrases: stop word   case-insensitive

between 1800 and 2000 from the corpus English   with smoothing of 3 . Search lots of books

(click on line/label for focus)

Fig. 11.4 "Stop word" ascendant as a term in the Google Books corpus.

read them. This puts the words of the stop list into a curious linguistic space: there but not here, text but not data for processing. In the stop list, a catalog of words of no interest, there is a kind of poetry. Georges Perec, a member of the French writers group OULIPO—founded in 1960 just as electronic stop lists were first being implemented—referred to the plane just below the level of traditional literary interest as the *infraordinary*, "that which is generally not taken note of, that which is not noticed, that which has no importance"[17] (fig. 11.5). For electronic text, one might further specify this idea: the stop list belongs to the domain of the *automatic infraordinary*. In this same category, one might include a diverse array of interstitial media artifacts just below the surface of language such as embedded source code, metadata, and failed words in optical character recognition (OCR) transcripts. From a formal point of view, the computer routines at issue here resemble OULIPO's experimental writing procedures, which apply mechanical strictures to language, as in Perec's novel *La disparition*, composed in a great tour de force of term-stopping without a single word containing the letter *e*, as well as in compositions such as Perec's list/object on the word *space* in figure 11.3, which at once evokes both the visual vocabulary of modern poetry and that of the computer-generated index.[18]

There is a special literality in places where language interacts with data processing systems to which ordinary users do not have access. In a search

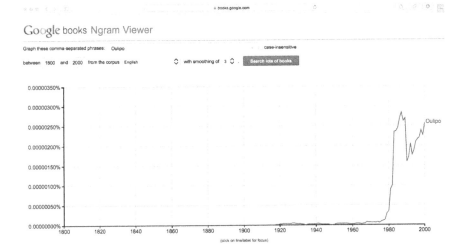

Google books Ngram Viewer

Graph these comma-separated phrases: Oulipo   case-insensitive

between 1800 and 2000 from the corpus English   ◇ with smoothing of 3 ◇ .   [Search lots of books]

(click on line/label for focus)

Fig. 11.5 "Oulipo" ascendant as a term in the Google Books corpus.

system employing a typical stop list, the user enters the character string *A Room with a View*; the system reads *room, view*. The system is nonetheless able to return information about the E. M. Forster novel, *A Room with a View*, because of the strong correlation of the word pair and documents referring to the novel, not because it searches the full expression. Between human language and the language returned by the computer, there is a mediating set of expressions, a tumble of words that have been filtered and normalized, a token language. Perhaps these words belong to an *infraextraordinary* plane in which everything is of heightened significance.

One of the interesting things about how people write to search engines is the attention they pay to choosing just the right word. In other linguistic contexts, writers use syntax, style, and context to convey meaning. Word choice is just one expressive tool among many. In the realm of the search line, writing is different. Search engine users effortlessly improvise in the blank box, creating new kinds of syntax, and, in a kind of language dance with the computer, dropping function words and other common terms themselves. In doing so, they seek to optimize the relationship between *recall* (the proportion of useful instances retrieved) and *precision* (the proportion of retrieved instances that are useful). These linguistic practices have helped intensify focus on the power of individual words in expression—not words in general, but individual words, alone and in combination. They

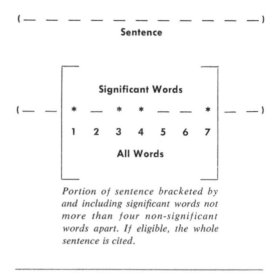

Sentence
Significant Words
* — * * — — *
1 2 3 4 5 6 7
All Words
Portion of sentence bracketed by and including significant words not more than four non-significant words apart. If eligible, the whole sentence is cited.

Figure 2 **Computation of significance factor.** *The square of the number of bracketed significant words (4) divided by the total number of bracketed words (7) = 2.3.*

Fig. 11.6 Example of text manipulation: computation of significance factor. Hans Peter Luhn, "The Automatic Creation of Literature Abstracts," *IBM Journal of Research and Development* 2, no. 2 (April 1958): 162.

have also intensified the need for a reflexive historiography of language processing from the humanities. Language, not just computer language, changes in our interaction with computers, as is so palpable in cases such as those of the hashtag and emoticon.

THE START OF THE STOP LIST

The most important of the early electronic stop lists appeared in the *IBM Journal of Research and Development* in 1960. "Keyword-in-Context Index for Technical Literature" was one in a series of influential articles in the 1950s and 60s by IBM engineer Hans Peter Luhn on automatic indexing and abstracting.[19] Here, Luhn proposed his Keyword-in-Context, or KWIC, indexing system. KWIC worked much like a traditional literary concordance. It presented the words of a text in alphabetical order surrounded by snippets of surrounding context to the left and the right along with a text location. It differed from a traditional concordance principally in that it was generated mechanically (fig. 11.7).

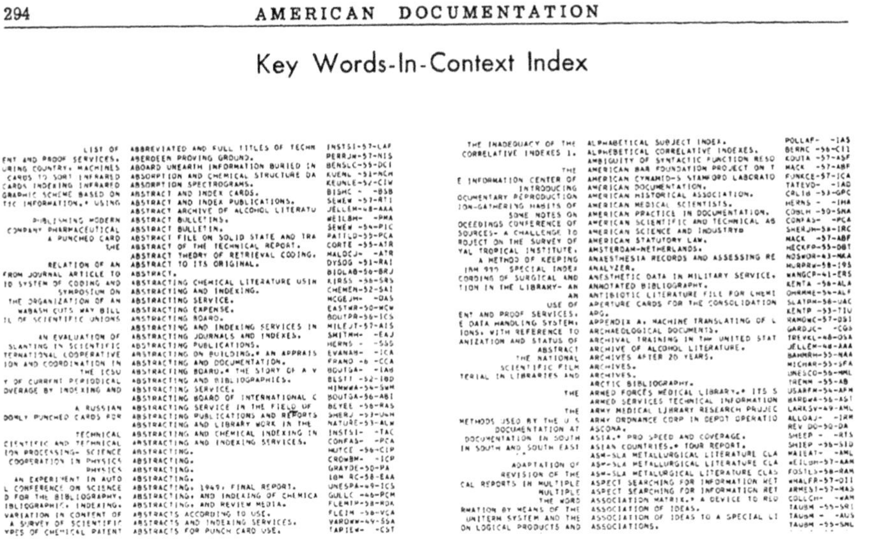

Key Words-In-Context Index

Fig. 11.7 KWIC index example. H. P. Luhn, "Keyword-in-Context Index for Technical Literature," *American Documentation* 11, no. 4 (October 1, 1960): 294.

In KWIC and his other systems, Luhn proposed mechanisms for identifying *significant words*—later *key words*, and then *keywords*—though in fact, his approach was not to directly determine which words were *key* but rather those that were *non-significant* and could therefore be set aside in processing. Luhn writes, "Keywords need only be defined as those which characterize a subject more than others. To derive them, rules have to be established for differentiating that which is significant from the non-significant. Since significance is difficult to predict, it is more practicable to isolate it by rejecting all obviously non-significant or 'common' words"[20] (fig. 11.6). Already in 1958, Luhn had suggested removing "common words such as pronouns, prepositions, and articles" at the start of an indexing routine.[21] Here, Luhn offered a list of sixteen common English-language articles, conjunctions, and prepositions: *a, an, and, as, at, by, for, from, if, in, of, on, or, the, to, with.*[22]

Luhn's originary stop list is about as infraordinary as one might imagine. With no verbs or nouns, not even a pronoun, it does not even offer sufficient building blocks for a simple sentence. At the same time, and as we have seen in the analyses of Busa, Weinrich, and James, the linguistic importance of these non-significant words should not be underestimated. *The* is the most common word in general English. Removing *the* alone makes a big dent in the word count of nearly any text. In a world of punch cards and "inches of line-printer pages," the effect could be measured in storage weight and volume, a fact of which organizations employing computers for big data applications of the time were acutely aware.[23] Paul Edwards' account of a thirty-ton punch card deck created at the National Weather Records Center in Asheville, North Carolina, in the late 1950s is only a particularly vivid example of how very impractical a large paper data set could become.[24] In the realm of text analysis, the stop list proved to be a very effective device for reducing the size of a data set. Early tests with a *Biological Abstracts* stop list of one thousand words showed that it reduced index entries by nearly half, and, as Luhn guessed, a mere fourteen of those one thousand stop words accounted for eighty percent of the difference.[25]

Luhn, of course, was not the first person to exclude words from an index. One could say that every index operates with an implicit stop list containing every term that does not rise to its level of attention. But the organizing principle of a traditional index is opposite that of a stop list: human indexers choose what terms to *include* as potentially significant to a search; stop lists *exclude* what is not. The first is applicable to any selective indexing process, the second only becomes sensible in the context of a complete index to all words from a text.

As it happens, the full-text index has a long history stretching to the rise of alphabetic indexes in the late Middle Ages and to the concurrent development of the Bible concordance.[26] And it is no accident that some of the very interesting early work on computerized stop lists was done by the Aquinas scholar Roberto Busa, steeped as he was in medieval scholarly tradition. In his 1971 article "Concordances" in the *Encyclopedia of Library and Information Sciences*, Busa makes the connection explicit, arguing that the electronic text created the condition for the generalization of medieval textual method. Busa writes:

> [T]he "computer era" caused a true explosion in the specialized, narrow field of concordances. But that field is not an isolated one. To the contrary it is only a portion of much larger waves of social development, which at present effect a true change of dimensions in the means of human communication: in Gutenberg's time we went from the handwritten book to the printed one; today we are moving from the printed book towards the "magnetic tape book" or, if you prefer, the "electronic library." In fact the sign and symbol of human concepts was up to now only a physical mark or spot, mostly ink, visible by human eye. But from now on, in addition to that, electronic impulses, perceptible only by means of an intermediate machine, will act as signals of the human mind and will. Concordances are one of the chapters of the rapidly growing world-wide "computer and the humanities" phenomenon affecting everything that may be called philology.[27]

Though a concordance is often defined as an index of every word in a work, until the nineteenth century, concordances rarely sought to achieve that degree of completeness.[28] Luhn's sixteen stop words were excluded from most early concordances, but such exclusions were not usually made explicit. Among the first published concordances to explicitly stop words is the fifteenth-century Hebrew scripture concordance, *Me'ir Nativ* (Illuminator of the Path), printed first a century later in 1523.[29] The *Me'ir Nativ* list, heavy on function words, resembles Luhn's, as does the rationale behind it.[30]

Early Latin concordances typically omitted words referred to as *indeclinables*, including prepositions, conjunctions, and interjections such as *amen*. During the Council of Basel in the 1430s, the theologian John of Ragusa initiated a controversy with the Bohemians over the biblical usage of the particles *nisi* and *per*. Later, when the council sent him to Constantinople, John engaged the Greeks over the true meaning of the particle *ex*. Finding common ground proved particularly difficult because the accepted reference for the Latin Bible, the 1247 concordance by the French Dominican cardinal Hugh of St-Cher, did not account for these words. Consequently, the council commissioned Spanish theologian John of Segovia to compose a concordance of indeclinables.[31]

In the age of print, theologians continued to produce and improve concordances, but the labor involved in indexing every word in a long work guaranteed that only a small number of books received such treatment. Most concordances were of the Bible. Most of these built directly on the work of predecessors, all the way back to Hugh of St-Cher. Eventually, a handful of other works received concordances too, including, during the eighteenth century, some secular texts such as the works of Shakespeare.[32] For the English language, Alexander Cruden's 1738 *Concordance to the Old and New Testament, or A Dictionary and Alphabetical Index to the Bible*, based on the King James Bible, marked an apogee, proving complete enough and intuitive enough to remain in print continuously to the present day.[33]

It is a notable feature of concordances that many contain highly detailed accounts of the indexing strategies of their predecessors. This makes the concordance particularly valuable for book history. It also directs attention to the concordance as an apparatus. We might think of the concordance as a *program* in two senses of the term, as a concerted plan and as a routine for processing literary data. For very good reasons, concordances look and feel like printouts. Cruden's *Concordance* offers a contextual line for each appearance of an indexed term. Each *word* in scripture produces a *line* in the concordance, and thus the concordance is often a larger work than the one indexed, even after the omission of function words and others.[34] Cruden's

Concordance is interesting, too, because of its relative independence from its source. One does not need to have a Bible handy in order to make use of it. The passages Cruden quotes are full enough to be read on their own. And this well-conceived feature of the text partly accounts for its remarkable durability.

In the nineteenth century, the Bible concordance finally achieved lexical completeness. In 1890, the American Methodist scholar James Strong published another enduring concordance to the King James Bible. It included every word, coding each as an expression of 8,674 root words from the Hebrew Old Testament and 5,624 roots from the Greek New Testament. Strong gave an identifying number to each of these roots, now referred to as "Strong's numbers."[35] But the value of the move from the comprehensiveness of Cruden to the completeness of Strong was far from settled. The 1906 report of the Dante Society, for example, singled out Strong for criticism. Completeness, it argued, is "a positive disadvantage, on account of the disproportionate amount of space required." Strong's *Concordance*, it stated with disapproval, "has more than four hundred thousand quotations in its main part, followed by over two hundred thousand references, without quotations, for some forty-seven particles."[36] It was a forest of words. Strong was aware of the difficulties posed by function words. In his concordance, these were separated into their own category, identified by citation only, without contextual quotation.

Other scholars of the period were unrepentant in their pursuit of completeness. In 1874, the American scholar Helen Kate Furness published a concordance to Shakespeare's poetry that proudly included "every word therein contained" not excepting the "minutest particulars."[37] Furness writes, "As it is impossible to limit the purposes for which the language of Shakespeare may be studied, or to say that the time will not come, if it has not already, when his use of every part of speech, down to the humblest conjunction, will be criticised with as much nicety as has been bestowed upon Greek and Latin authors, it seems to me that, in the selection of words to be recorded, no discretionary powers should be granted to the 'harmless drudge' compiling a Concordance. Within a year or two a German scholar has published a pamphlet of some fifty pages on Shakespeare's use of the auxiliary verb *to do*, and *Abbot's* Grammar shows with what success the study of Shakespeare's language in its minutest particulars may be pursued. I have therefore cited in the following pages every word in his Poems."[38]

Though concordances are designed to aid in the study of a source text, in practice, they produce corpora with their own logic and integrity. In the cases of Cruden, Strong, Furness, and others, the concordance texts have a structure and identity very much their own. They also demand reading practices different from those of scripture or imaginative literature, erstwhile practices of art, now become normative. Cruden's and Strong's sentences make one kind of corpus; Luhn's character-limited strings, another. Knowing this, one may observe that the concordance text itself demands a formal analysis equivalent to that which it performs.

A GRAND GAME OF SOLITAIRE

In the account given by Roberto Busa, there are two periods in the history of the concordance, the period from Hugh of St-Cher in the mid-thirteenth century to "the beginning of the computer age" in 1950, and the period since.[39] It is a bold proposition, but, as we have seen, Busa was in a privileged position to make it. His manually researched 1946 dissertation analyzed over ten thousand instances of the Latin preposition, *in*, across the fifteen million words of the collected works of Thomas Aquinas. Beginning in 1949, he had the opportunity to re-pose the question using an IBM mainframe computer.[40] Perhaps the first humanities scholar to use a computer in this way, Busa was struck by the power of the new tools. At the same time, he stressed their continuity with old scholarly techniques.[41] Applications to traditional objects of humanistic study such as the works of Aquinas and the then-recently disinterred Dead Sea Scrolls reinforced the point.[42]

In Busa's case, what began as a study of a rather uncommon thematic word, *praesens*, ended up as a study of a very common function word, *in*. And as such, it demanded exceptional scholarly patience. Busa writes, "According to the scholarly practices, I first searched through tables and subject indexes for the words of *praesens* and *praesentia*. I soon learned that such words in Thomas Aquinas are peripheral: his doctrine of presence is linked with the preposition *in*. My next step was to write out by hand 10,000 3″ × 5″ cards, each containing a sentence with the word *in* or a word connected with *in*. Grand games of solitaire followed."[43]

Five years later, everything had changed, and Busa was no longer playing solitaire. He had built up a substantial operation in data entry, correction, and analysis at Gallarate in northern Italy, and he was already releasing results from his "automatized concordance" to Aquinas, a founding work in the history of humanities computing, one that pulses with the future in a title mashed up from the technical languages of philology and programming, the *S. Thomae Aq. Hymnorum Ritualium Varia Specimina Concordantiarum. A First Example of a Word Index Automatically Compiled and Printed by IBM Punched Card Machines* (fig. 11.8). The full fifty-six volumes of

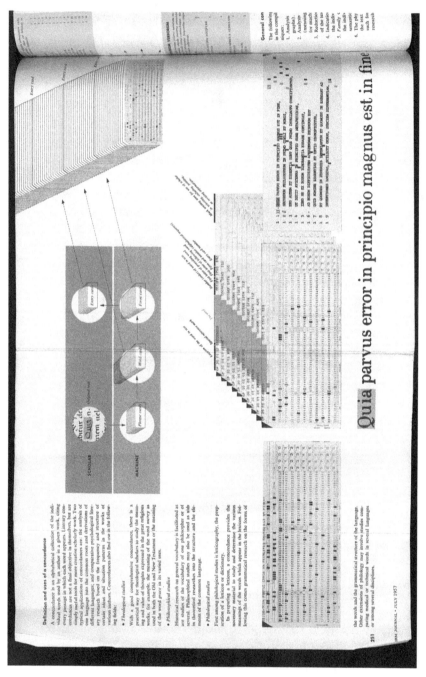

Fig. 11.8 Diagram explaining the use of punch cards for Busa's "automatized concordance" in Paul Tasman, "Literary Data Processing," *IBM Journal of Research and Development* 1, no. 3 (1957): 251–52.

Fig. 11.9 Punch card machine operators working on the *Index Thomisticus* in Gallarate, Italy (29 June 1967). Courtesy of the Busa Archive and the CIRCSE Research Centre, Università Cattolica del Sacro Cuore, Milan, Italy.

the *Index Thomisticus*, Busa's computer-generated concordance to Aquinas, took an additional quarter century to complete.

Busa's work set an enduring standard in humanities computing. At the time, however, his approach was best applied on a relatively small scale, on projects with a long time frame and substantial staffing, and, in the best case, all of the above. To even begin work on the *Index Thomisticus*, Busa had to build a custom Latin dictionary, the *Lexicon Electronicum Latinum*, to group inflected words and to code morphological categories for 150,000 Latin forms.[44] Establishing distinctions, entering and correcting text, and developing a system to process the data were hugely labor intensive. Over the course of twenty-five years, Busa wrote afterward, "The ratio of human work to machine time was more than 100:1. Computer hours were less than 10,000 while man hours were much more than one million."[45] A large portion of these "man hours" were, in fact, put in by women staffers performing data input, as Busa intentionally employed women, "because they were more careful than men."[46] Photographs of Busa's workshop in Gallarate— Busa called it his "school for punching cards"—powerfully convey the gendered dynamics of the digital workshop of the 1960s (figs. 11.9, 11.10).[47]

Fig. 11.10 Livia Canestraro, a punch card machine operator, working on the *Index Thomisticus* in Gallarate, Italy (29 June 1967). Courtesy of the Busa Archive and the CIRCSE Research Centre, Università Cattolica del Sacro Cuore, Milan, Italy.

It was another world. Not to say that the production of reliable electronic versions of literary texts is a trivial matter today. To the contrary, whether by keying or scanning, and whether employing hand coding or natural language processing for lemmatization and other grammatical taxonomy, properly digitizing print and manuscript remains a demanding task for humanities scholars. But the environment in which such work takes place now is very different from that of the mid-twentieth century. Today, so much of what we read comes to us through an electronic medium that we risk losing sight of the specific character of these different kinds of texts.

This is not only a question of materiality. Certainly the experience of handling and reading a paper book is different from that of an electronic book. But the differences run deeper. The "books" in a system such as Google Books are not really books at all. They may have been digitized from print books. They can be read as books. But they are always also electronic data. And, likely as not, it is on account of this other identity as data that we encounter them in the first place, having been directed by a search engine. Even if scanning is perfect, metadata pristine, and the paratextual apparatus of the book convincingly reproduced, the electronic corpus, is simply, *textually*, different from its source, having passed through layers of mediation and encoding, as a glance at any page of OCR text, even very high quality OCR text, will tell you (fig. 11.11).

Busa's original project on Aquinas was freestanding. The fifteen-million word electronic corpus he created pressed at the limits of what a research group could manage in its day, but it was infinitesimal in comparison to the text base that one takes for granted today. Consider the Google Books Ngram Viewer, which plots relative frequency over time for words in the Google Books corpus. In order to improve results, the designers of the Ngram Viewer excluded works in the Google Books corpus for which they had incomplete or faulty metadata. In 2011, that amounted to about three quarters of the objects in the corpus, or 15 million books. That is a considerable number by any standard, more than 10 percent of the estimated 130 million books ever published. Still, that left the Ngram Viewer a hefty 5 million books to index. Even book data is now big data, so big that Google can simply ignore 15 million books in order to improve its results.[48] The contrast with the completism of Busa's approach could not be sharper, nor could the implications for what we understand our library to be, greater. Google Books offers an exceptionally powerful finding tool but no catalog

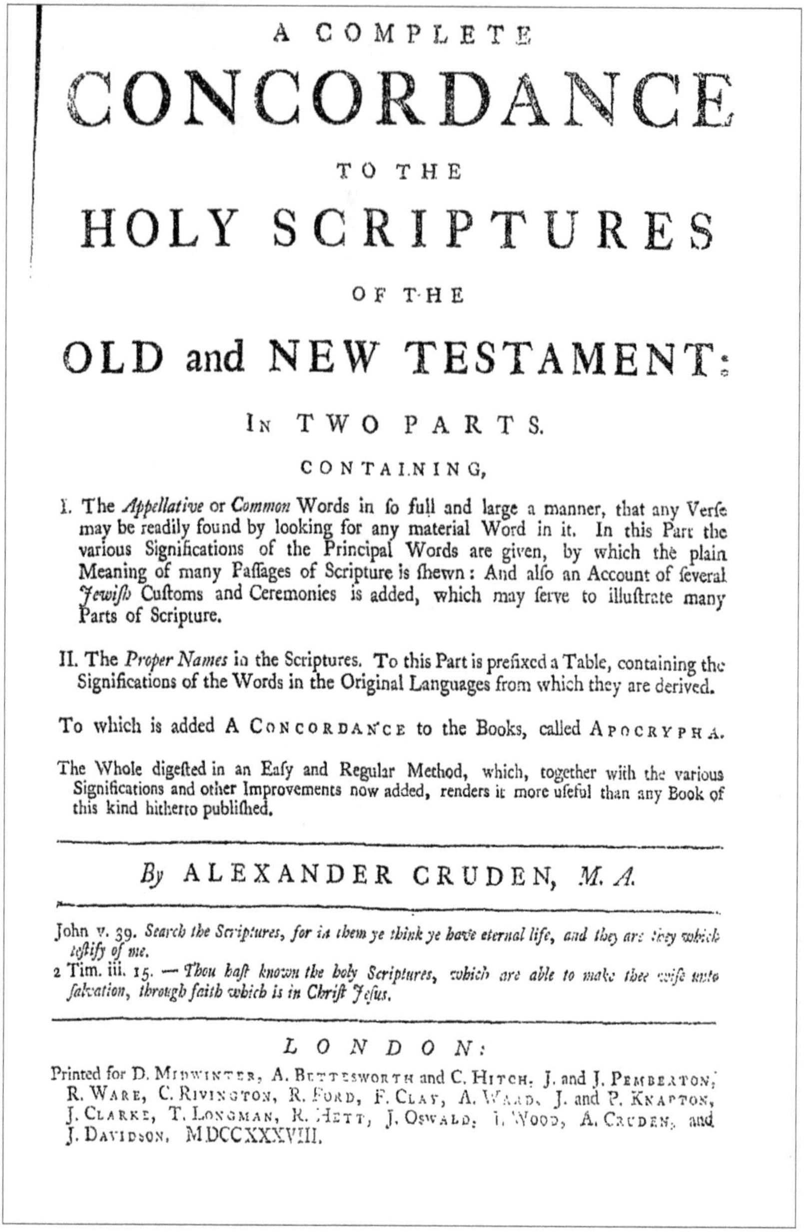

A COMPLETE

CONCORDANCE

TO THE

HOLY SCRIPTURES

OF THE

OLD and NEW TESTAMENT:

IN TWO PARTS.

CONTAINING,

I. The *Appellative* or *Common* Words in fo full and large a manner, that any Verfe may be readily found by looking for any material Word in it. In this Part the various Significations of the Principal Words are given, by which the plain Meaning of many Paffages of Scripture is fhewn : And alfo an Account of feveral *Jewifh* Cuftoms and Ceremonies is added, which may ferve to illuftrate many Parts of Scripture.

II. The *Proper Names* in the Scriptures. To this Part is prefixed a Table, containing the Significations of the Words in the Original Languages from which they are derived.

To which is added A CONCORDANCE to the Books, called APOCRYPHA.

The Whole digefted in an Eafy and Regular Method, which, together with the various Significations and other Improvements now added, renders it more ufeful than any Book of this kind hitherto publifhed.

By ALEXANDER CRUDEN, *M. A.*

John v. 39. *Search the Scriptures, for in them ye think ye have eternal life, and they are they which teftify of me.*
2 Tim. iii. 15. — *Thou haft known the holy Scriptures, which are able to make thee wife unto falvation, through faith which is in Chrift Jefus.*

L O N D O N:

Printed for D. MIDWINTER, A. BETTESWORTH and C. HITCH, J. and J. PEMBERTON, R. WARE, C. RIVINGTON, R. FORD, F. CLAY, A. WARD, J. and P. KNAPTON, J. CLARKE, T. LONGMAN, R. HETT, J. OSWALD, J. WOOD, A. CRUDEN, and J. DAVIDSON, MDCCXXXVIII.

Fig. 11.11 Title page of Alexander Cruden, *Concordance to the Holy Scriptures* (1738) (a) as page image from Eighteenth-Century Collections Online and (b) in Adobe basic OCR.

!

.b C 0 M P LET E
T 0 THE
VT u
0 F T·H E
OLD and NE TESTAI\IIEN "
I N T w 0 p A R T s.
C 0 NT A I.N IN G,

•

I. The Appellative or Commou Words in fo full and large a manner, that any Verf~
may be readily found by looking for any material Word in it. In this Parr the
various Significations of the Principal Words are given, by which the plain.
Meaning of many Faifages of Scripture is !hewn: And alfo an Account of feveral
ewijb Cuil:oms and Ceremonies is added, which may ferve to iilufl:r~te many
arts of Scripture.
II. The Proper Names i,1 the Scriptures. To this Part is prelixcd a Table, containing th<:
Significations of the \Vords in the Original Langnages from which they are derived.
To which is added A CnNcORDAN'CE to the Books, called APOCRYPHA.
The Whole digefted in an Eafy and Reguhr Method, which, together wiih the various
Significations and other Improvements now added, renders it more ufeful than ~ny Book
of
this kind hitherto publi0 1ed.

--------------------------------- ·-------------- -----

By A L E X A N D E R C R U D E N, .711. A.

·~----------------------------------~---------------------------...

John v. 39· Search tbe Sc;·ip:uru, for i,1 1hem)'e tbh:kye hat"" eternal life, c.ild lht) aro
:i:yy <iil.>:·d;
l<jlifj· of me.
2 Tim. iii. 15. 'I'hou h~f. kno;;;n tbe bo[v Scrip!rtrrs, ~uhich m c able to ma(·.: thee
·::•ifltii.~!c'
jah·atiOJt, tJ.,rougb faith wbicb is in Chrift Jc:fit.r.

L 0 N D 0 N:
Printed for D. Mrn•,:r::."""""" A. Br.':'T':swon. TH and C. HncH. j. and J. PH!BE.tToN;
R'\. , Äi~'f' :'~ RI'~ , (' • }-~. JV'~ "••·'-'" ro·:•,, .:"i,, .r. ::·v.h,., D , 1·~ , CL,·'\ . ~-)
1·•" -• ,,.,, ::,, ·.. "1) ., ~T, u'! lc\ T[J, J:.'r\.o.'l -Ar~,l .·,'-.">
J. CLARK!:, 1'. I..,o;.;-\.";M.AN, H.. >L~TT] J. o~v.: . .:..LD: i, '.Yoo...::·· Ji,
C:?.CD£N;, and.
J. D A\. lD~Q~,' J~.V 1. 'L) (. (",..X\. v..\ ~T Y•. IJ- •

Sacred Electronics

The five machines stood, rectangular, silver-green, silent. They were obviously not thinking about anything at all as Archbishop Giovanni Battista Montini of Milan raised his hand to bless them.

"It would seem at first sight," said the archbishop, "that automation, which transfers to machines operations that were previously reserved to man's genius and labor, so that machines think and remember and correct and control, would create a vaster difference between man and the contemplation of God. But this isn't so. It mustn't be so. By blessing these machines, we are causing a contract to be made and a current to run between the one pole, religion, and the other, technology . . . These machines become a modern means of contact between God and man."

So last week at the Jesuit philosophical institute known as the Aloysianum (for St. Aloysius Gonzaga) in Gallarate, near Milan, man put his electronic brains to work for the glory of God. The experiment began ten years ago, when a young Jesuit named Roberto Busa at Rome's Gregorian University chose an extraordinary project for his doctor's thesis in theology: sorting out the different shades of meaning of every word used by St. Thomas Aquinas. But when he found that Aquinas had written 13 million words, Busa sadly settled for an analysis of only one word—the various

J. Watson threw up his hands. "Even if you had time to waste for the rest of your life, you couldn't do a job like that," he said. "You seem to be more go-ahead and American than we are!"

But in seven years IBM technicians in

Bettmann Archive

St. Thomas Aquinas

Philosophy went on the punch cards.

Fig. 11.12 "Sacred Electronics," *Time*, December 31, 1956, 40.

INDEX READING

In the news media in the 1950s and '60s, IBM promoted Busa's project with panache. Decades before the first mention of the Singularity, an item in *Time* magazine, for example, under the heading "Sacred Electronics," stages emergent fantasies of the "scriptural brain" of the computer and its metaphysical and spiritual implications (fig. 11.12).

The five machines stood, rectangular, silver-green, silent. They were obviously not thinking about anything at all as Archbishop Giovanni Battista Montini of Milan raised his hand to bless them.

"It would seem at first sight," said the archbishop, "that automation, which transfers to machines operations that were previously reserved to man's genius and labor, so that machines think and remember and correct and control, would create a vaster difference between man and the contemplation of God. But this isn't so. It mustn't be so. By blessing these machines, we are causing a contract to be made and a current to run between the one pole, religion, and the other technology . . . These machines become a modern means of contact between God and man."[49]

Internally, IBM was also placing substantial effort behind technical text processing initiatives including Luhn's KWIC indexing system. The immediate goal of the KWIC (Keyword-in-Context) index, as the acronym suggests, was not the kind of comprehensive textual analysis that Busa's group was after, but quick and ready reference. The idea of the stop list was articulated explicitly, though not yet named, in the earliest documents on electronic text indexing and handling published by Luhn in the mid and late 1950s. Luhn was developing a system of "business intelligence" for IBM to automatically identify and distribute relevant technical documents to interested parties throughout an organization. His proposals, many of which were eventually realized, focused on enabling computing machines to act on texts. For Luhn, it was not a matter of teaching machines to read. He sought rather to group texts thematically and coordinate these groupings with the interests of individuals "profiled" in the organization.[50]

Unsurprisingly, the push to process and organize information that drove Luhn's project produced lively interest in developing new catalog and classification schemes and in retrofitting old ones to suit the demands of information technology.[51] Controlled vocabulary approaches including Mortimer Taube's Uniterm and Calvin Mooers's Zator systems relied on establishing key terms and conceptual relationships prior to data entry, but potential searches were constrained by initial choices.[52] Luhn sought to devise a system that would offer greater flexibility to future researchers. He also wanted his systems to be able to learn. Luhn's programs had multiple layers and included feedback cycles through which user choices would create and reinforce associations much as in present-day social media applications.

The aspect of Luhn's system most widely adopted during this period was the KWIC index. Though referred to as a "permuted index," KWIC was more properly a "rotated index." After removing stop words, it generated a complete alphabetized list of words in a given text surrounded by a string of characters of a determined length before and after that word at each appearance in the source text, as in a traditional concordance. On a typical KWIC index page, the alphabetical list of keywords would run top to bottom,

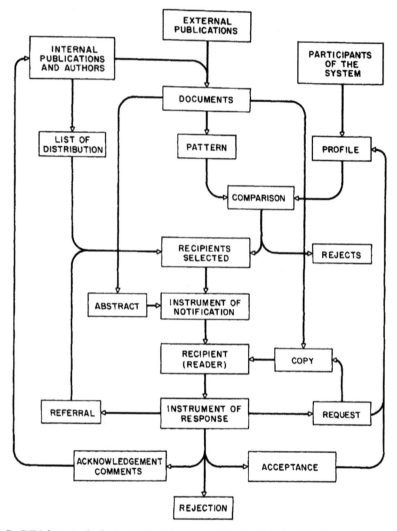

Fig. 11.13 Information feedback system for Luhn's index. H. P. Luhn, "Selective Dissemination of New Scientific Information with the Aid of Electronic Processing Equipment," *American Documentation* 12:2 (Apr. 1, 1961), 133.

down the center of a column, with contextual words to the left and right (fig. 11.13). The reader could then scan the instances of a given word, referring to the contextual information alongside it to determine which instances matched specific research concerns.

As Georges Perec noticed when he adapted it to a literary use in *Espèces d'espace*, a KWIC index is a new kind of text. Though made up of elements from a source text, the KWIC index has many characteristics different from it. In addition to being differently ordered and formatted, it is a different length than the source and has different word frequencies. It has a different statistical signature. In the traditional print concordance, these differences are unmistakable.[53] Not only would no one mistake Cruden's *Concordance* for the Bible, readers intuitively understand it as a work with a structure, approach, and point of view. More generally, as the linguist John Sinclair has argued, it is valuable to think of a concordance as "a text in itself."[54] This is particularly important in the case of computerized concordances in which processing procedures may be black-boxed and the specific character of the working text can too easily slip out of view.[55]

To the end user, the KWIC index proved a gregarious instrument. First, it offered a lot of choices. Standard indexes are typically much shorter than the works to which they refer. Not so, KWIC. Like a traditional print concordance, a KWIC index is typically longer than its source. Since the KWIC index is rotated, bits of the same passage are repeated in entries for different words. Though a funny kind of text, KWIC is quite engaging. It encourages a style of reading that is fleeting but constructive, that refers to itself, providing immediate content, but that also refers to a source text address where more context is always available. As it spread in the culture, KWIC and its daughter technologies modeled and propagated styles of what we might call *index reading*.

Removing stop words was crucial to making print versions of KWIC work. As a starting point, Luhn proposed a list of sixteen function words, but many other words immediately suggested themselves. In a medical journal, words such as *medicine, doctor*, and *illness*, were so common as to be of little use in indexing and were thus often put onto stop lists. Additionally words that appeared very *infrequently* in a text were often removed in order to highlight terms of greater significance.

It is important to note that "word" in this context has a technical meaning. In all of these programming contexts, "word" should be read as "token." A "word" as referred to above may be a character string bracketed by blank spaces. Or it may be a fixed number of alphabetic characters occurring in a row—six is common number—and a quick-and-dirty way to group words by stem in a language heavy with terminal declensions. Or a "word" may be a string with a predicted frequency profile. Regardless, the *idea* of words (collected in what the computational linguists refer to as a "bag of words") as the fundamental units of analysis is pervasive in this literature. Luhn writes,

Figure 1 **Word-frequency diagram.**
Abscissa represents individual words arranged in order of frequency.

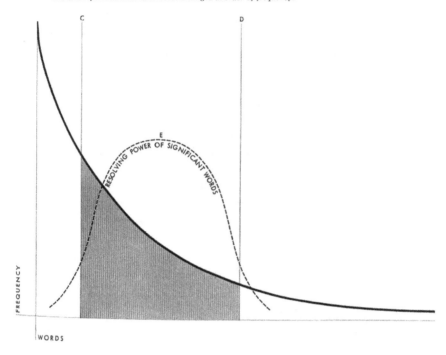

Fig. 11.14 Luhn's interval of significance. Lines C and D represent the high- and low-frequency cut-offs. The central range, in black, represents the "most useful range of words" (Luhn 1958, 161).

The presence in the region of highest frequency of many of the words previously described as too common to have the type of significance being sought would constitute "noise" in the system. This noise can be materially reduced by an elimination technique in which text words are compared with a stored common-word list. A simpler way might be to determine a high-frequency cutoff through statistical methods to establish "confidence limits" . . . Establishing optimum locations for [these] would be a matter of experience with appropriately large samples of published articles. It should even be possible to adjust these locations to alter the characteristics of the output.[56]

Though Luhn initially proposed an automatic procedure to distinguish keywords from non-significant words, Luhn later elaborated a role for human editors, and in practice, stop lists produced by editors quickly took over (fig. 11.14).[57]

Luhn's statistical mechanism itself was meant to be adjusted for par-

FRONT

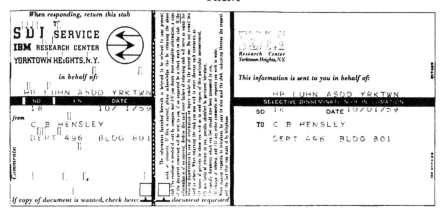

BACK

Keyword-in-Context Index for Technical Literature (KWIC Index)

Luhn, H. P.   Aug. 1959
IBM ASDD RC 127 16 pages   00018
    ABSTRACT. A distinction is made between bibliographical indexes for new and
past literature based on the willingness of the user to trade perfection for
currency. Indexes giving keywords in their context are proposed as suitable
for disseminating new information. These can be entirely machine-generated
and hence kept up-to-date with the current literature. A compatible coding
scheme to identify the indexed documents is also proposed. In it elements are
automatically extracted from the usual identifiers of the document so that the
coded identifier yields a maximum of information while remaining susceptible
to normal methods of ordering.

Keyword-in-Context Index for
Technical Literature (KWIC Index)

Luhn, H. P.   Aug. 1959
IBM ASDD RC 127   00018

keyword	context	index
technical	literature	bibliography
disseminate	information	machine
generate	code	identify
document	order	retrieval
derive	readable	record
intellectual	compile	update
publication	journal	communication
print	text	concordance
title	list	service
notation	name	letter
word	punch	tape
card	dictionary	sort

Fig. 11.15 Index card automatically generated by Luhn's system for mail by post. H. P. Luhn, "Selective Dissemination of New Scientific Information with the Aid of Electronic Processing Equipment," *American Documentation* 12:2 (Apr. 1, 1961): 133.

ticular cases. Boundaries of significance were to be set depending on the needs of a given project. In general, Luhn's account of human input into such processes was ambivalent. None of his proposed systems was autonomous. All were designed to "augment human intellect," to use the phrase from computing pioneer Douglas Engelbart.[58] In fact, even the most ambitious of Luhn's proposals, such as his system for "business intelligence," looped back constantly through mechanisms of user input. (In that particular case, the computer was to automatically print a feedback card to be sent by postal mail to a human reader for each article that it recommended [fig. 11.15].) Luhn writes, "One must keep in mind that machine products of the kind discussed here can never reach the level of perfection of which human beings are capable, and that there will always be residual effort left for human beings."[59] In the 1960s, these "residual efforts"—in another context, one might have called them "reading," "writing," and "editing"—turned out to be substantial.[60]

In the years immediately following Luhn's first proposals, users experimented widely with statistical parameters in order to produce results tailored to the very diverse projects in which KWIC was being implemented in industry, government, and scholarship. Some followed Luhn's approach, adjusting for "optimum locations." Others devised alternative statistical tests.[61] As we shall see, many organizations jettisoned the statistical approach altogether, but even in the simplest case, where Luhn's frequency curves were adopted, decisions about parameters were generally editorial.

ARCHIVES OF EXCLUDED LANGUAGE

By the early 1960s, KWIC was a success. In 1958, the National Science Foundation's Office of Science Information granted $150,000 to the Chemical Abstracts Service, allowing it to develop a keyword indexing scheme using the KWIC system. In 1960, seven thousand copies of the new index were distributed at the annual meeting of the American Chemical Society. The first efforts with KWIC in the American scientific societies were self-consciously experimental. The KWIC index was adopted by the American Institute of Biology for *Biological Abstracts* (*stopwords, forbidden words*, 1961), and in the next several years, the American Meteorological Society (*forbidden words*, 1962), the American Geological Institute (*words prevented from indexing*, 1964), and so on. In a 1966 retrospective on a half-decade of KWIC, Marguerite Fischer of the American College of Physicians counted more than thirty such ongoing efforts.[62] These first experiments offer layers of ethnographic detail on the ways in which editors and users interacted with early systems of "automatic" text processing.

The American scientific societies adopted KWIC because it offered both efficiency and functionality absent in traditionally indexed volumes. As Barbara Flood, an editor at *Biological Abstracts*, writes, by 1961 the "traditional subject index was falling further and further behind." KWIC seemed like a good solution. After a year of planning, a modified KWIC index became Biological Abstracts' Subject in Context or BASIC.[63] Like the other programs running on IBM mainframes at that time, BASIC existed in a hybrid world of electronics and paper. And, in the first instance, it was the physicality of the paper output devices that drove the design of the BASIC stop list. Needless to say, since the stop list continues to be widely employed today, the conversion to magnetic media altered but did eliminate the logistical concerns so urgent in the paper world.[64]

At the scientific societies, stop lists were generated by editors and editorial committees. Flood writes,

> I was given early runs of computer printouts to look through editorially. There were inches of line-printer pages with index entry words such as *of, the, in, and*. It might be of trivial interest that *of* was the most prevalent, perhaps because these were titles in biology. I chose to delete these words and began to compile a list of words that were going to be automatically stopped from printing.[65]

In 1961, Flood and her colleagues were already referring to this as a "stop list."

The relationship between electronics and paper is palpable in all of these first indexes, among which only one, the 1963 *Proceedings* of the American Documentation Institute, of which Luhn himself was president, was published in a magnetic format. Luhn offered the index to the *Proceedings* on magnetic tape for $100. Reportedly, no one took him up on this offer.[66] At *Biological Abstracts*, the process of selecting stop words was governed by committee. Flood writes,

> Members of the editorial department met often to discuss candidate stop list words generated from the print-outs to make sure that homographs (such as *a* in *Vitamin A* or *are* as a measure of area) were not overlooked. Thus, the Stop List was generated from frequency data with concurrence by committee. Multiple word terms such as *Rana esculenta* had to be evaluated as to whether the second word provided a significant index entry. Should the second word be added to the Stop List? Was it frequent enough? The cost of adding a word to the Stop List with resultant added computer time had to be compared to the cost per copy of printing the extra line.[67]

It is hard to imagine a better characterization of the organizational cyborg than "data with concurrence by committee." As Flood and Parkins describe it, the early efforts at *Biological Abstracts* produced valuable empirical insights. Editors tried out different word selections and different size lists and observed the changed output from the system.

Though the idea behind KWIC was to reduce human involvement in indexing procedures, at least at the level of design, the editors at publications such as *Biological Abstracts* were highly engaged. Editors developed a system of annotations made by hand, which they referred to as "slashing and dashing," to inform keypunch operators of words to be held together and treated as one, for example, a place name such as *New York*, and words to be split apart for indexing by roots, such as *autoimmune, neonatal, endoparasitism.*[68] The editors at the American Bar Association took the further step of "adding one or more words" to titles and subtitles to "reveal the contents of a thesis."[69] According to Phyllis Parkins, later director of *Biological Abstracts*, "It is quite possible . . . that even now it should no longer be said of all permuted-title indexes: 'The chief advantage is elimination of the creative work of human indexers,' . . . since we believe at present that a great advantage of the subject index to *Biological Abstracts* is its potential for using the creative work of skillful, biologically trained, decidedly human editors."[70] No more "harmless drudges."

Parkins's account of the work of human indexers has particular poignancy in this context. In the early literature on automatic indexing and abstracting in this period, there is a notable experimentalism and theoretical freedom. Beyond the larger background debates about controlled and uncontrolled vocabularies, machine translation, artificial intelligence, and natural language processing, these texts convey a very powerful sense of the development of a new professional, intellectual, and social space in the overlapping territory among fields of computer programming, scientific administration, data management, and librarianship. And it is hard not to notice how many of the voices and hands in this new region were those of women.

Phyllis Parkins's own story is a case in point. A biologist trained at Syracuse University, Parkins came to Philadelphia with her husband when he was hired as a physician at the University of Pennsylvania. She began work at *Biological Abstracts* as a subject editor in 1953 and moved up the ranks. When *Biological Abstracts* director Miles Conrad died unexpectedly in 1964, Parkins was named to an interim directorship of the entire operation. In 1965, over objections from individuals who preferred not to appoint a woman to the permanent position, Parkins became director and, later, CEO of the parent company, BIOSIS.[71] And Parkins was far from alone. Agatha Hughes has detailed the parallel story of Mary Elizabeth Stevens of the National Bureau of Standards, another key actor in the field.[72] And related observations may be made for many others including Mary Veilleux at the Central Intelligence Agency and Sally Dennis at IBM.[73]

In many applications, stop lists quickly grew in size and complexity. By 1963, the BASIC stop list comprised approximately 1,000 words. At the same time, the *Chemical Titles* stop list grew to 950 words, at which point editors began to reduce it, eventually trimming it back to only 328. In a note introducing its new KWIC format, *Geoscience Abstracts* posed the question

A
ABOUT
ABOVE
ABSENCE
ABSTRACT
ABSTRACTS
ACADEMICIAN
ACCOMPANIED
ACCOMPANYING
ACCOMPLISHED
ACCORDING
ACCOUNT
ACCOUNTING
ACHIEVEMENTS
ACIDS
ACROSS
ACT
ACTING
ACTION
ACTIONS
ACTUAL
ADAPTED
ADDITIONAL
ADDRESS
ADJACENT
ADVANTAGE
ADVANTAGES
AFFECTED
AFFECTING
AFTER
AGAINST
AGE
AGES
AHEAD
AID
ALONE
ALONG
ALSO
AMONG
AN
AND
ANNIVERSARY
APPEAL
APPEARANCE
APPEARING
APPENDIX
APPLICABILITY
APPLICATION
APPLICATIONS
APPLIED
APPROACH
APPROPRIATE
APPROXIMATE
APPROXIMATELY
APPROXIMATION
APPROXIMATIONS
ARBITRARILY
ARE
AREA
AREAS
ARGUMENT
AROUND
ARRANGED
AS
ASPECTS
ASSOCIATED
ASSUMED
ASSUMING
AT
ATTAINABLE
ATTEMPT
B
BASED
BASIC
BE
BEFORE
BEGINNING
BEHAVIOR

BEHAVIOUR
BEINGS
BELOW
BENEATH
BENEFITS
BETTER
BETWEEN
BEV
BEYOND
BIOL
BOT
BOTH
BRIEF
BY
C
C/S
CAN
CARRIED
CASE
CASES
CAUSE
CAUSED
CAUSES
CERTAIN
CHANGES
CHARACTER
CHARACTERISTICS
CHARACTERIZATION
CHEM
CHOOSING
CLOSE
CM
COMBINED
COMMENT
COMMENTS
COMPARED
COMPARING
COMPARISONS
COMPLEMENTARY
COMPLETED
CONCEPT
CONCERNING
CONCLUDING
CONCLUSIONS
CONDITION
CONDITIONS
CONDUCTED
CONNECTED
CONNECTING
CONNECTION
CONNECTIONS
CONSEQUENCES
CONSIDERABLE
CONSIDERATION
CONSIDERATIONS
CONSIDERED
CONSIDERING
CONSISTING
CONTAINING
CONTRIBUTION
CONTRIBUTIONS
CORRESPONDING
COURSE
CRITERIA
CRITERION
DEDICATED
DEDUCED
DEDUCTIONS
DEG
DEGREE
DELETED
DEPENDENCE
DEPOSIT
DEPOSITS
DERIVED
DESCRIPTIVE
DESIGNATIONS
DETAIL
DETERMINATION

Fig. 11.16 First of four page of stop list ("words prevented from indexing"), *Geoscience Abstracts*, 1964.

of the stop list to its readership: "In selecting keywords for alphabetization in the permuted index, the machine memory is programmed by a stop list to ignore words of no subject interest, such as *the*, *studies*, *of*, etc. . . . Users are particularly asked to note that the inclusion of an occasional unsuitable indexing word is inevitable; it is not possible to predict what words of no subject interest will be used in future titles."[74] In fact, the stop list provided in the appendix to *Geoscience Abstracts* 1964 is considerably more elaborate than this notice suggests (fig. 11.16). Though the list is sober and comprehensible in its general outlines, it produces some vertigo as categories emerge and recede in waves. The stop list is a "negative dictionary," so it makes sense that it provides relatively little solid ground. Yet it is a pleasure to read the *Geoscience Abstracts* list and to imagine the curious geological language that remains, without articles, prepositions, and pronouns, without *studies*, *science*, and—of course—without *rocks*.

We may wonder at any number of curious features of the *Geoscience Abstracts* list. It seems quite reasonable that it includes the word *ordinary* since it also includes *unusual*. But then it is odd to find *peninsula* without *island* or any other topographic designation; *sea*, *ocean*, and *water* but not *land*, *continent*, *earth*, or *soil*; *novel* but not *old*, or for that matter, *new*. The inclusion of *etc.* on the list gives it a particularly Borgesian quality, one that also throws us back to the miscellaneous political usage of *stop list* in the early twentieth century as "a list of persons, *etc.*, deprived of particular rights."[75]

Of course, unlike Borges, *Geoscience Abstracts* is not in any fundamental way concerned with the category of *etc.* To the contrary, the inclusion of *etc.* in its stop list highlights the fact that systems such as KWIC formalize disciplinary protocols of reading. As in the stories of Barbara Flood and Phyllis Parkins, examination of the list as a whole clarifies how the architecture of linguistic selection, both human and machine, determines what falls into areas of "interest" and "no subject interest," in the corporatized textual universe. It highlights the role of "preprocessing," in larger indexing and abstracting processes and finally in our encounter with texts. The persistence of the stop list, and its high placement in the procedural hierarchy of modern software, reinforces the importance of these processes.

Despite the statistical framing in presentations such as that of Luhn, in practice, the stop lists of the 1960s seem mostly to have been constructed through a kind of kitchen science, adding and subtracting to taste, attending to the results and adjusting again. The introductory material to the *Geoscience Abstracts* asking readers for suggestions, says exactly this. In this single gesture, *Geoscience Abstracts* helps clarify the place of that curious self-cancelling artifact, the stop list, in the history of disciplines as well as in the development of information practice in the twentieth century.

What has become of the stop list? In some applications, nothing. As the example of Wordle suggests, in simple frequency calculations, stop lists continue to be useful tools. But in sophisticated search and data mining operations, more powerful computers and new methods of economizing searches have diminished the importance of the stop list as it was originally employed.

Yet the lessons of the stop list are, if anything, more pressing today than ever. A recent flurry of critical work has highlighted the ways in which the embodiment of literature in electronic corpora is changing how readers today encounter texts. From Franco Moretti's *distant reading* to the *surface reading* of Stephen Best, Sharon Marcus, and Heather Love, to the *database aesthetics* of Lev Manovich and Geoffrey Bowker, numerous critics have pointed out, and many have promoted, a departure from the close reading practices that were once the bread and butter of textual scholarship.[76]

In these frameworks, all of which pertain in one way or another to the long tradition of *index reading*, boundaries between works are loosened, conceptions of textuality are broadened, and books themselves tend to be seen as an aspect of the corpus rather than the corpus as an assemblage of books. Moreover, one need not ascend to the heights of criticism to notice this phenomenon. Unassuming objects such as a word cloud that appears alongside a newspaper transcript of the State of the Union address demonstrate how techniques of reading in electronic corpora have become embedded in our everyday experience and expectation of text.

At the same time, it is not clear that we yet have a grasp on just what such text is, how it is created, and—importantly—how it endures. A transcript of the State of the Union speech is not the same as a speech. A digital version is not the same as print. And the systems through which we have access to text in electronic form do not all operate according to common rules. The Internet may feel like a universal archive. In fact, it is more like a textual archipelago. Moreover, the archive of this "archive" is everywhere wanting.[77]

The question of the stop list is not precisely new. Viewed broadly, it encompasses a long history of indexing extending to the Middle Ages, at least. Even viewed narrowly, it extends more than half a century to the work of Roberto Busa and his contemporaries at the dawn of the age of digitized text. In this context, the stop list is of special interest as a mechanism fundamental to electronic "systems of statements" and as an example of the kind of para-literary object that we should be especially attentive to archiving today.

1. Michel Foucault, *The Archeology of Knowledge*, trans. Alan Sheridan (New York: Pantheon, [1969] 1972), 128.

2. Daniel Rosenberg, "Early Modern Information Overload," *Journal of the History of Ideas* 64, no. 1 (January 1, 2003): 1–9; Daniel Rosenberg, "An Eighteenth-Century Time Machine: The *Encyclopedia* of Diderot and d'Alembert," in Daniel Gordon, ed., *Postmodernism and the Enlightenment* (New York: Routledge, 2001), 45–66.

3. http://www.gutenberg.org/ebooks/1400; http://www.wordle.net/. Accessed June 2013.

4. Roberto Busa, "The Annals of Humanities Computing: The Index Thomisticus," *Computers and the Humanities* 14 (1980): 83–90.

5. Harald Weinrich, "The Textual Function of the French Article," in *Literary Style: A Symposium*, ed. Seymour Chatman (New York: Oxford University Press, 1971), 227.

6. Franco Moretti, *Distant Reading* (London: Verso, 2013).

7. William James, *Principles of Psychology*, 2 vols. (New York: Holt, 1890), 1:238; also cited in Willard McCarty, "Beyond the Word: Modelling Literary Context," Canadian Symposium on Text Analysis (Fredricton, NB, 2006). None of this has been lost on contemporary psychologists or data miners. See, for example, J. W. Pennebaker, *The Secret Life of Pronouns: What Our Words Say about Us* (New York: Bloomsbury Press, 2011).

8. Barbara J. Flood, "Historical Note: The Start of a Stop List at Biological Abstracts," *Journal of the American Society for Information Science* 50, no. 12 (October 1999): 1066.

9. Richard L. Venezky, "Computer-Aided Humanities Research at the University of Wisconsin," *Computers and the Humanities* 3, no. 3 (January 1, 1969): 129–38.

10. Antonette DiPaolo Healey and Richard L. Venezky, eds., *Microfiche Concordance to Old English: The List of Texts and Index of Editions* (Toronto: University of Toronto, 1980).

11. Richard L. Venezky, *Microfiche Concordance to Old English: The High-Frequency Words*. (Newark, DE: University of Delaware, 1983).

12. "Librarians Stop Abortion Stop-Listing," *American Libraries* 39, no. 5 (May 2008): 23.

13. David Bamman, Brendan O'Connor, and Noah Smith, "Censorship and Deletion Practices in Chinese Social Media," *First Monday* 17, no. 3 (March 5, 2012), http://journals.uic.edu/ojs/index.php/fm/article/view/3943/3169, accessed August 2016; J. R. Crandall et al., "ConceptDoppler: A Weather Tracker for Internet Censorship," *Proceedings of the 14th ACM Conference on Computer and Communications Security*, 2007, 352–65; J. R. Crandall and Phillipp Winter, "The Great Firewall of China: How It Blocks Tor and Why It Is Hard to Pinpoint," *Login* 37, no. 6 (December 2012): 42–50.

14. David Thies, "Stop Words Are Dead! Did I Miss Another Memo?," *SEO Fast Start*, January 9, 2008, http://www.seofaststart.com/stop-words-are-dead/, accessed August 2016; Bill Slawski, "New Google Approach to Indexing and Stopwords," *SEO by the Sea*, January 17, 2008, http://www.seobythesea.com/2008/01/new-google-approach-to-indexing-and-stopwords/.

15. Google drew a contrast between its new approach and older techniques relying on stop lists. "Typically, given a query, the performance bottleneck is the time it takes to decode the occurrences . . . of the most frequently occurring term, especially if this term is a so-called stop-word such as *the*. The present invention allows the search engine to simply look at the documents that contain the least popular term . . . . This allows the search engine to decode other terms in the document to see if any of the other terms

match more popular terms in the query. If so, the search engine can simply advance to that position." Document Compression Scheme that Supports Searching and Partial Decompression, US Patent 7319994 (assigned to Google Inc., January 15, 2008).

16. See Matthew Jones, chapter 12 in this volume.

17. Georges Perec, *An Attempt at Exhausting a Place in Paris*, trans. Marc Lowenthal (Cambridge, MA: Wakefield Press, 2010), 3.

18. Georges Perec, *La disparition* (Paris: Gallimard, 1989); *Espèces d'espaces* (Paris: Editions Galilée, 1974).

19. H. P. Luhn, "A New Method of Recording and Searching Information," *American Documentation* 4, no. 1 (1953): 14; "A Statistical Approach to Mechanized Encoding and Searching of Literary Information," *IBM Journal of Research and Development* 1, no. 4 (October 1957): 309–17; "The Automatic Creation of Literature Abstracts," *IBM Journal of Research and Development* 2, no. 2 (April 1958): 159–65; "A Business Intelligence System," *IBM Journal of Research and Development* 2, no. 4 (October 1958): 314–19; "Keyword-in-Context Index for Technical Literature," *American Documentation* 11, no. 4 (1960): 288; "Selective Dissemination of New Scientific Information with the Aid of Electronic Processing Equipment," *American Documentation* 12, no. 2 (1961): 131.

20. Luhn, "Keyword-in-Context," 289.

21. Luhn, "Automatic Creation of Literature Abstracts," 162.

22. Luhn, "Keyword-in-Context," 291.

23. Flood, "Historical Note," 1066.

24. Paul N. Edwards, *A Vast Machine: Computer Models, Climate Data, and the Politics of Global Warming* (Cambridge, MA: MIT Press, 2010), 205.

25. Phyllis V. Parkins, "Approaches to Vocabulary Management in Permuted Title Indexing of Biological Abstracts," in *Automation and Scientific Communication* (Washington, DC, 1963), 27–28.

26. Ivan Illich, *In the Vineyard of the Text: A Commentary to Hugh's Didascalicon* (Chicago: University of Chicago Press, 1996).

27. Roberto Busa, "Concordances," in *Encyclopedia Library and Information Science*, vol. 5 (New York: Marcel Dekker, 1971), 592–604, 598–99.

28. Busa, "Concordances," 596.

29. Bella Haas Weinberg, "Predecessors of Scientific Indexing Structures in the Domain of Religion," in *History and Heritage of Scientific and Technological Information Systems*, ed. W. Boyd Rayward and Mary Ellen Bowden (Medford, NJ: Information Today, 2004), 129.

30. "The author uses the Hebrew equivalent of the term function words to characterize the stopwords. The author admits to having modeled his Hebrew concordance on a Latin one, but from the wording of the passage on the stoplist in the (unpaginated) introduction, I infer that this stoplist was an original compilation. The author wanted to complete the concordance in a reasonable time, and so he omitted frequent nonsubstantive words." Weinberg, "Predecessors of Scientific Indexing Structures," 129.

31. Weinberg, "Predecessors of Scientific Indexing Structures," 129; J. J. Herzog, P. Schaff, A. Hauck, C. C. Sherman, G. W. Gilmore, and S. M. Jackson, *The New Schaff-Herzog Encyclopedia of Religious Knowledge* (New York: Funk & Wagnalls, 1909), vol. 1, 522.

32. Andrew Becket, *A Concordance to Shakespeare* (London: Robinson, 1787); Alexander Cruden, *A Verbal Index to Milton's Paradise Lost* (London: 1741).

33. Alexander Cruden, *A Complete Concordance to the Holy Scriptures of the Old and New Testament: In Two Parts Containing the Appellative or Common Words . . . , the Proper Names . . . , to Which Is Added a Concordance to the Books Called Apocrypha* (London: Printed for W. Owen and 18 others, 1738).

34. See Busa's discussion of the distinctive characteristics of the concordance as text. Busa, "Concordances," 600.

35. James Strong, *The Exhaustive Concordance of the Bible: Showing Every Word of the Text of the Common English Version of the Canonical Books, and Every Occurrence* (London: Hodder & Stoughton, 1890).

36. Kenneth McKenzie, "Means and End in Making a Concordance, with Special Reference to Dante and Petrarch," in *25th Annual Report of the Dante Society* (Boston: Ginn & Co, 1907), 18–46.

37. Helen Kate Furness, *Concordance to Shakespeare's Poems: An Index to Every Word Therein Contained* (Philadelphia: J. B. Lippincott, 1874), 3. Furness's volume complemented the New Variorum editions of Shakespeare published by her husband, Horace Howard Furness, and later her son, Horace Howard Furness, Jr., beginning in these same years.

38. Furness, *Concordance*, 3.

39. Busa, "Concordances," 596.

40. Busa, "The Annals of Humanities Computing," 84.

41. Busa, "Concordances," 592–604.

42. Paul Tasman, "Literary Data Processing," *IBM Journal of Research and Development* 1, no. 3 (1957): 249–56; *Indexing the Dead Sea Scrolls by Electronic Literary Data Processing Methods* (White Plains, NY: IBM, 1958).

43. Busa, "The Annals of Humanities Computing," 83.

44. Ibid., 86.

45. Ibid., 87.

46. Melissa Terras, "For Ada Lovelace Day: Father Busa's Female Punch Card Operatives," Melissa Terras Blog, http://melissaterras.blogspot.com/2013/10/for-ada-lovelace-day-father-busas.html, accessed April 2014; Julianne Nyhan, Andrew Flynn, and Anne Welsh, "Oral History and the Hidden Histories Project: Towards Histories of Computing in the Humanities," *Literary and Linguistic Computing*, first published online on July 30, 2013, at http://dsh.oxfordjournals.org/content/early/2015/01/11/llc.fqt044]

47. On the more general subject, see Jennifer S. Light, "When Computers Were Women," *Technology and Culture* 40, no. 3 (July 1999): 455–83.

48. Erez Aiden and Jean-Baptiste Michel, *Uncharted: Big Data as a Lens on Human Culture* (New York: Riverhead, 2013).

49. "Sacred Electronics," *Time*, December 31, 1956, 40. From June 1963 to his death in August 1978, Montini would be Pope Paul VI.

50. Luhn, "Business Intelligence System," 314–19.

51. Mary Elizabeth Stevens, *Automatic Indexing: A State-of-the-Art Report*, National Bureau of Standards Monograph 91 (Washington: US Government Printing Office, 1965).

52. Gerald Salton, "Historical Note: The Past Thirty Years in Information Retrieval," *Journal of the American Society for Information Science* 38, no. 5 (1987): 375; C. P. Bourne and T. B. Hahn, *A History of Online Information Services: 1963–1976* (Cambridge, MA: MIT Press, 2003); Jana Varlejs, "The Technical Report and Its Impact on Post–World War II Informa-

tion Systems: Two Case Studies," in *The History and Heritage of Scientific and Technological Information Systems: Proceedings of the 2002 Conference of the American Society for Information Science and Technology*, ed. W. Boyd Rayward and Mary Ellen Bowden. Information Today, for ASIS&T and Chemical Heritage Foundation, 2004, 89–99.

53. Busa, "Concordances," 600.

54. John McHardy Sinclair, *Corpus, Concordance, Collocation* (Oxford: Oxford University Press, 1991), 34.

55. Frank Pasquale, *The Black Box Society: The Secret Algorithms That Control Money and Information* (Cambridge, MA: Harvard University Press, 2015); Tarleton Gillespie, "The Relevance of Algorithms," in Gillespie, Pablo Boczkowski, and Kirsten Foot, eds., *Media Technologies: Essays on Communication, Materiality, and Society* (Cambridge, MA: MIT Press, 2014), 167–94.

56. Luhn, "Literature Abstracts," 160.

57. Luhn, "Keyword-in-Context," 288.

58. Douglas C. Engelbart, "Augmenting Human Intellect: A Conceptual Framework," SRI Summary Report AFOSR-3223, Air Force Office of Scientific Research (October 1962); Charles P. Bourne and Douglas C. Engelbart, "Facets of the Technical Information Problem," *SRI Journal* 2, no. 1 (1958). The latter report stresses the importance of information economies in view of the very large number of scientific and technical indexers employed by the Soviet Union.

59. Luhn, "Keyword-in-Context," 295.

60. Charles F. Balz and Richard H. Stanwood, *Some Applications of the KWIC Indexing System* (Oswego, NY: IBM, 1962); Saul Herner, *Deep Subject Indexing by Manual Permutation Methods* (Washington, DC: Herner and Co., 1963); Fischer, "The KWIC Index Concept," 57.

61. For example, Mary Elizabeth Stevens, Vincent E. Giuliano, and Laurence B. Heilprin, *Statistical Association Methods for Mechanized Documentation: Symposium Proceedings Washington 1964*, National Bureau of Standards Miscellaneous Publications 269 (Washington, DC: US Government Printing Office, 1965).

62. Marguerite Fischer, "The KWIC Index Concept: A Retrospective View," *American Documentation* 17, no. 2 (1966): 57. By 1971, Busa believed there were too many such projects in humanities fields to count them any longer. Busa, "Concordances," 598.

63. Flood, "Historical Note," 1066.

64. See esp. Markus Krajewski, *Paper Machines: About Cards and Catalogs, 1548–1929*, trans. Peter Krapp (Cambridge, MA: MIT Press, 2011). As Paul Edwards recounts, new paper formats such as microcard proved resilient, even in face of new magnetic and electronic media. Edwards, *A Vast Machine*, 205.

65. Flood, "Historical Note," 1066.

66. C. P. Bourne, "40 Years of Database Distribution and Use: An Overview and Observation," NFAIS Miles Conrad Lecture (1999), accessed June 17, 2013, http://nfais.org/1999-miles-conrad-lecture, accessed June 2013. H. P. Luhn and P. C. Janaske, eds., *Automation and Scientific Communication: Short Papers Contributed to the Theme Sessions of the 26th Annual Meeting of the American Documentation Institute at Chicago, Pick-Congress Hotel, October 6-11, 1963* (American Documentation Institute, 1963), preface. Busa reports the first humanities concordance in magnet format to be J. W. Ellison's 1957 *Nelson's Complete Concordance of the Revised Standard Version Bible*. Busa, "Concordances," 597.

67. Flood, "Historical Note," 1066.

68. Parkins, "Approaches to Vocabulary Management," 27–28.

69. Glenn Greenwood, *Index to Legal Theses and Research Projects* (Chicago: American Bar Foundation, 1962), 1.

70. Parkins, "Approaches to Vocabulary Management," 28; William Campbell Steere, Hazel A. Philson, and Phyllis V. Parkins, *Biological Abstracts/BIOSIS: The First Fifty Years, the Evolution of a Major Science Information Service* (New York: Plenum Press, 1976).

71. Steere et al., *Biological Abstracts/BIOSIS*, 124–27; George Tomezsko, *Fully Occupied Years: The Rise and Fall of a Company Called BIOSIS* (Philadelphia: Xlibris, 2006), 81.

72. Agatha C. Hughes, *Systems, Experts, and Computers: The Systems Approach in Management and Engineering, World War II and After* (Cambridge, MA: MIT Press, 2000), 197–99.

73. M. P. Veilleux, "Permuted Title Word Indexing Procedures for a Man/Machine System," in *Machine Indexing: Progress and Problems* (Washington, DC: American University, 1962), 77–111; S. F. Dennis, "Construction of a Thesaurus Automatically," in Stevens, Giuliano, and Heilprin, *Statistical Association Methods for Mechanized Documentation*, 61–155.

74. *Geoscience Abstracts* (Washington: American Geological Institute, 1964), 3.

75. "These ambiguities, redundancies and deficiencies remind us of those which doctor Franz Kuhn attributes to a certain Chinese encyclopaedia entitled 'Celestial Empire of benevolent Knowledge'. In its remote pages it is written that the animals are divided into: (a) belonging to the emperor, (b) embalmed, (c) tame, (d) sucking pigs, (e) sirens, (f) fabulous, (g) stray dogs, (h) included in the present classification, (i) frenzied, (j) innumerable, (k) drawn with a very fine camelhair brush, (l) *et cetera*, (m) having just broken the water pitcher, (n) that from a long way off look like flies." Jorge Luis Borges, "The Analytical Language of John Wilkins," in *Other Inquisitions, 1937–1952*, trans. Ruth L. C. Simms (Austin: University of Texas Press, 1965), 103, emphasis added.

76. Moretti, *Distant Reading*; Steven Best and Sharon Marcus, "Surface Reading: An Introduction," *Representations* 108, no. 1 (November 1, 2009): 1–21; Heather Love, "Close but Not Deep: Literary Ethics and the Descriptive Turn," *New Literary History* 41, no. 2 (Spring 2010): 371–91; Lev Manovich, "The Database as Symbolic Form," *Convergence* 5, no. 2 (June 1999): 80–99; Geoffrey C. Bowker, *Memory Practices in the Sciences* (Cambridge, MA: MIT Press, 2005).

77. On the fantasy/melancholy of the total electronic archive, see Daniel Rosenberg, "Electronic Memory," in Daniel Rosenberg and Susan Harding, eds., *Histories of the Future* (Durham: Duke University Press, 2005), 123–52; Wendy Chun, "The Enduring Ephemeral, or the Future is a Memory," *Critical Inquiry* 35, no. 1 (Autumn 2008): 148–71; chapters by Matthew Jones (chapter 12), Rebecca Lemov (chapter 10), and Bruno Strasser (chapter 7) in this volume.

TWELVE

Querying the Archive: Data Mining from Apriori to PageRank

Matthew L. Jones

In 1998, amid the blossoming of large-scale corporate, government, and academic "data warehouses," Usama Fayyad was worried.

> If I were to draw on a historical analogy of where we stand today with regards to digital information manipulation, navigation, and exploitation, I find myself thinking of Ancient Egypt. We can build large impressive structures. We have demonstrated abilities at the grandest of scales in being able to capture data and construct huge data warehouses. However, our ability to navigate the digital stores and truly make use of their contents, or to understand how they can be exploited effectively is still fairly primitive. A large data store today, in practice, is not very far from being a grand, write-only, data tomb.[1]

Fayyad, then at Microsoft, previously at the Jet Propulsion Laboratory, explained that he hoped "the recent flurry in activity in data mining and KDD [knowledge discovery in databases] will advance us a little towards bringing some life into our data pyramids."[2] "Big data" there was aplenty. Ready access to that data and its significance, not so much. Excavating this entombed data, Fayyad argued, required richer forms of querying, especially transforming techniques drawn from statistics and the machine learning branch of artificial intelligence. In the same year, two Stanford computer science graduate students were hard at work pushing the latest in data mining techniques to apply to a far more anarchic archive, not a data pyramid of well-organized corporate data, but a jumble of noncurated, interlinked pages: the World Wide Web. Their end product: the Google search engine.

This chapter stresses the centrality of the database community of academic computer science and industry, the community entrusted with figuring out how to secure digital archives, in the creation of data sciences of the 2000s. Database practitioners could never forget the scale of data, understood not as something intangible but as something physical existing on slow hard drives, something incapable of being resident in memory, something requiring time to move from place to place and from drives to processors. Large amounts of data broke things easily, as Fayyad noted in 1998.

> A typical statistical package assumes small data sets and low dimensionality. For example, suppose you want to do some simple database segmentation by running [. . .] a basic simple method for clustering data. Let's say you have managed to lay your hands on an implementation in some statistical library. The first operation the routine will execute is "load data." In most settings, this operation will also be the last as the process hits its memory limits and comes crashing down.[3]

Adapting statistical and artificial tools to very large databases required dramatic modifications to the tools, epistemic shifts, and transformations in the epistemic values and protocols surrounding their use.

This chapter focuses on two centers of tremendous activity in mining databases, just down the road from each other: the data mining group at Stanford and IBM's Almaden research center in San José. While the technical means for contending with the scale of the archive remained underdetermined, the creators of data mining offered powerful technological determinist narratives holding that contending with great volumes of data requires the development of new algorithms, the loosening of traditional account of statistical rigor, the creation of new epistemic virtues, and the creation of new experts.[4] The challenging materiality of corporate and government digital archives came to justify fundamental transformations in practices and values: new tools for automation that demanded heightened skills in data cleaning and "munging" and alternative forms of algorithmic judgment.[5] The sheer scale of data was held to demand—and to justify— new forms of scientific knowledge, at times in conflict with long-held views of statistical rigor.

The "data warehousing" guru Ralph Kimball explained in 1995 that makers of large databases in the 1980s and 1990s had prioritized perfecting storage over improving access to data.[6] Building large databases that could accurately, durably, and securely record large numbers of transactions and other forms of cleaned data had dominated database research and practical implementation. In the 1970s and 1980s, database researchers had focused squarely upon putting stuff into databases.[7] Central to the developments of database practice was an insistence that changes to the database be "atomic": any set of connected changes to a database must be entirely completed or not completed at all. If you transfer $100 from one account to the other, the database system must either alter both accounts or alter neither. Researchers had to solve the problem for large-scale systems, where hardware, software, and power were presumed to fail with great regularity—as they do.[8]

However necessary their efforts were, Kimball explained, "we came perilously close to forgetting why we bought relational databases in the first place."[9] Seeking meaningful information from massive databases had been insufficiently prioritized. Kimball proclaimed the situation to be dire: "To be blunt, today's systems are very good at transaction processing and pretty horrible at querying." The time to create new forms of access was now.

> Fortunately, the chief executives in most companies have long memories. They remembered the promise that we would be able to "slice and dice" all of our data. These executives have noticed that we have almost succeeded in storing all the corporate data in relational databases. They also haven't forgotten that they have spent several billion dollars. From their point of view, it is now time to get all that data out.[10]

By the mid-1990s database practitioners in industry and academia alike were focusing attention upon new techniques for querying increasingly large databases.

Data mining was initially one of the many efforts to improve querying.[11] For database management researchers, data mining—or as they originally called it, "database mining"—was just such a better way of getting at the contents of databases. A primary driver of IBM's data mining effort, Rakesh Agrawal, explained in 2003: "I'm a database person, so my view of data mining has been that it is essentially a richer form of querying. We want to be able to ask richer questions than we could conveniently ask earlier."[12] In 2004, the Microsoft database researcher Jim Gray explained, "We are slowly climbing the value chain from data to information to knowledge to wisdom. Data mining is our first step into the knowledge domain."[13]

KDD

Data mining, or, as it was more formally dubbed in the 1990s, knowledge discovery in databases (KDD), is the activity of creating nontrivial knowledge suitable for action from databases of vast size and dimensionality.[14] Data miners never speak of an *information* overload. Their mantra is "we're data rich, but information poor." For them, information, taken to be largely synonymous with knowledge, comprises interesting, nontrivial patterns in data. Their task was to create such "interesting," "actionable" patterns from vast quantities of data in practically computable ways.

Underlying the practice of data mining is a *critique* of artificial reason—a recognition of the limits of the abilities of human beings and machines when faced with vast amounts of data, followed by the creation of tools more suited to limited human abilities, limited computing memories and speeds, and relatively short-term real-world goals for investigation. Data mining concerns databases of very large size—millions or billions of records, usually with elements of high dimensionality (meaning that every record typically comprises a large number of elements). For each record in a retail database, a data mining operation might seek unexpected relationships among the item purchased, the store's zip code, the purchaser's zip code, variety of credit card, time of day, date of birth, other items purchased at the same time, even every item viewed, or the history of every previous item purchased or returned. Performing reasonably fast analyses of high-dimensional, messy real-world data is central to the identity and purpose of data mining, in contrast to its predecessor fields such as statistics and machine learning. The technical and social solutions of data mining involve converting theoretical algorithms into everyday practices and high-dimensional data into actionable knowledge.

According to KDD advocates, traditional scientific approaches to data—and the traditional competencies of scientists—simply could not keep up with the volume of data and multidimensionality possible thanks to computers. Something else is needed, something less pure—because it deals with vast impurities of dynamic data, nearly always from a particular business, governmental, or scientific research goal. A now canonical programmatic piece explains, "Scientists can reformulate and rerun their experiments should they find that the initial design was inadequate. Database managers rarely have the luxury of redesigning their data fields and recollecting the data."[15] Establishing the legitimacy of KDD meant demonstrating that lack of luxury and showing its techniques to be productive and meaningful in dealing with such challenging data.

The contrast with statistics and machine learning, however polemically overdrawn, is philosophical, methodological, and institutional; data mining involves a different scientific "way of life" with far different epistemological virtues from its antecedent disciplines. The contrast is also historical. As a field, data mining had statistical concerns and used numerous statistical techniques but did not emerge from a statistical culture; it emerged less from a culture self-consciously attempting to be a branch of mathematics than one uniting the practices of machine learning and of the management and use of large-scale databases for corporations and scientific researchers.

The sweet smell of positivism wafted over data mining from the start. Through data unwedded to theory, data mining promises to overcome wonted ways of dividing the world:

> With traditional statistical modeling, an analyst would pose a question such as: 'Are higher-income people prone to be more loyal to a warehouse club than those with lower income levels?' and the hypothesis would either be supported or unsupported. Data mining, on the other hand, potentially would provide more insight by pointing out other factors contributing to store loyalty that the analyst would not otherwise have been able to consider testing.[16]

Science studies practitioners are not the only ones likely to question such claims. One of the great figures of AI, John McCarthy, wrote in a presentation composed after spending time in IBM's database laboratory in San José, that positivistic data mining was "BAD PHILOSOPHY and INADEQUATE COMPUTER SCIENCE."[17]

COMPUTATIONAL CONSTRAINTS AND DATABASE CULTURE

The notes for a lecture course on data mining at Stanford in 2000 detail the various communities involved in data mining: statistics, artificial intelligence, "visualization researchers," and "databases." Like the IBM researchers, the prominent Stanford database researcher Jeffrey Ullman explained,

> We'll be taking this [database] approach, of course, concentrating on the challenges that appear when the data is large and the computations complex. In a sense, data mining can be thought of as algorithms for executing very complex queries on non-main-memory data.[18]

These concerns, in turn, encouraged an environment in which creative database practitioners adapted analytical processes from elsewhere in computer science and computational statistics to allow them to work with much bigger realms of data.

While futurists and politicians opined endlessly about the utopian possibilities of virtual spaces, cyber-realms, and dystopian visions of alternate reality matrices, database practitioners had the essential job of working within the constraints of materialized systems to enable the preservation, storage, and quick access to data. Disk access times, efficient use of memory, component failure, and distributed and parallel computing are central concerns in nearly every database paper and the maintenance of working databases.[19]

Sophisticated statistical and machine learning algorithms are typically devised for sets of data that can easily fit in memory or that require a relatively small use of slower disk access. Adapting such algorithms to huge quantities of data that cannot be held in memory is nontrivial: different epistemic values and metrics for gauging difficulty and success are brought into play. Many of the developments of greatest significance to data mining were efforts to choose among the tradeoffs necessary to make statistical and machine learning algorithms scale.

One constraint is of the utmost importance for database miners. The database cannot be entirely stored in a computer's memory: "the number of times we need to read each datum is often the best measure of the running time of the algorithm."[20] A generation of practitioners devoted tremendous effort to reworking algorithms to reduce disk access time and the use of memory. The algorithms of big data, deeply indebted to an earlier generation of machine learning tools, matter just because they were made to scale. And whatever the increase of processor speed, successes in parallelizing processing, and the availability of memory, the size and dimensionality of data keep these values central for large data mining operations.

Concerns with computational efficiency and the constraints of disk access times were no mere surface phenomenon: they were deeply inscribed into the algorithms themselves and their implementations in real, faulty machines.

ASSOCIATION MINING: APRIORI AND QUEST

Suppose you run a large grocery store chain. For merchandising purposes, you'd like to know which items customers tend to purchase together, so that you could optimally place those items in your store to maximize revenue.

You have an enormous database that collects all the items purchased together in every transaction. The "market-basket problem" is to find items that tend to be purchased together. More formally, the problem asks us to discern from our database a series of "association rules" where the presence of some set of items suggests a highly likelihood of another, so that $\{X_1, X_2, X_3, X_4, \ldots, X_n\} \to Y$ with a high degree of probability. Most famously, researchers working on a data set from the Osco drug store chain discovered that $\{diapers\} \to beer$, that is, there's a fairly high probability that someone buying diapers is likely to buy beer. Note that the converse, $\{beer\} \to diapers$, has a much lower probability.[21]

Researchers in the Quest group at IBM's Almaden research lab in San José developed an algorithm called Apriori to discover just such association rules to deal with the market-basket problem. The researchers in Quest cast their investigations in terms of how "database technology should be enhanced," notably, "classification, queries on sequences, successive query refinements, query language extensions and optimization as some of the key technical issues in this area." Already in 1991 the group had experienced some success with "an algorithm for discovering rules (patterns) in a historical database of customer transactions."[22] The latter would eventually become the Aprioi association algorithm, published in the flagship journal of database management in 1993.[23]

Soon dubbed one of the "top ten data mining algorithms," Apriori spurred a broad range of researchers in industry and academia to improve it, apply it to new domains, and to discern new uses for it. Of little interest within the more rarified domains of artificial intelligence, machine learning, and statistics, such association algorithms accounted for a sizeable percentage of early successes of the data mining community, quickly moving from the drug store to genomics and beyond.

Association algorithms produced rules in great number, indeed in huge numbers. This turned out to be a boon. In an interview, Agrawal explained,

> When we started doing data mining, we were concerned that we were generating too many rules, but the companies we worked with said, "this is great, this is what exactly what we want!" The prevailing mode of decision making was that somebody would make a hypothesis, test if the hypothesis was correct, and repeat the process. Once they had data mining tools, the decision-making process changed. Now they could use a data mining algorithm to generate all rules, and then debate which of them were valuable.[24]

Traditional statistical notions of significance had little place here. Instead, the focus was on the value of "interestingness."[25]

The IBM Quest team worked from the "perspective of database mining as the confluence of machine learning techniques and the performance emphasis of database technology." The confluence demanded that the two fields influence one another: "Unfortunately the database systems of today offer little functionality to support such 'mining' applications. At the same time, statistical and machine learning techniques usually perform poorly when applied to very large data sets."[26] Most data mining algorithms were drawn from statistics or the branch of artificial intelligence called machine learning, typically housed within computer science departments. Even as they shared an interest in algorithms and their use on sets of data, practitioners of the three diverse fields had—and have—radically different goals and values when implementing and transforming these basic algorithms. No database practitioner would discuss an algorithm without a focus on scale. For a database practitioner, scale means reworking an algorithm to deal with large volumes of data on real computers taking nontrivial time to perform various operations.

Computational statisticians, machine learning specialists, and database engineers have produced a large number of different algorithms to perform classification, decision trees, neural nets, clustering algorithms, various forms of regression, "support vector machines," and so forth. Algorithms requiring both data and human classification of some of that data, called "training data," are examples of "supervised learning." Even more radical, in many ways, were algorithms that simply worked on the data without requiring a "domain-specific" expert to classify some cases. Such approaches are known as "unsupervised learning," and from them a million positivistic promises to funders have sprung. No human "trains" the computer to make certain associations or classifications by giving it human produced classifications of the data.

Database researchers could not simply import algorithms as described in statistical papers or machine learning textbooks; scaling algorithms usually required radical reworking, both of code and of values at stake. An early paper on the topic explains that approaches in machine learning and statistics "do not adequately consider the case that the dataset can be too large to fit in main memory. In particular, they do not recognize that the problem must be viewed in terms of how to work with a [sic] limited resources (e.g., memory that is typically much smaller than the size of the dataset) to do the clustering as accurately as possible while keeping I/O [input/output] costs low."[27] Recognizing such constraints changes the problem to be solved:

We adopt the problem definition [of clustering] used in Statistics, but with an additional database-oriented constraint: The amount of memory available is limited (typically, much smaller than the data set size) and we want to minimize the time required for I/O. A related point is that it is desirable to be able to take into account the amount of time that a user is willing to wait for the results of the clustering algorithm.[28]

Providing a solution to this problem demands a radically different kind of algorithm with a choice of statistical trade-offs that made analyzing larger sets of data tractable. Distance-based approaches to clustering, for example, "assume that all data points are given in advance and can be scanned frequently. They totally or partially ignore the fact that not all data points in the dataset are equally important with respect to the clustering purpose, and that data points which are close and dense should be considered collectively instead of individually."[29] From the database perspective, not all the data points could be treated the same, and so they *should* not be. The algorithm creates something called a "CF tree," a condensation of the total data that balances the value of accounting for every data vector with the need to fit within memory and reduce scans of the disk. The authors explain the balance of the values. The algorithm is

1. fast because (a) no I/O operations are needed and (b) the problem of clustering the original data is reduced to a smaller problem of clustering the subclusters in the leaf entries;
2. accurate because (a) a lot of outliers are eliminated and (b) the remaining data is reflected with the finest granularity *that can be achieved given the available memory*.[30]

Further developing this approach to extending clustering algorithms to large datasets, the (then) Microsoft researcher Usama Fayyad and several collaborators offered a yet more demanding set of *"Data Mining Desiderata"*:

1. **One scan:** The algorithm requires at most one database scan with early termination highly desirable.
2. **Anytime algorithm:** The algorithm is always able to provide a "best" answer at anytime during its computation (i.e. it exhibits "online, anytime" behavior).
3. **Interruptible and Incremental:** The algorithm is suspendable, stoppable and *resumable*. Incremental progress can be saved for continued computation later, possibly on new data.
4. **Limited RAM Requirement:** The algorithm works within the confines

of a limited memory (RAM) buffer, allocated by the user, insuring good behavior when run as a server process.

5. **Forward-only cursor:** The algorithm has the ability to operate with a forward-only cursor over a view of the database.[31]

Always attending to input/output and memory constraints, these practitioners underscored the value of considering different provisional runs of an algorithm. Rather than waiting for an algorithm to complete and issue a final result, a provisional result can always be rendered. The result of nearly two decades of work now figures centrally in courses on mining massive data sets, including the descendant of the Stanford CS course described above.

In the late 1990s, IBM's Almaden lab in San José hosted an ongoing seminar series that brought in academic researchers, industrial researchers, and IBM's own employees, and more generally served as a center of sociability for the local data mining community.[32] Many of the transformative papers presented there would become standard works in scaling statistical and machine learning algorithms for use in real machines with large data sets.

One Wednesday morning in November 1997, a Stanford computer science graduate student came south to Almaden to speak on the topic of "Mining the Web":

A new project at Stanford is the WebBase project. The goals are to collect a large amount of data from the Web and to make it available for research. While the project is relatively new (several months), it has already produced some interesting results.

The speaker, Sergey Brin, was the organizing force of a data mining group, MIDAS—**MI**ning **Da**ta **A**t **S**tanford—which had the support of several faculty members, each a pioneer in database management. At its regular meeting, the MIDAS group discussed the state of the field, from algorithms to ethics: "Topics range from admistrative [*sic*] issues and grant proposals to conference-style presentations by students and visitors."[33] For his talk at IBM, Brin promised to range widely over their work on the web.

I will talk about some of the things we have discovered with this data and some algorithms that have been developed including link analysis, quality filtering, searching and phrase detection.

The project would soon bear much algorithmic fruit.[34] And, before long, many billions of dollars. The webpage for the Stanford Data Mining group notes, "The most impressive and useful demo is the super search engine, called Google, built by Larry Page and Sergey Brin."[35]

THE GENEALOGY OF PAGERANK

Numerous computer science communities were underprepared in the 1990s to contend with the vast, expanding, and decidedly noncurated World Wide Web. A generation of work dedicated to producing reliable, consistent corporate databases and data warehouses had not created tools nearly adequate for the Web. The field of Information Retrieval, dedicated precisely to the creation of tools for search and extraction, had created numerous techniques for indexing and querying in what had previously been considered large databases, such as library catalogs or indices of journals.[36] Those search and indexing tools were designed for highly standardized, curated, centralized collections of text or other data such as a collection of journals with their metadata; they struggled with both the nonstandard and unstructured quality of Web pages and their number.[37] Like many machine-learning algorithms, information retrieval algorithms did not scale easily to the number of pages in the Web. Early search engines tended to index the words in documents, which proved disappointing and was easily manipulated simply by adding long lists of popular words to a webpage. Web-search positivism of the simplest kind had failed. By the mid-1990s, search seemed to many an unpromising approach to the Web, and the major industry players focused increasingly on curated portals, exemplified by the approach of Yahoo. Search came to dominate after 2000, with the gradual, then exponential, rise of Google. (This short chapter must skip over Jon Kleinberg's simultaneous invention of a similar approach to ranking search results using indices of authorities.)[38]

In 1998, Brin, in the database group at Stanford, and his fellow graduate student Larry Page, in the human-computer interactions group, attempted to devise newer association algorithms for a generalized market-basket problem. Rather than looking at items in consumers' baskets, they looked for associations within documents on the Web. Their approach, called "dynamic data mining," did not "exhaustively explore the space of all possible association rules"—as the web was far too big to do so:

> When standard market basket data analysis is applied to data sets other than market baskets, producing useful output in a reasonable amount of time is very difficult. For example, consider a data set with tens of millions of items and an average of 200 items per basket. . . . A traditional algorithm could not compute the large itemsets in the lifetime of the universe.[39]

Just as machine learning algorithms had to change to deal with the scale of early database mining, association mining algorithms had to change to deal with the scale of the early Web. Combining his long-standing interests in mathematics with the values and concerns of a database group, Brin undertook, in his earliest publication, to modify classical statistical algorithms for "nearest neighbor search" to contend with "large metric spaces"—that is, to be able to deal with very large databases with high-dimensional vectors.[40] In their adaptation of such an approach to commercial databases, Brin and Page exemplified the drive of database practitioners to minimize disk and memory usage.

> To make an entire pass over the data for every set of candidates is prohibitively expensive. . . . Instead of making mining a finite, closed-ended process which produces a well defined, complete set of items, we make it a continuous process, generating continually improving sets of itemsets . . . . the process takes advantage of intermediate counts to form estimates of itemsets' occurrence so there is no need to wait until the end of a pass to estimate an itemset's weight.[41]

They likewise recognized the value of designing algorithms capable of yielding provisional answers as they continued to scan the data.

Brin and Page, along with their collaborators, argued that the scale of the Web, which made it so challenging, simultaneously made it deeply promising:

> We take advantage of one central idea: the Web provides its own metadata through its link structure, anchor text [the visible name of the link], and partially redundant content. This is because a substantial portion *of* the Web is *about* the Web. To take full advantage of this data would require human intelligence or more. However, simple techniques that focus on a small subset of the potentially useful data can succeed due to the scale of the Web.[42]

Based within a database community deeply interested in transforming existing statistical and machine learning techniques, Brin and his collaborators were prepared not just to deal with scale, but to make it into a central resource for discovery. Fundamentally, they realized that the scale of the Web included vast human efforts to classify and categorize the Web in billions of piecemeal ways. Rather than creating any form of artificial intelligence capable of classifying the Web itself, they created a mechanism for leveraging human judgment at great scale.

Brin and Page's greatest breakthrough in mining the Web came in adapting an everyday academic practice into algorithmic form most fruitful at vast scales. Following an insight of Page's, they adapted the idea of counting high-quality citations to gauge the authority or value of academic work. Web pages could be "ranked" as more or less authoritative by counting citations, that is, links to pages. More authoritative pages are those that have been linked to by other authoritative pages. The total numbers of links to a page counted far less than the authority of the pages linking to that page. They called the result PageRank, and they soon made it central to a new search engine, Google.

Google search emerged from within a culture fusing database values and practice with machine learning. Much scholarship on Google search has focused tightly on the clever development and implementation of PageRank as a problem in linear algebra and computation of Markov chains.[43] Brin and Page recognized from the start the need for structuring databases capable of implementing the beautiful mathematics on fallible and limited machines.

> Google's data structures are optimized so that a large document collection can be crawled, indexed, and searched with little cost. Although, CPUs and bulk input output rates have improved dramatically over the years, a disk seek still requires about 10 ms to complete. Google is designed to avoid disk seeks whenever possible, and this has had a considerable influence on the design of the data structures.[44]

A process for leveraging human judgment at mass scale, PageRank had to be materialized in a creatively designed set of databases. PageRank and its instantiation within commodity hardware in time led to the development of new architectures for distributed databases and distributed analytic processing, called BigTable and MapReduce respectively. These developments of these approaches figure centrally in the distributed storage and processing of big data today.

In a 2009 piece, "The Unreasonable Effectiveness of Data," three Google affiliates premised the power of huge data sets on the dispersed human curation of the data available "in the wild."[45] Like the IBM database researchers happy to find lots of interesting results in the data, the Google researchers praise the discovery of the multitude of rules found, say, in understanding translation. Memorization, not first principles, have allowed the advances in machine translation, they argue:

> Instead of assuming that general patterns are more effective than memorizing specific phrases, today's translation models introduce general rules only when they improve translation over just memorizing particular phrases (for instance,

in rules for dates and numbers). Similar observations have been made in every other application of machine learning to Web data.

The authors craft from this a conception of linguistics itself, as a new form of big data natural history:

> For those who were hoping that a small number of general rules could explain language, it is worth noting that language is inherently complex, with hundreds of thousands of vocabulary words and a vast variety of grammatical construc-tions. Every day, new words are coined and old usages are modified. This suggests that we can't reduce what we want to say to the free combination of a few abstract primitives.[46]

In their celebration of drawing upon vast arrays of data from the Web, the Google researchers explicitly condemned several generations of work trying to build up computational linguistics and other fields through the arduous process within artificial intelligence of creating generalized "ontologies." The granularity of knowledge—of language and many other domains—becomes evident far better through leveraging the human curation of information. Data mining has thus yielded a new positivism, grounded less in the data themselves than in a billion classifications taken together. Truly this was a new science grounded in empirical specificity of the archive, with the help of its plentiful, often contradictory, cacophonous finding aids.

VOLUME WILL NEVER WIN, FOR VOLUME IS OUR FRIEND

Brin and Page weren't the only people recognizing the dangers and poten-tial of volume in the late 1990s. A 1996 interview in the highly classified house magazine of the US National Security Agency with a deputy director turned to the question of the volume of world communication to be spied upon:

> Let me add to all of that the third biggest challenge facing us, and that is vol-ume. And I could just end the sentence there and everything is said. [Paragraph redacted.]
>
> That gives you some idea of the daunting challenge volume presents, forcing us to look for new technologies.

However great a challenge, volume, the deputy director underscored, would not destroy signals intelligence: "Volume will never win, the reason being that volume is not the only way the world is constructed."[47] By 2006, a top-secret email "Volume Is Our Friend" suggests a newfound confidence in the agency's ability to contend with data overload: indeed, the enabling quality central to the celebration of big data elsewhere. The bigger the volume, the better.

NOTES

1. Usama Fayyad, "Mining Databases: Towards Algorithms for Knowledge Discovery," *Bulletin of the Technical Committee on Data Engineering* 21, no. 1 (1998): 48.

2. Ibid.

3. Usama Fayyad, "Taming the Giants and the Monsters: Mining Large Databases for Nuggets of Knowledge," *Database Programming and Design* 11, no. 3 (1998).

4. For accounts of earlier history of data, see David Sepkoski, "Towards 'a Natural History of Data': Evolving Practices and Epistemologies of Data in Paleontology, 1800–2000," *Journal of the History of Biology* 46, no. 3 (August 2013): 401–44, doi:10.1007/s10739-012-9336-6; Bruno J. Strasser, "Data-Driven Sciences: From Wonder Cabinets to Electronic Databases," *Studies in History and Philosophy of Science Part C: Studies in History and Philosophy of Biological and Biomedical Sciences* 43, no. 1 (March 2012): 85–87, doi:10.1016/j.shpsc.2011.10.009; Paul Edwards, *A Vast Machine: Computer Models, Climate Data, and the Politics of Global Warming* (Cambridge: MIT Press, 2010). See also Sepkoski (chapter 2) and Strasser (chapter 7) in this volume.

5. For data preparation, see Vladimir Janković (chapter 9) and Bruno Strasser (chapter 7) in this volume.

6. Ralph Kimball, "The Database Market Splits," http://www.kimballgroup.com/1995/09/01/the-database-market-splits/, accessed June 18, 2013.

7. The primary academic histories of database systems are Thomas J. Bergin and Thomas Haigh, "The Commercialization of Database Management Systems, 1969–1983," *Annals of the History of Computing, IEEE* 31, no. 4 (2009): 26–41; Thomas Haigh, "How Data Got Its Base: Information Storage Software in the 1950s and 1960s," *Annals of the History of Computing, IEEE* 31, no. 4 (2009): 6–25; Avi Silberschatz, Michael Stonebraker, and Jeffrey D. Ullman, "Database Systems: Achievements and Opportunities," *ACM Sigmod Record* 19, no. 4 (1990): 6–22; more generally, see Richard L. Nolan, "Information Technology Management Since 1960," in *Nation Transformed by Information: How Information Has Shaped the United States from Colonial Times to the Present*, ed. Alfred Dupont Chandler and James W. Cortada (New York: Oxford University Press, 2000), 217–56; and for the very recent history, see Jeff Hammerbacher, "Information Platforms and the Rise of the Data Scientist," in *Beautiful Data: The Stories Behind Elegant Data Solutions*, ed. Toby Segaran and Jeff Hammerbacher (Beijing, China, and Sebastopol, CA: O'Reilly Media, 2009), 73–84. Fine studies of the profound effects of database design on scientific cultures include Hallam Stevens, *Life Out of Sequence: A Data-Driven History of Bioinformatics* (Chicago: University of Chicago Press, 2013), 139–41, 161, 168–69; Sabina Leonelli and Rachel A. Ankeny, "Re-Thinking Organisms: The Impact of Databases on

Model Organism Biology," *Studies in History and Philosophy of Science Part C: Studies in History and Philosophy of Biological and Biomedical Sciences* 43, no. 1 (March 2012): 29–36, doi:10.1016/j.shpsc.2011.10.003. See also Geoffrey C. Bowker, *Memory Practices in the Sciences* (Cambridge, MA: MIT Press, 2005).

8. Silberschatz, Stonebraker, and Ullman, "Database Systems," 11.

9. Kimball, "The Database Market Splits."

10. Ibid.

11. For a good sense of the array of research projects in addition to data mining, see Jim Gray, "The Next Database Revolution," in *Proceedings of the 2004 ACM SIGMOD International Conference on Management of Data*, 2004, 1–4, http://dl.acm.org/citation.cfm?id=1007570.

12. Marianne Winslett and Rakesh Agrawal, "Rakesh Agrawal Speaks Out on Where the Data Mining Field Is Going, Where It Came From, How to Choose Problems and Open Up New Fields, Our Responsibilities to Society as Technologists, What Industry Owes Academia, and More," 2003, http://www.sigmod.org/publications/interview/pdf/D15.rakesh-final-final.pdf.

13. Gray, "The Next Database Revolution."

14. While "data mining" and "KDD" are usually used interchangeably, the term "KDD" was introduced to clarify that the most ideational algorithmic processes of processing data are only a small subset of wide ranges of processes needed to produce "knowledge" from large databases; see Usama M. Fayyad, Gregory Piatetsky-Shapiro, and Padhraic Smyth, "From Data Mining to Knowledge Discovery: An Overview," in *Advances in Knowledge Discovery and Data Mining*, ed. Usama M. Fayyad et al. (Menlo Park, CA: AAAI/MIT Press, 1996), 1–34.

15. William J. Frawley, Gregory Piatetsky-Shapiro, and Christopher J. Matheus, "Knowledge Discovery in Databases: An Overview," in *Knowledge Discovery in Databases*, ed. Gregory Piatetsky-Shapiro (Cambridge, MA: AAAI/MIT Press, 1991), 8.

16. Shawn Thelen, Sandra Mottner, and Barry Berman, "Data Mining: On the Trail to Marketing Gold," *Business Horizons* 47 (2004): 26, doi:10.1016/j.bushor.2004.09.005.

17. John McCarthy Papers (SC0524) Department of Special Collections and University Archives, Stanford University Libraries, Stanford, CA; Accession 2012-055, box 11, folder 11. He published a version of this paper: John McCarthy, "Phenomenal Data Mining: From Data to Phenomena," *ACM SIGKDD Explorations Newsletter* 1, no. 2 (2000): 24–29.

18. Jeffrey D. Ullman, "CS 345 Data Mining Lecture Notes" (unpublished, 2000), 1, https://web.archive.org/web/20030623052236/http://www-db.stanford.edu/~ullman/mining/overview.pdf.

19. For the "materiality" of data and its significance, see especially Edwards, *A Vast Machine*, e.g. at 83–85, 97.

20. Ullman, "CS 345 Data Mining Lecture Notes," 3.

21. Dan Power, "Origins of Beer & Diapers," *DSS News* 3, no. 23 (2002), http://www.dssresources.com/newsletters/66.php; Ronny Kohavi, "Origin of 'Diapers and Beer,'" July 6, 2000, http://www.kdnuggets.com/news/2000/n14/8i.html.

22. Laura M. Haas and Patricia G. Selinger, "Database Research at the IBM Almaden Research Center," *SIGMOD Rec.* 20, no. 3 (September 1991): 97,

doi:10.1145/126482.126493. Research on Quest enabled by Wayback machine caches of IBM Quest website from the mid to late 1990s.

23. Rakesh Agrawal, Tomasz Imielinski, and Arun Swami, "Mining Association Rules between Sets of Items in Large Databases," *Proceedings of the 1993 ACM SIGMOD International Conference on Management of Data*, 1993, 207–16, doi:10.1145/170035.170072.

24. Winslett and Agrawal, "Rakesh Agrawal Speaks Out."

25. Debate around "interestingness" continues around association mining. For a mid-1990s perspective, see the important Avi Silberschatz and Alexander Tuzhilin, "What Makes Patterns Interesting in Knowledge Discovery Systems," *IEEE Transactions on Knowledge and Data Engineering* 8, no. 6 (1996): 970–74, doi:http://doi.ieeecomputersociety.org/10.1109/69.553165.

26. R. Agrawal, T. Imielinski, and A. Swami, "Database Mining: A Performance Perspective," *IEEE Transactions on Knowledge and Data Engineering* 5 (1993): 914.

27. Tian Zhang, Raghu Ramakrishnan, and Miron Livny, "BIRCH: An Efficient Data Clustering Method for Very Large Databases," in *Proceedings of the 1996 ACM SIGMOD International Conference on Management of Data, Montreal, Quebec, Canada, June 4–6, 1996*, ed. H. V. Jagadish and Inderpal Singh Mumick (ACM Press, 1996), 104.

28. Ibid., 103.

29. Ibid., 104.

30. Ibid., 107.

31. Paul S. Bradley, Usama M. Fayyad, and Cory Reina, *Scaling EM (Expectation-Maximization) Clustering to Large Databases*, Technical Report MSR-TR-98-35 (Microsoft Research, 1998), 2.

32. The list of seminars as of the end of 1998 are available at http://web.archive.org/web/19990116232602/http://www.almaden.ibm.com/cs/quest/seminars.html and http://web.archive.org/web/19980210042739/http://www.almaden.ibm.com/cs/quest/seminars-hist.html.

33. The webpage for MIDAS is preserved at http://infolab.stanford.edu/midas/; a listserv of the data mining group can be found on Yahoo e-groups. See Jeffrey D. Ullman, "The MIDAS Data-Mining Project at Stanford," in *Database Engineering and Applications, 1999, IDEAS'99, International Symposium Proceedings*, 1999, 460–64, http://ieeexplore.ieee.org/xpls/abs_all.jsp?arnumber=787298.

34. A printed version of this material appeared as Sergey Brin, Rajeev Motwani, and Terry Winograd, "What Can You Do with a Web in Your Pocket," *Data Engineering Bulletin* 21 (1998): 37–47.

35. http://infolab.stanford.edu/midas/.

36. For the early history of some of these tools, see Daniel Rosenberg (chapter 11) in this volume.

37. Thomas Haigh, "The Web's Missing Links: Search Engines and Portals," in *The Internet and American Business*, ed. William Aspray and Paul Ceruzzi (Cambridge, MA: MIT Press, 2008), 160–61. S. Brin and L. Page, "The Anatomy of a Large-Scale Hypertextual Web Search Engine," in *Seventh International World-Wide Web Conference (WWW 1998)*, 1998, http://ilpubs.stanford.edu:8090/361/, §3.1. "Most of the research on information retrieval systems is on small well controlled homogeneous collections such as collections of scientific papers or news stories on a related topic."

38. Jon M. Kleinberg, "Authoritative Sources in a Hyperlinked Environment," *Journal of the Association of Computing Machinery* 46, no. 5 (September 1999): 604–32, doi:10.1145/324133.324140.

39. Sergey Brin and Lawrence Page, *Dynamic Data Mining: Exploring Large Rule Spaces by Sampling*, Technical Report (Stanford InfoLab, November 1999), 2, http://ilpubs .stanford.edu:8090/424/.

40. Sergey Brin, "Near Neighbor Search in Large Metric Spaces," *21st International Conference on Very Large Data Bases (VLDB 1995)*, 1995, http://ilpubs.stanford.edu:8090 /113/. The historical section of Brin's paper offers a rich multiple genealogy of such efforts.

41. Brin and Page, *Dynamic Data Mining*, 7.

42. Brin, Motwani, and Winograd, "What Can You Do with a Web in Your Pocket," 2.

43. Amy N. Langville and Carl D. Meyer, *Google's PageRank and Beyond: The Science of Search Engine Rankings* (Princeton: Princeton University Press, 2006).

44. Brin and Page, "The Anatomy of a Large-Scale Hypertextual Web Search Engine," §4.2.

45. For the human transformation of archival materials, see Rebecca Lemov (chapter 10) and Liba Taub (chapter 4) in this volume.

46. A. Halevy, P. Norvig, and F. Pereira, "The Unreasonable Effectiveness of Data," *Intelligent Systems, IEEE* 24, no. 2 (April 2009): 12, doi:10.1109/MIS.2009.36.

47. (redacted author name), "Confronting the Intelligence Future (U) An Interview with William P. Crowell, NSA's Deputy Director (U)," *Cryptolog* 22, no. 2 (1996): 1–5.

The Time of the Archive

Lorraine Daston

Every scientific archive has a timeline. As long as the history of life on earth, or as short as the meteorological havoc wrought by the latest El Niño; reaching deep into astronomical past and future; spanning generations of natural philosophers since Thales and physicians since Hippocrates—the archive is where the scientific past, present, and future converge. More specifically, the archive is the physical expression of how present science creates a usable past for future science.

The word "physical" should be chiseled in stone. Because archival metaphors are rampant—as they always are when a major shift in documentary medium, whether printing in the sixteenth century or digitalization in the twenty-first, awakens the imagination to new opportunities and risks—the literal fact of the archive sometimes turns ghostly. The chapters in this volume remind us of the very palpable, fragile, expensive, and often recalcitrant stuff that archives are made of: the yellowing pages of a nineteenth-century medical journal; the glass photographic plates carefully stored in a French observatory; the electricity-gobbling servers of the US National Climate Program; the frozen blood samples taken from Havasupai tribe; the balky hard-drives of today's data miners. Cast in the image of the humans who create them, archives are mortals with aspirations to immortality.

No archive is forever—not even the natural archives of the earth's fossil record or the human genome. But some archives have been very long-lived: astronomers still make use of ancient Mesopotamian and Chinese observations; historians are still mining Herodotus and Strabo for details about the ancient world; no one knows when a freak storm or earthquake recorded in a medieval chronicle might suddenly become relevant for science right now. Longevity is no accident: a chain of individuals and institutions links Babylonian cuneiform tablets with NASA's *Five Millennium* canons and the fossils piled up in a seventeenth-century cabinet of curiosities with a twentieth-century digital database. At each stage of transmission, key information about the original context in which the archive was compiled might be lost; standards for precision, reliability, and relevance also have their history. Without scholars and scientists, copyists, printers, proofreaders, curators, librarians, archivists, programmers, and the institutions that at every step support them, from monastic scriptoria to the modern university library, the chain would break. Nor is what and who gets into an archive accidental: just because the costs of curating and preserving its contents are so high, absorbing fortunes and lifetimes, selectivity is paramount and access is jealously guarded. Whether in eighteenth-century astronomy or late twentieth-century genetics, it takes titanic acts of disciplinary will to create, sustain, and make public a scientific archive. An archive is a discipline's wager on its own longevity.

It's a bet riskier than any casino gamble. Will the discipline survive with enough resources, human and material, to tend its archive over generations, centuries, perhaps millennia? Will the archive survive the inevitable changes in instruments, techniques, and media? Above all, will the archive contain what the future disciplinary community wants to know? These questions are still more acute for personal archives like that of the quantified self movement, which is perhaps why some are donated to more enduring institutions like universities in the hopes that future social scientists will convert them into disciplinary archives.

Time is the archive's enemy. Under the right conditions, a medieval parchment codex may last for centuries, nineteenth-century acidic paper crumbles in decades, a compact disk deteriorates in thirty years, and does anyone remember the floppy disk? Voluminous social science archives of the mid-twentieth century are embalmed in the once futuristic medium of microfiche. In the first flush of excitement over the possibilities of astrophotography in the late nineteenth century, the revelations of spectroscopy and the rise of astrophysics were not so much as a gleam in the astronomer's eye. Civilizations rise, civilizations fall, libraries burn and fall to ruins. Less grandly, government grants dry up, the brightest students want MBA's rather than Ph.D.'s, and disciplines wither. But time is also the archive's friend. At any moment, a new research question—Is this really a brand new disease? Were there mass extinctions? Does dark matter exist?—can rouse a Sleeping Beauty archive from its slumbers. The very fact that the archive was in all likelihood not originally compiled to answer these questions only increases the value of its evidence: serendipity is the best guarantee of lack of bias.

Against the odds, scientific archives bet on continuity. More than that: they enforce continuity. Without continuity of practices, the archive would not just slumber from time to time; it would sink into a coma. Stable practices of collecting, selecting, canonizing, scrubbing, and ordering data insure that the contents of archives are commensurable and retrievable. At key junctures—e.g., when new instruments or techniques enable a leap in precision, as in the case of early modern European astronomy—older contents may have to be corrected and edited, but they are rarely expunged from the archive altogether. Even when a media revolution translates manuscripts into print, print into microfilm, microfilm into digital documents, the older archive material is not discarded. Every copying process, including high-quality digital scans, introduces error (as proofreaders of astronomical tables or genetic sequences or even Google books know to their woe). Every migration between media loses metadata; every material repository of the archive may embed valuable clues about the original context in which information was collected, whether in the chemistry of the ink that fixes an approximate date or the arrangement of specimens in drawers and vitrines that hints at taxonomic assumptions. New finding aids are often deliberately parasitic upon old ones: the venerable methods used to compile Biblical concordances and subject indices underpin digital indexing tools; the Google PageRank algorithm mimics the time-tested citation practices of scholars and scientists. These practices are as much a part of the scientific archives as their contents.

Archival practices are surprisingly uniform across disciplines, evidence of both intertwined histories that antedate modern classifications of knowledge and resilience in the face of periodic media ruptures. An index is an index is an index is an index, whether you're a paleontologist or a medievalist, and it's as useful online as it is on paper. Familiar divisions among the natural sciences, the social sciences, and the humanities begin to teeter when the focus shifts from subject matter to practices, and that is one reason why this volume has not respected such conventional boundaries. The same holds in spades for current developments in more applied data mining and management, whether for the insurance industry or the National Security Agency, in which the line between tools developed for research and for commerce has become so faint as to vanish.

As historians of science have pointed out in other contexts, practices have a chronology that is all their own, slower and steadier than the allegretto of new empirical results or even the andante progression of new theories. Once tabular display of data or laboratory experiment or statistical significance tests or other practices entrench themselves in a discipline, they endure. Indeed, they discipline the disciplines, as the essential handiwork that every advanced student masters in seminars to certify competence as a practitioner. The continuity of archival practices is simply the extreme point on the spectrum of scientific practices passed on from generation to generation.

Like other scientific practices, archival practices have their history, partly told in these pages, but it is of necessity a longer, slower history. Accelerate the tempo, and the continuity of the archive would be endangered. The attentive reader of this volume will have noticed passages in which the past and present blur together in the description of archival practices: databases before the computer, data mining before algorithms, medieval concordances that resemble printouts. Historically tuned ears will register such conflations of past and present as anachronisms, and the wider the chronological gap, the more jarring the discord. When gap widens to chasm, anachronism tips into comedy à la Monty Python: ancient Romans in togas checking their cell phones. But perception of historical anachronism presupposes a timeline measured on a human scale, upon which points separated by more than a few generations belong to different epochs governed by different rules. The time of archives ticks more slowly. It stretches human time into the far past and the far future. Continuity demands that new practices mirror old ones, however faster, cheaper, and more efficient new media may make them. Archival practices deliberately cultivate anachronism: pasts, presents, and futures merge.

This volume began with the observation that we are currently in the midst of an archival moment. It would perhaps be more accurate to say that we are in a moment of archival anxiety, compounded of hope and fear. New digital media rekindle old dreams of bigger, better archives of everything (and everyone); they also revive old nightmares of too much to know and the irretrievable loss of what we once knew. Dreams and nightmares are two sides of the same coin: technological time has speeded up; can the time of archives keep pace? What is to be gained and lost by tinkering with the clock? The cautious conservatism of archival practices must be balanced against the restless progressivism of technology and science: no scientific archive can stand still, any more than it can reinvent itself tabula rasa. The balancing act is delicate—and precarious. We have been here before. The long history of science in the archives is punctuated by such moments of anxiety, when human time collides with archive time.

Lorraine Daston
Max Planck Institute for the History of
    Science
14195 Berlin
Germany

Cathy Gere
Department of History
University of California—San Diego
La Jolla, CA 92093

Florence Hsia
Department of the History of Science
University of Wisconsin—Madison
Madison, WI 53706

Vladimir Janković
Faculty of Life Sciences
The University of Manchester
Manchester M13 9PL
United Kingdom

Matthew L. Jones
Department of History
Columbia University
New York, NY 10027

Rebecca Lemov
Department of the History of Science
Harvard University
Cambridge MA 02138

Suzanne Marchand
Department of Comparative Literature
Louisiana State University
Baton Rouge, LA 70803

J. Andrew Mendelsohn
Department of History
Queen Mary, University of London
London E1 4NS
United Kingdom

Daniel Rosenberg
Robert D. Clark Honors College
University of Oregon
Eugene, OR 97403

David Sepkoski
Max Planck Institute for the History of
    Science
14195 Berlin
Germany

Bruno J. Strasser
Section of Biology
University of Geneva
1211 Geneva 4
Switzerland

Liba Taub
Whipple Museum Department of History
   and Philosophy of Science
University of Cambridge
Cambridge CB2 3RH
United Kingdom

Published sources only: for archival sources consulted, please see citations in the individual papers.

Abbattista, Guido. "The English *Universal History*: Publishing, Historiography, and Authorship in an European Project (1736–1790)." *Storia della Storiografia* 39 (2001): 100–105.

Aber, James S. "Torbern Olaf Bergman." http://academic.emporia.edu/aberjame/histgeol /bergman/bergman.htm.

Académie royale des sciences. *Recueil d'observations faites en plusieurs voyages par ordre de sa majesté, pour perfectionner l'astronomie et la géographie.* Paris: De l'Imprimerie Royale, 1693.

Ackerknecht, Erwin H. *A Short History of Medicine.* New York: Ronald, 1955.

Adams, Mark B., ed. *The Evolution of Theodosius Dobzhansky: Essays on His Life and Thought in Russia and America.* Princeton: Princeton University Press, 1994.

Adams, Robert. "Cases of Diseases of the Heart, Accompanied with Pathological Observations." *Dublin Hospital Reports* 4 (1827): 353–453.

Agrawal, Rakesh, Tomasz Imielinski, and Arun Swami. "Database Mining: A Performance Perspective." *IEEE Transactions on Knowledge and Data Engineering* 5 (1993): 914.

Agrawal, Rakesh, Tomasz Imielinski, and Arun Swami. "Mining Association Rules between Sets of Items in Large Databases." *Proceedings of the 1993 ACM SIGMOD International Conference on Management of Data*, 1993, 207–16. doi:10.1145/170035.170072.

Aiden, Erez, and Jean-Baptiste Michel. *Uncharted: Big Data as a Lens on Human Culture.* New York: Riverhead Books, 2013.

Algra, Keimpe A. "Chrysippus, Carneades, Cicero: The Ethical *Divisiones* in Cicero's *Lucullus*." In *Assent and Argument: Studies in Cicero's Academic Books*, proceedings of the 7th Symposium Hellenisticum, Utrecht, August 21–25, 1995, edited by Brad Inwood and Jaap Mansfeld, 107–39. Leiden: Brill, 1997.

Allen, David Elliston. "Amateurs and Professionals." In *The Cambridge History of Science: The Modern Biological and Earth Sciences*, edited by Peter J. Bowler and John Pickstone, 15–33. Cambridge: Cambridge University Press, 2009.

Allen, David Elliston. *The Naturalist in Britain: A Social History*. London: A. Lane, 1976.

Alsted, Johann Heinrich. *Thesaurus chronologiae*. 2nd ed. Herborn: [Corvinus Erben], 1628.

Althoff, Jochen. "Aristoteles als Medizindoxograph." In *Ancient Histories of Medicine: Essays in Medical Doxography and Historiography in Classical Antiquity*, edited by Philip J. van der Eijk, 57–94. Leiden: Brill, 1999.

Alvarez, Walter. *T. Rex and the Crater of Doom*. Princeton: Princeton University Press, 1997.

Amad, Paula. *Counter-Archive: Film, the Everyday, and Albert Kahn's Archives de la Planète*. New York: Columbia University Press, 2010.

Anderson, Warwick. "The Case of the Archive." *Critical Inquiry* 39, no. 3 (2013): 532–47.

Ankeny, Rachel A. "Using Cases to Establish Novel Diagnoses: Creating Generic Facts by Making Particular Facts Travel Together." In *How Well Do Facts Travel? The Dissemination of Reliable Knowledge*, edited by Peter Howlett and Mary S. Morgan, 252–72. Cambridge: Cambridge University Press, 2011.

Anonymous. "Crystallography Protein Data Bank." *Journal of Molecular Biology* 78 (1971): 587.

Anonymous. "More Bang for Your Byte." *Scientific Data* 1 (2014), accessed July 10, 2014. doi:10.1038/sdata.2014.10.

Anonymous. "Triumph der Symbolik." *Hermes, oder kritisches Jahrbuch der Literatur* 26 (1826): 344–55.

App, Urs. *The Birth of Orientalism*. College Station: University of Pennsylvania Press, 2010.

Aristotle. *The Complete Works of Aristotle*. Edited by J. Barnes. 2 vols. Princeton: Princeton University Press, 1984.

Aronova, Elena, Christine von Oertzen, and David Sepkoski, eds. *Data Histories* Special issue of *Osiris* 32 (2017), forthcoming.

Aronova, Elena, Karen S. Baker, and Naomi Oreskes. "Big Science and Big Data in Biology: From the International Geophysical Year through the International Biological Program to the Long Term Ecological Research (LTER) Network 1957–Present." *Historical Studies in the Natural Sciences* 40 (2010): 183–224.

Assmann, Jan. *Cultural Memory and Early Civilization: Writing, Remembrance, and Political Imagination*. Cambridge: Cambridge University Press, 2011.

Atmospheric Climate Data: Problems and Promises. Washington DC: National Academy Press, 1986.

Aubin, David, Charlotte Bigg, and H. Otto Sibum, eds. *The Heavens on Earth: Observatories and Astronomy in Nineteenth-century Science and Culture*. Durham: Duke University Press, 2010.

Bailly, Jean-Sylvain. *Histoire de l'astronomie ancienne, depuis son origine jusqu'à l'établissement de l'école d'Alexandrie*. Paris: Frères Debure, 1775.

Bainbridge, John. *Procli sphaera: Ptolemaei de hypothesibus planetarum liber singularis . . . cui accesit ejusdem Ptolemaei canon regnorum*. London: Excudebat Guilielmus Iones, 1620.

Baird, Davis, Alfred Nordmann, and Joachim Schummer, eds. *Discovering the Nanoscale*. Amsterdam: IOS Press, 2004.

Baltussen, Han. "Peripatetic Dialectic." In *Theophrastus: His Psychological, Doxographical, and Scientific Writings*, edited by W. W. Fortenbaugh and D. Gutas. New Brunswick, NJ: Transaction Publishers, 1992.

Balty-Fontaine, J. "Pour une edition nouvelle du 'Liber Aristotelis de inundatione Nili.'" *Chronique d'Egypte* 34 (1959): 95–102.

Balz, Charles F., and Richard H. Stanwood. *Some Applications of the KWIC Indexing System.* Oswego, NY: IBM, 1962.

Bamman, David, Brendan O'Connor, and Noah Smith. "Censorship and Deletion Practices in Chinese Social Media." *First Monday* 17, no. 3 (March 5, 2012). http://journals.uic.edu/ojs/index.php/fm/article/view/3943/3169.

Barnes, Jonathan. "Aristotle and the Methods of Ethics." *Revue Internationale de Philosophie* 133–34 (1980): 490–511.

Barretus, Lucius [Albert Curtz], ed. *Historia coelestis ex libris commentariis manuscriptis observationum vicennalium viri generosi Tichonis Brahe.* Augsburg: Apud Simonem Utzschneiderum, [1666].

Barrow, Mark. "The Specimen Dealer: Entrepreneurial Natural History in America's Gilded Age." *Journal of the History of Biology* 33 (2000): 493–534.

Bartholin, Erasmus. *Specimen recognitionis nuper editarum observationum astronomicarum Tychonis Brahe.* Copenhagen: prostat apud Danielem Paulli: literis Henrici Gödiani, 1668.

Beatty, John. "Dobzhansky's Worldview." In *The Evolution of Theodosius Dobzhansky: Essays on His Life and Thought in Russia and America,* edited by Mark B. Adams, 195–218. Princeton: Princeton University Press, 1994.

Beaulieu, Paul-Alain. "The Astronomers of the Esagil Temple in the Fourth Century BC." In *If a Man Builds a Joyful House,* edited by Erle Leichty and Ann K. Guinan, 5–22. Leiden: Brill, 2006.

Becker, Peter, and William Clark, eds. *Little Tools of Knowledge: Historical Essays on Academic and Bureaucratic Practices.* Ann Arbor: University of Michigan Press, 2001.

Becket, Andrew. *A Concordance to Shakespeare.* London: Robinson, 1787.

Behm, Johannes. *Chronologica manuductio, & deductio annorum.* Frankfurt an der Oder: Impensis Johannis Thymii, 1620.

Behrens, Rudolf, and Carsten Zelle, eds. *Der ärztliche Fallbericht: epistemische Grundlagen und textuelle Strukturen dargestellter Beobachtung.* Wiesbaden: Harrassowitz Verlag, 2012.

Behrisch, Lars. "Zu viele Informationen! Die Aggregierung des Wissens in der frühen Neuzeit." In *Information in der frühen Neuzeit: Status, Bestände, Strategien,* edited by Arndt Brendecke, Markus Friedrich, and Susanne Friedrich, 455–73. Berlin: LIT Verlag, 2008.

Bell, Gordon. "Foreword." *The Fourth Paradigm: Data-Intensive Scientific Discovery.* Edited by Tony Hey, Stewart Tansley, and Kristin Tolle, xi–xv. Redmond, WA: Microsoft Research, 2009.

Bell, Gordon, and Jim Gemmell. *Your Life, Uploaded: The Digital Way to Better Memory, Health, and Productivity.* New York: Plume, 2010.

Bell, Gordon, Tony Hey, and Alex Szalay. "Beyond the Data Deluge." *Science* 323, no. 5919 (2009): 1297–98.

Bensaude-Vincent, Bernadette. *L'opinion publique et la science.* Paris: La Découverte, 2013.

Benson, Etienne. "One Infrastructure, Many Global Visions: The Commercialization and Diversification of Argos, a Satellite-based Environmental Surveillance System." *Social Studies of Science,* published online October 3, 2012. doi: 10.1177/0306312712457851.

Benton, Michael. *The Fossil Record 2.* London: Chapman & Hall, 1993.

Benton, Michael. "The History of Life: Large Databases in Palaeontology." In *Numerical*

Palaeobiology: Computer-based Modelling and Analysis of Fossils and Their Distributions, edited by D. A. T. Harper, 249–83. New York: John Wiley & Sons, 1999.

Bergin, Thomas J., and Thomas Haigh. "The Commercialization of Database Management Systems, 1969–1983." *Annals of the History of Computing, IEEE* 31, no. 4 (2009): 26–41.

Bergman, Torbern Olof. *Physical Description of the Earth (Physisk beskrifning ofvert jordklotet)*. Uppsala: 1766. Quoted in James S. Aber, "Torbern Olaf [sic] Bergman," http://academic.emporia.edu/aberjame/histgeol/bergman/bergman.htm.

Bernal, Martin. *Black Athena: The Afroasiatic Roots of Classical Civilization.* Vol. 1: *The Fabrication of Ancient Greece, 1785–1985.* New Brunswick, NJ: Rutgers University Press, 1987.

Berner, Eta S., ed. *Clinical Decision Support Systems.* New York: Springer, 2007.

Berriman, G. Bruce, and Steven L. Groom. "How Will Astronomy Archives Survive the Data Tsunami?" *Queue* 9, no. 10 (2011): 20–27.

Berville, Saint-Albin. "Éloge de Rollin." In *Oeuvres complètes de Rollin, nouvelle edition, accompagnée d'observations et d'éclaircissements historiques, par M. Letronne.* Vol. 1: *Histoire ancienne.* Paris: Firmin Didot Frères, 1820, xii.

Best, Stephen, and Sharon Marcus. "Surface Reading: An Introduction." *Representations* 108, no. 1 (November 1, 2009): 1–21.

Bianchini, Francesco. *Astronomiae, ac geographicae observationes selectae Romae*, edited by Eustachio Manfredi. Verona: Typis Dyonisii Ramanzini, 1737.

Bigg, Charlotte. "Photography and Labour History of Astrometry: The Carte du Ciel." In *The Role of Visual Representations in Astronomy: History and Research Practice: Contributions to a Colloquium Held at Göttingen in 1999*, edited by Klaus Hentschel and Axel D. Wittmann, 90–106. Thun: Verlag Harri Deutsch, 2000.

Blair, Ann. *Too Much to Know: Managing Scholarly Information before the Modern Age.* New Haven: Yale University Press, 2010.

Blair, Ann, and Jennifer Mulligan, eds. *Toward a Cultural History of the Archives. Archival Science* 7, special issue, 2007.

Bleek, Wilhelm. *Friedrich Christoph Dahlmann: Eine Biographie.* Munich: CH Beck, 2010.

Blok, Josine. "Quest for a Scientific Mythology: F. Creuzer and K. O. Müller on History and Myth." *History and Theory*, suppl. 33 (1994): 26–52.

Blumenbach, Johann Friedrich. "Beyträge zur Naturgeschichte der Vorwelt." *Magazin für das Neueste aus der Physik und Naturgeschichte* 4 (1790): 1–2.

Bolkestein, Hendrik. *Adversaria critica et exegetica ad Plutarchi Quaestionum Convivalium librum primum et secundum.* Amsterdam: H. J. Paris, 1946.

Bonet, Théophile. *Sepulchretum sive Anatomia practica ex cadaveribus morbo denati.* Geneva: Chouet, 1679.

Bonneau, Danielle. "Liber Aristotelis De inundatione Nili." *Etudes de Papyrologie* 9 (1971): 1–33.

Borges, Jorge Luis. "The Analytical Language of John Wilkins." In *Other Inquisitions, 1937–1952*, translated by Ruth L. C. Simms, 101–5. Austin: University of Texas Press, 1965.

Borri, Cristoforo. *Collecta astronomica: ex doctrina P. Christophori Borri, mediolanensis, ex Societate Iesu.* Lisbon: Apud Matthiam Rodrigues, 1631.

Borne, Kirk. "Virtual Observatories, Data Mining, and Astroinformatics." In *Planets, Stars, and Stellar Systems*, edited by Terry D. Oswalt and Howard E. Bond, 2: 403–43. Dordrecht: Springer, 2013.

Boudway, Ira. "Is Chris Dancy the Most Quantified Self in America? One Man's Project to Keep Track of Everything He Sees, Does, Thinks, Eats." *Business Week*, June 5, 2014. http://www.businessweek.com/articles/2014-06-05/is-chris-dancy-the-most -quantified-self-in-america

Bouguer, Pierre. *Justification des memoires de l'Académie royale des sciences de 1744: et du livre de la figure de la terre, déterminée par les observations faites au Pérou, sur plusieurs faits qui concernent les opérations des académiciens*. Paris: Chez Charles-Antoine Jombert, 1752.

Bourne, Charles P. "40 Years of Database Distribution and Use: An Overview and Observation." NFAIS 1999 Miles Conrad Lecture. http://nfais.org/1999-miles-conrad -lecture. Accessed June 17, 2013.

Bourne, Charles P., and Douglas C. Engelbart. "Facets of the Technical Information Problem." *SRI Journal 2*, no. 1 (1958).

Bourne, Charles P., and T. B. Hahn. *A History of Online Information Services: 1963–1976*. Cambridge, MA: MIT Press, 2003.

Bowen, Alan C. *Simplicius on the Planets and Their Motions: In Defense of a Heresy*. Leiden: Brill, 2013.

Bowker, Geoffrey C. *Memory Practices in the Sciences*. Cambridge, MA: MIT Press, 2005.

Brack-Bernsen, Lis. *Zur Entstehung der babylonischen Mondtheorie: Beobachtung und theoretische Berechnung von Mondphasen*. Stuttgart: F. Steiner, 1997.

Bradley, James. *Astronomical Observations, Made at the Royal Observatory at Greenwich, from the Year MDCCL to the Year MDCCLXII*. Edited by Thomas Hornby and Abram Robertson. Oxford: At the Clarendon Press, 1798–1805.

Bradley, Paul S., Usama M. Fayyad, and Cory Reina. *Scaling EM (Expectation-Maximization) Clustering to Large Databases*, Technical Report MSR-TR-98-35. Microsoft Research, 1998.

Brahe, Tycho. *Astronomiæ instauratæ mechanica*. Wandesburg: [Philippi de Ohr], 1598.

Brembs, Björn, Katherine Button, and Marcus Munafò. "Deep Impact: Unintended Consequences of Journal Rank." *Frontiers in Human Neuroscience 7* (2013). doi:10.3389/ fnhum.2013.00291.

Brendecke, Arndt, Markus Friedrich, and Susanne Friedrich, eds. *Information in der Frühen Neuzeit: Status, Bestände, Strategien*. Berlin: LIT Verlag, 2008.

Brin, Sergey. "Near Neighbor Search in Large Metric Spaces." In *VLDB '95 Proceedings of the 21st International Conference on Very Large Data Bases (VLDB 1995)*, 1995. http:// ilpubs.stanford.edu:8090/113/.

Brin, Sergey, Rajeev Motwani, and Terry Winograd. "What Can You Do with a Web in Your Pocket." *Data Engineering Bulletin* 21 (1998): 37–47.

Brin, Sergey, and Lawrence Page. "The Anatomy of a Large-Scale Hypertextual Web Search Engine." In *Seventh International World-Wide Web Conference (WWW 1998)*, 1998, http://ilpubs.stanford.edu:8090/361/.

Brin, Sergey, and Lawrence Page. *Dynamic Data Mining: Exploring Large Rule Spaces by Sampling*, Technical Report (Stanford InfoLab, November 1999), http://ilpubs.stanford .edu:8090/424/.

Britton, John Phillips. *Models and Precision: The Quality of Ptolemy's Observations and Parameters*. New York: Garland, 1992.

Broman, Thomas H. "J. C. Reil and the 'Journalization' of Physiology." In *The Literary Structure of Scientific Argument: Historical Studies*, edited by Peter Dear, 13–42. Philadelphia: University of Pennsylvania Press, 1991.

Broman, Thomas H. *The Transformation of German Academic Medicine, 1750–1820*. Cambridge: Cambridge University Press, 1996.

Bronn, H. G. *Index Palaeontologicus, oder, Übersicht der bis jetzt bekannten fossilen Organismen*. Stuttgart: E. Schweizerbart, 1848.

Bronn, H. G. *Italiens Tertiär-Gebilde und deren organische Einschlüsse*. Heidelberg: K. Groos, 1831.

Bronn, H. G. *Lethaea Geognostica, oder Abbildungen und Beschreibungen der für die Gebirgs-Formationen bezeichnendsten Versteinerungen*. Stuttgart: E. Schweizerbart, 1835.

Bronn, H. G. *Untersuchungen über die Entwicklungsgesetze der organischen Welt während der Bildungs-zeit unserer Erdoberfläche*. Stuttgart: E. Schweizerbart, 1858.

Brosius, Maria, ed. *Ancient Archives and Archival Traditions: Concepts of Record-Keeping in the Ancient World*. Oxford: Oxford University Press, [2003] 2011.

Brosius, Maria. "Ancient Archives and Concepts of Record-Keeping: An Introduction." In *Ancient Archives and Archival Traditions*, edited by Maria Brosius, 1–16. Oxford: Oxford University Press, [2003] 2011.

Brown, David. "What Shaped Our Corpuses of Astral and Mathematical Cuneiform Texts?" In *Looking at It from Asia: The Processes That Shaped the Sources of History of Science*, edited by Florence Bretelle-Establet, 277–303. Dordrecht: Springer, 2010.

Browne, Janet. *Charles Darwin: Voyaging*. New York: Knopf, 1995.

Bruch, Rüdiger vom. "Mommsen und Harnack: Die Geburt von Big Science aus den Geisteswissenschaften." In *Theodor Mommsen: Wissenschaft und Politik im 19. Jahrhundert*, edited by Alexander Demandt et al., 121–41. Berlin: Walter de Gruyter, 2005.

Buchwald, Jed Z. "Discrepant Measurements and Experimental Knowledge in the Early Modern Era." *Archives for History of Exact Sciences* 60, no. 6 (2006): 565–649.

Buchwald, Jed Z., and Mordechai Feingold. *Newton and the Origins of Civilization*. Princeton: Princeton University Press, 2012.

Bünting, Heinrich. *Chronologia hoc est, omnium temporum et annorum series*. Zerbst: Impressum typis Bonaventurae Fabri, 1590.

Burke, Peter. "Reflections on the Information State." In *Information in der Frühen Neuzeit: Status, Bestände, Strategien*, edited by Arndt Brendecke, Markus Friedrich, and Susanne Friedrich, 51–64. Berlin: LIT Verlag, 2008.

Burke, Peter. "A Survey of the Popularity of Ancient Historians, 1450–1700." *History and Theory* 5, no. 2 (1966): 135–52.

Burnet, John. *Early Greek Philosophy*. London: A & C Black, 1920.

Burnett, William. "Case of Epilepsy Attended with Remarkable Slowness of the Pulse." *Medical-Chirurgical Transactions* 13 (1825): 202–11.

Burnyeat, Myles. "The Origins of Non-deductive Inference." In *Science and Speculation: Studies in Hellenistic Theory and Practice*, edited by J. Barnes, J. Brunschwig, M. Burnyeat, and M. Schofield, 193–283. Cambridge: Cambridge University Press, 1982.

Burtin, François-Xavier. "Réponse a la question physique, proposé par la société de teyler, sur les révolutions générales, qu'a subies la surface de la terre, et sur l'ancienneté de notre globe." In *Algemeene Omkeeringen*, 1–242. Haarlem, 1789.

Busa, Roberto. "The Annals of Humanities Computing: The Index Thomisticus." *Computers and the Humanities* 14 (1980): 83–90.

Busa, Roberto. "Concordances." In *Encyclopedia Library and Information Science*, 5: 592–604. New York: Marcel Dekker, 1971.

Busa, Roberto, Robert S. Casey, James W. Perry, and Madeleine M. Berry. "The Use of Punched Cards in Linguistic Analysis." In *Punched Cards: Their Application to Science and Industry*, edited by Robert S. Casey, 2nd ed., 357–73. London: Reinhold, 1958.

Bush, Vannevar. "As We May Think." *Atlantic Monthly*, July 1945.

Butrica, Andrew J. "Redefining Celestial Mechanics in the Space Age: Astrodynamics, Deep-space Navigation, and the Pursuit of Accuracy." In *Exploring the Solar System: The History and Science of Planetary Exploration*, edited by Roger D. Launius, 105–27. New York: Palgrave Macmillan, 2012.

Butson, Keith D., and Warren L. Hatch. *Selective Guide to Climatic Data Sources*. US Department of Commerce: Washington, DC, 1979.

Buttmann, Philipp. "Ueber die philosophische Deutung der griechischen Gottheiten, inbesondere von Apollon und Artemis" (1803). In *Mythologus, oder gesammelte Abhandlungen über die Sagen des Alterthums*, edited by Philipp Buttmann, vol. 1, 1–21. Berlin: 'Mylius'sche Buchhandlung, 1828.

Calvisius, Seth. *Opus chronologicum*. Frankfurt an der Oder: Impensis Johannis Thymii, 1620.

Caplan, Jane, and John Torpey. *Documenting Individual Identity: The Development of State Practices in the Modern World*. Princeton: Princeton University Press, 2001.

Cassini, Jacques. *Elemens d'astronomie*. Paris: De l'Imprimerie royale, 1740.

Cassini, Jacques. *Tables astronomiques du soleil, de la lune, des planets, des etoiles fixes, et des satellites de Jupiter et de Saturne: avec l'explication & l'usage de ces mêmes tables*. Paris: De l'Imprimerie royale, 1740.

Cassini, Jean-Dominique. *Mémoires pour servir à l'histoire des sciences et a celle de l'Observatoire royal de Paris, suivis de la vie de J.-D. Cassini, écrite par lui-même, et des éloges de plusieurs académiciens morts pendant la révolution*. Paris: Chez Bleuet, 1810.

Cassini de Thury, César-François. *La meridienne de l'observatoire royal de Paris, vérifiée dans toute l'étendue du royaume par de nouvelles observations*. Paris: Chez Hippolyte-Louis Guerin, & Jacques Guerin, 1744.

Cavalli-Sforza, Luigi Luca, and Walter F. Bodmer, *The Genetics of Human Populations*. San Francisco: W. H. Freeman, 1971.

Cavalli-Sforza, Luigi Luca, Paolo Menozzi, and Alberto Piazza. *The History and Geography of Human Genes*. Princeton: Princeton University Press, 1994.

Chabert, Joseph Bernard. *Voyage fait par ordre du roi en 1750 et 1751, dans l'Amérique septentrionale, pour rectifier les cartes des côtes de l'Acadie, de l'Isle royale & de l'Isle de Terre-Neuve; et pour en fixer les principaux points par des observations astronomiques*. Paris: De l'Imprimerie royale, 1753.

Chadarevian, Soraya de. *Designs for Life: Molecular Biology after World War II*. Cambridge: Cambridge University Press, 2002.

Chadarevian, Soraya de. "Genetic Evidence and Interpretation in History." *BioSocieties* 5, no. 3 (2010): 301–5.

Changnon, Stanley A. "The Past and Future of Climate Related Services in the United States." *Journal of Service Climatology* 1 (2007): 1–7.

Charmantier, Isabelle, and Staffan Müller-Wille. "Carl Linnaeus's Botanical Paper Slips (1767–1773)." *Intellectual History Review* 24 (2014): 1–24.

Chinnici, Ileana. *La Carte du Ciel: Correspondence, inédite conservée dans les Archives de l'Observatoire de Paris*. Paris: Observatoire de Paris, 1999.

Chinnici, Ileana. "La Carte du Ciel: Genèse, Déroulement et Issues." In *La Carte du Ciel: Histoire et actualité d'un projet scientifique international*, edited by Jérôme Lamy, 19–43. Paris: Observatoire de Paris, 2008.

Chun, Wendy. "The Enduring Ephemeral, or the Future Is a Memory." *Critical Inquiry* 35, no. 1 (Autumn 2008): 148–71.

Church, George M., Yuan Gao, and Sriram Kosuri. "Next Generation Digital Information Storage in DNA." *Science* 337, no. 6102 (September 28, 2012): 1628.

Churchill, Elizabeth, and Jeff Ubois. "Designing for Digital Archives." *Interactions* (March–April 2008). doi 10.1145/1340961.1340964.

Cicero, Marcus Tullius. *De Fato*. Translated by H. Rackham. Loeb Classical Library 349. Cambridge, MA: Harvard University Press, [1942] 1992.

Clement, Tanya, Wendy Hagenmaier, and Jennie Levine Knies. "Toward a Notion of the Archive of the Future: Impressions of Practice by Librarians, Archivists, and Digital Humanities Scholars." *Library Quarterly* 83 (2013): 112–30.

Climatological Data: Pennsylvania, January 1984. NOAA: Arlington, VA, 1984.

CODATA-ICSTI Task Group on Data Citation Standards and Practices. "Out of Cite, Out of Mind: the Current State of Practice, Policy, and Technology for the Citation of Data." *Data Science Journal* 12, no. 0 (2013), CIDCR1–CIDCR75.

Coletti, Margaret H., and Howard L. Bleich. "Medical Subject Headings Used to Search the Biomedical Literature." *Journal of the American Medical Informatics Association* 8 (2001): 317–23.

Colwell, Rita R., David G. Swartz, and Michael Terrell MacDonell. *Biomolecular Data: A Resource in Transition*. Oxford: Oxford University Press, 1989.

Committee on Atmospheric Sciences. "The Atmospheric Sciences: National Objectives for the 1980s." *Bulletin of the American Meteorological Society* 62 (1981): 226–31.

Comte, Auguste. *Cours de philosophie positive*. 6 vols. Paris: Bachelier, 1830–42.

"Confronting the Intelligence Future (U) An Interview with William P. Crowell, NSA's Deputy Director (U)" (redacted). Author's name redacted. *Cryptolog* 22, no. 2 (1996): 1–5.

Conway, Eric. "Drowning in Data: Satellite Oceanography and Information Overload in the Earth Sciences." *Historical Studies in the Physical and Biological Sciences* 37 (2006): 127–51.

Cook, Alan H. *Edmond Halley: Charting the Heavens and the Seas*. Oxford: Clarendon Press, 1998.

Copernicus, Nicolaus. *De revolutionibus orbium coelestium*. Nuremberg: apud Ioh. Petreius, 1543.

Corbin, Brenda G., and Donna J. Coletti. "Digitization of Historical Astronomical Literature." *Vistas in Astronomy* 39, no. 2 (1995): 161–65.

Costello, Mark J. "Motivating Online Publication of Data." *Bioscience* 59, no. 5 (2009): 418–27.

Craigie, David. "Case I. Case of Disease of the Spleen, in which Death Took Place in Consequence of the Presence of Purulent Matter in the Blood." *Edinburgh Medical and Surgical Journal* 64 (1845): 400–413.

Craik, Elisabeth. "Horizontal Transmission in the Hippocratic Tradition." *Mnemosyne* 59 (2006): 334–47.

Crandall, Jedidiah R., and Philipp Winter. "The Great Firewall of China: How It Blocks Tor and Why It Is Hard to Pinpoint." *Login* 37, no. 6 (December 2012): 42–50.

Crandall, Jedidiah R., Daniel Zinn, Earl Barr, and Rich East. "Concept Doppler: A Weather Tracker for Internet Censorship." *Proceedings of the 14th ACM Conference on Computer and Communications Security*, (2007): 352–65.

Creuzer, Friedrich. *Herodot und Thucydides: Versuch einer nähern Würdigung ihrer historischen Grundsätze mit Rücksicht auf Lucians Schrift Wie man Geschichte schreiben müsse*. Leipzig: Neue Akademische Buchhandlung Marburg, 1803.

Crowe, Michael, Thomas Reek, and Robert Mattingly. "Operational Automated Graphics at the NCDC." *Bulletin of the American Meteorological Society* 69 (1988): 28–38.

Cruden, Alexander. *A Complete Concordance to the Holy Scriptures of the Old and New Testament: In Two Parts Containing the Appellative or Common Words . . . , the Proper Names . . . , to Which Is Added a Concordance to the Books Called Apocrypha.* London: Printed for W. Owen and 18 others, 1738.

Cruden, Alexander. *A Verbal Index to Milton's "Paradise Lost."* London: Sold by W. Innys, and D. Browne, 1741.

Csiszar, Alex. "Seriality and the Search for Order: Scientific Print and Its Problems during the Late Nineteenth Century." *History of Science* 48 (2010): 399–434.

Cuvier, Georges. "Preliminary Discourse" to *Recherches sur les ossemens fossiles de quadrupèdes* (1812). In *Georges Cuvier, Fossil Bones, and Geological Catastrophes,* edited by Martin J. S. Rudwick, 183–252. Chicago: University of Chicago Press, 1997.

Dalbello, Marija. "Digitality, Epistolarity, and Reconstituted Letter Archives." *Information Research* 18 (2013). http://www.informationr.net/ir/18-3/colis/paperC26.html #Williams77. Accessed July 1, 2014.

Darnton, Robert. *The Business of Enlightenment: A Publishing History of the Encyclopédie, 1775–1800.* Cambridge: Harvard University Press, [1979] 2009.

Darnton, Robert. "Philosophers Trim the Tree of Knowledge: The Epistemological Strategy of the Encyclopédie." In *The Great Cat Massacre and Other Episodes in French Cultural History,* edited by Robert Darnton, 91–208. New York: Basic Books, [1984] 2009.

Darwin, Charles. *On the Origin of Species.* London: J. Murray, 1859.

Daston, Lorraine, ed. *Biographies of Scientific Objects.* Chicago: University of Chicago Press, 2000.

Daston, Lorraine. "Enlightenment Calculations." *Critical Inquiry* 21 (1994): 182–202.

Daston, Lorraine. "Objectivity and the Escape from Perspective." *Social Studies of Science* 22 (1992): 597–618.

Daston, Lorraine. "The Sciences of the Archive." *Osiris* 27, no. 1 (January 1, 2012): 156–87.

Daston, Lorraine. "Why Statistics Tend Not Only to Describe the World but to Change It." Essay review of: Alain Desrosières, *The Politics of Large Numbers: A History of Statistical Reasoning. London Review of Books* 22, no. 8 (2000): 35–36.

Daston, Lorraine, and Peter Galison. *Objectivity.* New York: Zone Books, 2007.

Daston, Lorraine, and Elizabeth Lunbeck, eds. *Histories of Scientific Observation.* Chicago: University of Chicago Press, 2011.

Daston, Lorraine, and Katharine Park. *Wonders and the Order of Nature 1150–1750.* New York: Zone Books, 2001.

Dawid, Igor B. "Editorial Submission of Sequences." *PNAS* 86 (1989): 407.

Day, Ronald. *Indexing It All: The Subject in the Age of Documentation, Information, and Data.* Cambridge, MA: MIT Press, 2014.

De Lacy, Phillip. Introduction, *Galen on the Doctrines of Hippocrates and Plato,* edited, translated, and with commentary by Phillip De Lacy. Berlin: Akademie-Verlag, 1984.

De Leemans, Pieter, and Michèle Goyens. *Aristotle's Problemata in Different Times and Tongues.* Leuven: Leuven University Press, 2006.

De Wreede, Liesbeth C. "Willebrord Snellius (1580–1626): A Humanist Reshaping the Mathematical Sciences." Ph.D. thesis, Utrecht University, 2007.

Dean, Gabrielle. "Disciplinarity and Disorder." *Archive Journal* 1 (2011). http://www .archivejournal.net/issue/1/archives-remixed/the-archeology-of-archival-practice/. Accessed June 9, 2013.

Dean-Jones, Lesley. "Literacy and the Charlatan in Ancient Greek Medicine." In *Written*

Texts and the Rise of Literate Culture in Ancient Greece, edited by Harvey Yunis and Hermann Diels, 97–121. New York: Cambridge University Press, 2003.

Débarbat, Suzanne, J. A. Eddy, H. K. Eichhorn, and A. R. Upgren, eds., *Mapping the Sky: Past Heritage and Future Directions.* Dordrecht: Kluwer, 1988.

Delambre, Jean-Baptiste Joseph. *Histoire de l'astronomie ancienne.* Paris: Mme Ve Courcier, 1817.

Demandt, Alexander, Andreas Goltz, and Heinrich Schlange-Schöningen, eds. *Theodor Mommsen: Wissenschaft und Politik im 19. Jahrhundert.* Berlin: Walter de Gruyter, 2005.

Dennis, Sally F. "Construction of a Thesaurus Automatically." In *Statistical Association Methods for Mechanized Documentation,* Symposium Proceedings Washington 1964, edited by Mary Elizabeth Stevens, Vincent E. Giuliano, and Laurence B. Heilprin, 61–73. Washington, DC: US Government Printing Office, 1965.

Depuydt, Leo. "'More Valuable Than All Gold': Ptolemy's Royal Canon and Babylonian Chronology." *Journal of Cuneiform Studies* 47 (1995): 97–117.

Derrida, Jacques. *Archive Fever: A Freudian Impression.* Translated by Eric Prenowitz. Chicago: University of Chicago Press, 1995.

Derrida, Jacques. *Mal d'archive: Une impression freudienne.* Paris: Galilée, 1995.

Deshayes, Gérard P. *Description des coquilles fossiles des environs de Paris.* Paris: L'auteur, chez Bechet jeune, 1824.

Dever, Maryanne. "Provocations on the Pleasures of Archived Paper." *Archives and Manuscripts* 41 (2013): 173–82.

De Vos, Abraham M., et al. "Three-dimensional Structure of an Oncogene Protein: Catalytic Domain of Human c-H-ras p21." *Science* 239 (1988): 888–93.

Dib, Lina. "The Forgetting Dis-Ease: Making Time Matter." *Differences: A Journal of Feminist Cultural Studies* 25, no. 3 (2012): 42–73.

Diels, Hermann. *Doxographi Graeci.* Berlin: G. Reimer, 1879.

Diodorus. *Diodorus of Sicily.* Vol. 1. Translated by Charles Henry Oldfather. Cambridge, MA: Harvard University Press, 1933.

Diogenes Laertius. *Lives of Eminent Philosophers.* 2 vols. Cambridge: Harvard University Press, 1925.

DiPaolo Healey, Antoinette, and Richard L. Venezky, eds. *Microfiche Concordance to Old English: The List of Texts and Index of Editions.* Toronto: University of Toronto, 1980.

Djorgovski, S. George, et al. "Sky Surveys." In *Planets, Stars, and Stellar Systems,* edited by Terry D. Oswalt and Howard E. Bond, 2: 223–81. Dordrecht: Springer, 2013.

Dobzhansky, Theodosius. *Genetics and the Origin of Species.* New York: Columbia University Press, [1937] 1951.

Dodge, Martin, and Rob Kitchin. "The Ethics of Forgetting in an Age of Pervasive Computing." Working Paper of Center for Advanced Spatial Analysis, 2005. http://www.casa.ucl.ac.uk/working_papers/paper92.pdf.

Doel, Ron. "Constituting the Postwar Earth Sciences: The Military Influence on the Environmental Sciences in the USA after 1945." *Social Studies of Science* 33 (2003): 635–66.

Doody, Aude. *Pliny's Encyclopedia: The Reception of the Natural History.* Cambridge: Cambridge University Press, 2010.

Drabiak-Syed, Katherine. "Lessons from Havasupai Tribe versus Arizona State University Board of Regents: Recognizing Group, Cultural, and Dignitary Harms as Legitimate Risks Warranting Integration into Research Practice." *Journal of Health and Biomedical Law* 6 (2010): 175–225.

Dreyer, John L. E., ed. *Tychonis Brahe Dani Opera omnia.* Copenhagen: in Libraria Gyldendaliana, 1913–29.

Dror, Otniel. "Seeing the Blush: Feeling Emotions." In *Histories of Scientific Observation*, edited by Lorraine Daston and Elizabeth Lunbeck, 326–48. Chicago: University of Chicago Press, 2011.

Droysen, Johann G. *Historik*, vol. 1: *Rekonstruktion der ersten Fassung der Vorlesungen* (1857). Stuttgart: Friedrich Fromann Verlag, 1977.

Dubs, Homer H. "A Canon of Lunar Eclipses for Anyang and China, –1400 to –1000." *Harvard Journal of Asiatic Studies* 10, no. 2 (1947): 162–78.

Dubs, Homer H., et al. *The History of the Former Han Dynasty.* Baltimore: Waverly Press Inc., 1938–55.

Duncker, Max. *Griechische Geschichte bis zum Tode des Perikles.* Vol. 1. Leipzig: Duncker & Humblot, 1888.

Dunthorne, Richard. "A Letter from the Rev. Mr. Richard Dunthorne to the Reverend Mr. Richard Mason F. R. S. and Keeper of the Wood-Wardian Museum at Cambridge, concerning the Acceleration of the Moon." *Philosophical Transactions* 46 (1749): 162–72.

Düring, Ingemar. *Aristotle in the Ancient Biographical Tradition.* Göteborg: Elanders Boktryckeri Aktiebolag, 1957.

Dvorsky, George. "The Russian Eccentric Who Wants to Make Surrogates a Reality." *io9,* June 3, 2013. http://io9.com/the-russian-eccentric-who-wants-to-make-surrogates-a -re-511064240. Accessed June 6, 2013.

Dyer, Geoff. *Zona.* New York: Vintage, 2012.

"Early Modern Information Overload." Special issue of *Journal of the History of Ideas* 64, no. 1 (2003).

Eddy, Amos. "The Economic Impact of Climate." Vols 1–3, Proceedings of Two Workshops on the Structure of Economic Models, Oklahoma Climatological Survey, University of Oklahoma, 1980.

"Education: Book on a Card?" *Time,* September 4, 1944.

Edwards, Paul N. *A Vast Machine: Computer Models, Climate Data, and the Politics of Global Warming.* Cambridge, MA: MIT Press, 2010.

Edwards, Paul, Steven J. Jackson, Geoffrey Bowker, and Cory P. Knobel. "Understanding Infrastructure: Dynamics, Tensions and Design, NSF Report of a Workshop on 'History and Theory of Infrastructure: Lessons for New Scientific Cyberinfrastructures.'" January 2007. http://www.si.umich.edu/cyber-infrastructure /UnderstandingInfrastructure_FinalReport25jan07.pdf.

Eichhorn, Guenther, et al. "Current and Future Holdings of the Historical Literature in the ADS." *Library and Information Services in Astronomy IV (LISA IV): Emerging and Preserving: Providing Astronomical Information in the Digital Age*, edited by Brenda G. Corbin et al., 145–52. Washington, DC: US Naval Observatory, 2003.

Eijk, Philip J. van der, ed. *Ancient Histories of Medicine: Essays in Medical Doxography and Historiography in Classical Antiquity.* Leiden: Brill, 1999.

Eijk, Philip J. van der. "Historical Awareness, Historiography, and Doxography in Greek and Roman Medicine." In *Ancient Histories of Medicine: Essays in Medical Doxography and Historiography in Classical Antiquity*, edited by Philip J. van der Eijk, 1–31. Leiden: Brill, 1999.

Eldredge, Niles, and Stephen Jay Gould. "Punctuated Equilibria: An Alternative to Phyletic Gradualism." In *Models in Paleobiology* edited by Thomas J.M. Schopf, 82–115. San Francisco: Freeman, Cooper & Co., 1972.

Endersby, Jim. *Imperial Nature: Joseph Hooker and the Practices of Victorian Science*. Chicago: University of Chicago Press, 2008.

Engelbart, Douglas C. "Augmenting Human Intellect: A Conceptual Framework." SRI Summary Report AFOSR-3223, Air Force Office of Scientific Research, October 1962.

Engerman, David. "The Rise and Fall of Wartime Social Science: Harvard's Refugee Interview Project, 1950–54." In *Cold War Social Science*, edited by Hamilton Cravens and Mark Solovey, 25–43. New York: Palgrave, 2013.

Environmental Data Management at NOAA: Archiving, Stewardship, and Access. Washington, DC: National Academy Press, 2007.

Epstein, Steven. *Impure Science: AIDS, Activism, and the Politics of Knowledge*. Berkeley: University of California Press, 1996.

Eskildsen, Kaspar. "Leopold Ranke's Archival Turn: Location and Evidence in Modern Historiography." *Modern Intellectual History* 5 (2008): 425–53.

Espenak, Fred, and Jean Meeus. NASA Technical Publication TP-2006-214141, *Five Millennium Canon of Solar Eclipses: −1999 to +3000 (2000 BCE to 3000 CE)*, Washington, DC: National Aeronautics and Space Flight Administration, 2006.

Espenak, Fred, and Jean Meeus. NASA Technical Publication TP-2009-214172, *Five Millennium Canon of Lunar Eclipses: −1999 to +3000 (2000 BCE to 3000 CE)*. Greenbelt, MD: NASA, Goddard Space Flight Center, 2009.

Fabian, Johannes. *Time and the Other: How Anthropology Makes Its Object*. New York: Columbia University Press, [1983] 2002.

Fan, Fa-ti. *British Naturalists in Qing China: Science, Empire, and Cultural Encounter*. Cambridge, MA: Harvard University Press, 2003.

Farge, Arlette. *The Allure of the Archives*. Translated by Thomas Scott-Railton. New Haven: Yale University Press, [1989] 2013.

Fassin, Didier. "Les économies morales revisitées." *Annales: Histoire, Sciences Sociales* 64, no. 6 (2009): 1237–66.

Fayyad, Usama. "Mining Databases: Towards Algorithms for Knowledge Discovery." *Bulletin of the Technical Committee on Data Engineering* 21, no. 1 (1998): 39–48.

Fayyad, Usama. "Taming the Giants and the Monsters: Mining Large Databases for Nuggets of Knowledge." *Database Programming and Design* 11, no. 3 (1998).

Fayyad, Usama M., Gregory Piatetsky-Shapiro, and Padhraic Smyth. "From Data Mining to Knowledge Discovery: An Overview." In *Advances in Knowledge Discovery and Data Mining*, edited by Usama M. Fayyad et al., 1–34. Menlo Park, CA: AAAI/MIT Press, 1996.

Fehling, Detlev. *Die Quellenangaben bei Herodot: Studien zur Erzählkunst Herodots*. Berlin: Walter de Gruyter, 1971.

Feuillée, Louis. *Journal des observations physiques, mathematiques et botaniques, faites par l'ordre du roy sur les côtes orientales de 'l'Amerique meridionale, & dans les Indes occidentales, depuis l'année 1707 jusques en 1712*. Paris: Chez Pierre Giffart: Jean Mariette, 1714–25.

Findlen, Paula. *Possessing Nature: Museums, Collecting, and Scientific Culture in Early Modern Italy*. Berkeley: University of California Press, 1994.

Finley, Klint. "The Quantified Man: How an Obsolete Tech Guy Rebuilt Himself for the Future." *Wired*, February 22, 2013, http://www.wired.com/2013/02/quantified-work/all/.

[Firmin Didot, Ambroise]. *Projets et rapports relatifs d'un recueil général d'épigraphie latine.* [Paris?]: [Firmin Didot?], [1843?].

Fischer, Katharina. "Der Begriff der '(wissenschaftlichen) Abhandlung' in der griechischen Antike—eine Untersuchung des Wortes πραγματεία." In *Antike Naturwissenschaft und ihre Rezeption* 23, 93–114. Trier: Wissenschaftlicher Verlag, 2013.

Fischer, Marguerite. "The KWIC Index Concept: A Retrospective View." *American Documentation* 17, no. 2 (1966): 57–70.

Fisher, John. "Conjectures and Reputations: The Composition and Reception of James Bradley's Paper on the Aberration of Light with Some Reference to a Third Unpublished Version." *British Journal for the History of Science* 43, no. 1 (2010): 19–48.

Fitzgibbon, Andrew, and Ehud Reiter. "'Memories for Life': Managing Information over a Human Lifetime." 22, 2003, Grand Challenge proposal, published by UK Computing Research Committee (UKCRC). http://homepages.abdn.ac.uk/e.reiter/pages/papers/memories.pdf.

Flammarion, Camille. "Le Congrès astronomique pour la photographie du ciel." *Astronomie* 6 (1887): 161–69.

Flammarion, Camille. "La photographie céleste à l'Observatoire de Paris." *Revue d'Astronomie Populaire* 5 (1886): 42–57.

Flamsteed, John. *The Correspondence of John Flamsteed*, edited by Eric G. Forbes, Lesley Murdin, and Frances Willmoth. Bristol, UK: Institute of Physics Pub., 1995.

Flamsteed, John. *Historia coelestis Britannica.* London: Typis H. Meere, 1725.

Flamsteed, John. *Historiae coelestis libri duo.* London: Typis J. Matthews, 1712.

Flamsteed, John. *The preface to John Flamsteed's Historia coelestis Britannica: or British catalogue of the heavens (1725),* translated by Alison Dione Johnson, edited by Allan Chapman. London: Trustees of the National Maritime Museum, 1982.

Flanders, Julia. "The Productive Unease of 21st-century Digital Scholarship." In *The Digitial Humanities: A Reader,* edited by Melissa Terras, Julianne Nyhan, and Edward Vanhoutte, 205–18. Burlington, VT: Ashgate, 2013.

Flashar, Hellmut. *Aristotle. Problemata physica.* Berlin: Akademie-Verlag, 1975.

Flood, Barbara J. "Historical Note: The Start of a Stop List at Biological Abstracts." *Journal of the American Society for Information Science* 50, no. 12 (October 1999): 1066.

Fontana, Francesco. *Novae coelestium terrestriumq[ue] rerum observationes, et fortasse hactenus non vulgatae.* Naples: apud Gaffarum, 1646.

Fortenbaugh, William W., and Dimitri Gutas, eds. *Theophrastus: His Psychological, Doxographical, and Scientific Writings.* New Brunswick, NJ: Transaction Publishers, 1992.

Foster, Hal. "Exhibitionists." *London Review of Books.* 4 June 2015: 13–14.

Foucault, Michel. *The Archaeology of Knowledge.* Translated by A. M. Sheridan Smith. New York: Pantheon, 1972.

Foucault, Michel. *L'Archéologie du savoir.* Paris: Gallimard, 1969.

Foucault, Michel. *The Hermeneutics of the Subject: Lectures at the Collège de France 1981–1982.* New York: Picador, 2005.

Foucault, Michel. "The Historical *a priori* and the Archive." In *The Archive,* edited by Charles Merewether, 28–29. Whitechapel: London, 2006.

Foucault, Michel. *History of Sexuality.* Vol. 3, *The Care of the Self.* New York: Vintage, 1988.

Foucault, Michel. *The Order of Things: An Archeology of the Human Sciences.* New York: Pantheon, 1970.

Foucault, Michel. *Technologies of the Self: A Seminar with Michel Foucault.* Amherst: University of Massachusetts Press, 1988.

Fournier, Georges. *Hydrographie, contenant la theorie et la practique de toutes les parties de la navigation*. Paris: Chez Michel Soly, 1643.

Fox, Herbert. "Remarks on the Presentation of Microscopical Preparations Made from Some of the Original Tissue Described by Thomas Hodgkin, 1832." *Annals of Medical History* 8 (1926): 370–74.

Frawley, William J., Gregory Piatetsky-Shapiro, and Christopher J. Matheus. "Knowledge Discovery in Databases: An Overview." In *Knowledge Discovery in Databases*, edited by Gregory Piatetsky-Shapiro, 1–27. Cambridge, MA: AAAI/MIT Press, 1991.

Freeland, C. A. "Scientific Explanation and Empirical Data in Aristotle's *Meteorology*." *Oxford Studies in Ancient Philosophy* 8 (1990): 67–102.

Fresneau, Alain. "La détection de la matière interstellaire sur les plaques photographiques de la Carte du Ciel." In *La Carte du Ciel, Histoire et actualité d'un projet scientifique international*, edited by Jérôme Lamy, 155–67. Paris: Observatoire de Paris, 2008.

Freud, Sigmund. "The uncanny." In *The Standard Edition of the Complete Psychological Works of Sigmund Freud*, Vol. XVII (1917–1919), 217–56, edited and translated by James Strachey, in collaboration with Anna Freud, assisted by Alix Strachey [ed; orig. 1919]. London: Hogarth 1955.

Friedrich, Markus. "Archiv und Verwaltung im frühneuzeitlichen Europa: Das Beispiel der Gesellschaft Jesu." *Zeitschrift für Historische Forschung* 35 (2008): 369–403.

Friedrich, Markus. *Die Geburt des Archivs: Eine Wissensgeschichte*. Munich: Oldenbourg Verlag, 2013.

Fuller, Buckminster. *Critical Path*. New York: Macmillan, 1981.

Fuller, Buckminster. Oregon Lecture #9, 324, 12 July 1962.

Furness, Helen Kate. *A Concordance to Shakespeare's Poems: An Index to Every Word therein Contained*. Philadelphia: J. B. Lippincott, 1874.

Galen. *On the Doctrines of Hippocrates and Plato*. Edited, translated, and with a commentary by Phillip De Lacy. Berlin: Akademie-Verlag, 1984.

Galison, Peter, and Lorraine Daston. "Scientific Coordination as Ethos and Epistemology." In *Instruments in Art and Science: On the Architectonics of Cultural Boundaries in the 17th Century*, edited by Helmar Schramm, Ludger Schwarte, and Jan Lazardzig, 296–333. Berlin: Walter de Gruyter, 2008.

Garcia-Sancho, Miguel. *Computing, and the History of Molecular Sequencing*. New York: Palgrave Macmillan, 2012.

Gardey, Delphine. *Écrire, calculer, classer: Comment une révolution de papier a transformé les sociétés contemporaines (1800–1940)*. Paris: Éditions de la Découverte, 2008.

Garfield, Eugene. "Citation Indexes for Science: A New Dimension in Documentation through Association of Ideas." *Science* 122, no. 3159 (July 15, 1955): 108–11.

Garg, Amit X., et al. "Effects of Computerized Clinical Decision Support Systems on Practitioner Performance and Patient Outcomes: A Systematic Review." *Journal of the American Medical Association* 293 (2005): 1223–38.

Gassendi, Pierre. *Epistolica exercitatio. . . . Cum appendice aliquot observationum coelestium*. Paris: Apud Sebastianum Cramoisy, 1630.

Gassendi, Pierre. *Opera omnia in sex tomos divisa*. Vol. 4. Lyon: Sumptibus Laurentii Anisson, & Ioannis Baptistae Devenet, 1658.

Geda, Carolyn L. "Social Science Data Archives." *American Archivist* 42 (1979): 158–66.

Geller, Mark J. "Babylonian Astronomical Diaries and Corrections of Diodorus." *Bulletin of the School of Oriental and African Studies* 53, no. 1 (1990): 1–7.

Gent, Robert H. van, and Albert Van Helden. "Lunar, Solar, and Planetary Representations to 1650." In *The History of Cartography*, vol. 3, edited by David Woodward, 123–34. Chicago: University of Chicago Press, 2007.

Geoscience Abstracts. Washington: American Geological Institute, 1964.

Gera, Dov, and Wayne Horowitz. "Antiochus IV in Life and Death: Evidence from the Babylonian Astronomical Diaries." *Journal of the American Oriental Society* 117, no. 2 (1997): 240–52.

Gérard, Alexandre. *Des perforations spontanées de l'estomac*. Paris: Impr. de Gillé fils, 1803.

Gerbezius, M. "Pulsus mira inconstantia." *Miscellanea curiosa, sive Ephemeridum medico-physicarum Germanicum Academiae Caesareo-Leopoldinae Naturae* 10 (1692): 115–18.

Gere, Cathy. "Inscribing Nature: Archaeological Metaphors and the Formation of New Sciences." *Public Archaeology* 2, no. 4 (2002): 195–208.

Gere, Cathy, and Bronwyn Parry. "The Flesh Made Word: Banking the Body in the Age of Information." *BioSocieties* 1, no. 1 (2006): 41–54.

Gibson, John R. "Case of Fits with Very Slow Pulse" (letter to the editor). *London Medical Gazette* 23 (1839): 123–26, 155–60.

Giddens, Anthony. *Modernity and Self-Identity: Self and Society in the Late Modern Age*. Stanford: Stanford University Press, 1991.

Gillespie, Tarleton. "The Relevance of Algorithms." In *Media Technologies: Essays on Communication, Materiality, and Society*, edited by Tarleton Gillespie, Pablo Boczkowski, and Kirsten Foot, 167–94. Cambridge: MA: MIT Press, 2014.

Gillies, John. *The History of Ancient Greece*, vol. 3 (of 4). Basel: J. J. Tourneisen and J. L. Grand, 1790.

Gingerich, Owen. *Astrophysics and Twentieth-Century Astronomy to 1950*. Cambridge: Cambridge University Press, 1984.

Gingerich, Owen, and Barbara L. Welther. *Planetary, Lunar, and Solar Positions, New and Full Moons, A.D. 1650–1805*. Philadelphia: American Philosophical Society, 1983.

Ginsburg, Carlo. "Clues: Roots of an Evidential Paradigm." In *Clues, Myths and the Historical Method*, 96–125. Baltimore: Johns Hopkins University Press, 1980.

Ginzel, Friedrich Karl. *Spezieller Kanon der Sonnen- und Mondfinsternisse für das Ländergebiet der klassischen Altertumswissenschaften und den Zeitraum von 900 vor Chr. bis 600 nach Chr.* Berlin: Mayer & Müller, 1899.

Ginzel obituary. *Observatory* 49 (1926): 348.

Giraud Soulavie, Jean-Louis. *Histoire naturelle de la France méridionale, ou recherches sur la minéralogie du vivarais.* [Nismes]: Belle, 1780–84.

Gitelman, Lisa, ed. *Raw Data Is an Oxymoron*. Cambridge, MA: MIT Press, 2013.

Glenn, Jacob. "NSF Data Management Plan" (2013). http://www.lib.umich.edu/research -data-services/nsf-data-management-plans. Accessed July 11, 2014.

Godefroy, Frédéric Eugène. *Dictionnaire de l'ancienne langue française et de tous ses dialectes du 9e au 15e siècle*. Paris: F. Vieweg, 1881.

Goldhill, Simon. *The Invention of Prose*. Oxford: Oxford University Press, 2002.

Goldstein, Bernard R. "What's New in Ptolemy's *Almagest*?" *Nuncius* 22 (2007): 261–85.

Goldstein, Bernard R., and Alan C. Bowen. "The Introduction of Dated Observations and Precise Measurement in Greek Astronomy." *Archive for History of Exact Sciences* 43, no. 2 (1991): 93–132.

Goldstein, Bernard R., and Alan C. Bowen. "The Role of Observations in Ptolemy's Lunar Theories." In *Ancient Astronomy and Celestial Divination*, edited by N. M. Swerdlow, 341–56. Cambridge, MA: MIT Press, 1999.

Goldstine, Herman H. *New and Full Moons 1001 B.C. to A.D. 1651*. Philadelphia: American Philosophical Society, 1973.

Golinski, Jan. *British Weather and the Climate of Enlightenment*. Chicago: University of Chicago Press, 2007.

Göransson, Tryggve. *Albinus, Alcinous, Arius Didymus*. Göteborg: Acta Universitatis Gothoburgensis, 1995

Gould, Stephen Jay. *The Structure of Evolutionary Theory*. Cambridge, MA: Belknap Press, 2002.

Gould, Stephen Jay. *Wonderful Life: The Burgess Shale and the Nature of History*. New York: W. W. Norton, 1989.

Goüye, Thomas, ed. *Observations physiques et mathematiques pour servir à l'histoire naturelle, & à la perfection de l'astronomie & de la geographie: envoyées de Siam à l'Academie Royale des Sciences à Paris, par les Peres Jesuites François*. Paris: Chez la Veuve d'Edme Martin, Jean Boudot, & Estienne Martin, 1688.

Goüye, Thomas, ed. *Observations physiques et mathematiques, pour servir a l'histoire naturelle & à la perfection de l'astronomie & de la geographie: envoyées des Indes et de la Chine à l'Académie Royale des Sciences à Paris, par les Peres Jesuites*. Paris: De l'Imprimerie Royale, 1692.

Government Data Center: Meeting Increasing Demands. Washington, DC: National Research Council, 2003.

Gowers, William R. "Splenic Leucocythaemia." In *A System of Medicine*, edited by John Russell Reynolds, 216–305, vol. 5 (of 5). Philadelphia: J. B. Lippincott, 1868–79.

Grafton, Anthony. *The Footnote: A Curious History*. Cambridge, MA: Harvard University Press, 1997.

Grafton, Anthony. "From Apotheosis to Analysis: Some Late Renaissance Histories of Classical Astronomy." In *History and the Disciplines: The Reclassification of Knowledge in Early Modern Europe*, edited by Donald R. Kelley, 261–76. Rochester, NY: University of Rochester Press, 1997.

Grafton, Anthony. *Joseph Scaliger: A Study in the History of Classical Scholarship* (2 vols.). Oxford: Clarendon Press, 1983–93.

Grafton, Anthony. "Libraries and Lecture Halls." In *The Cambridge History of Science*. Vol. 3: *Early Modern Science*, edited by Katharine Park and Lorraine Daston, 238–50. Cambridge: Cambridge University Press, 2006.

Grafton, Anthony. "Some Uses of Eclipses in Early Modern Chronology." *Journal of the History of Ideas* 64, no. 2 (2003): 213–29.

Grafton, Anthony. *What Was History? The Art of History in Early Modern Europe*. New York: Cambridge University Press, 2007.

Grafton, Anthony, Glenn W. Most, and James E. G. Zetzel. "Introduction." In F. A. Wolf, *Prolegomena to Homer (1795)*, edited by Anthony Grafton, Glenn W. Most, and James E. G. Zetzel, 3–35. Princeton: Princeton University Press, 1985.

Grattan-Guiness, Ivor. "Work for the Hairdressers: The Production of Prony's Logarithmic and Trigonometric Tables." *Annals of the History of Computing* 12 (1990): 177–85.

Gray, Jim. "The Next Database Revolution." In *Proceedings of the 2004 ACM SIGMOD International Conference on Management of Data*, 2004, 1–4. http://dl.acm.org/citation .cfm?id=1007570.

Greenberg, John Leonard. *The Problem of the Earth's Shape from Newton to Clairaut: The Rise of Mathematical Science in Eighteenth-Century Paris and the Fall of "Normal" Science*. Cambridge: Cambridge University Press, 1995.

Greenberg, Stephen J., and Patricia E. Gallagher. "The Great Contribution: Index Medicus, Index-Catalogue, and IndexCat." *Journal of the Medical Library Association* 97 (2009): 108–13.

Greene, Gretchen, Brian McLean, and Barry Lasker. "Development of the Astronomical Image Archive and Catalog Database for Production of GSC-II." *Future Generation Computer Systems* 16, no. 1 (1999): 29–38.

Greenhalgh, Trisha. *How to Read a Paper: The Basics of Evidence-Based Medicine*. 5th ed. Chichester: Wiley-Blackwell, 2014.

Greenwood, Glenn. *Index to Legal Theses and Research Projects*. Chicago: American Bar Foundation, 1962.

Grier, David Alan. *When Computers Were Human*. Princeton: Princeton University Press, 2006.

Grote, George. "Grecian Legends and Early History" (1843). In *The Minor Works of George Grote*, edited by Alexander Bain, 73–143. London: John Murray, 1873.

Grote, George. *A History of Greece*. 4th ed., vol. 1. Bristol: Thoemmes Press, [1872] 2000.

Gurrin, Cathal. C. Gurrin at QS, vimeo video. http://vimeo.com/32054542.

Guttman, N., C. Karl, T. Reek, and V. Shuler. "Measuring the Performance of Data Validators." *Bulletin of the American Meteorological Society* 69 (1988): 1448–52.

Haas, Laura M., and Patricia G. Selinger. "Database Research at the IBM Almaden Research Center." *SIGMOD Rec.* 20, no. 3 (September 1991): 92–98. doi:10.1145/126482.126493.

Hafner, Katie. "Doctor vs. Computer for the Right Diagnosis." *International Herald Tribune*, December 5, 2012.

Haigh, Thomas. "How Data Got Its Base: Information Storage Software in the 1950s and 1960s." *Annals of the History of Computing, IEEE* 31, no. 4 (2009): 6–25.

Haigh, Thomas. "The Web's Missing Links: Search Engines and Portals." In *The Internet and American Business*, edited by William Aspray and Paul Ceruzzi, 159–200. Cambridge, MA: MIT Press, 2008. http://ieeexplore.ieee.org.ezproxy.cul.columbia.edu /ebooks/6267214/6270064.pdf?bkn=6267214.

Halevy, A., P. Norvig, and F. Pereira. "The Unreasonable Effectiveness of Data." *Intelligent Systems, IEEE* 24, no. 2 (April 2009): 8–12. doi:10.1109/MIS.2009.36.

Hallerstein, Augustin. *Observationes astronomicae ab anno 1717 ad annum 1752*, edited by Maximilian Hell. Vienna: Typis Joannis Thomae, 1768.

Halley, Edmond. *Catalogus stellarum australium sive supplementum catalogi tychonici*. London: Typis Thomae James, 1679.

Halliday, W. R. *The Greek Questions of Plutarch*. Oxford: Oxford University Press, [1928] 1995.

Hamilton, Walter C. "The Revolution in Crystallography." *Science* 169, no. 941 (July 10, 1970): 133–41.

Hammerbacher, Jeff. "Information Platforms and the Rise of the Data Scientist." In *Beautiful Data: The Stories behind Elegant Data Solutions*, edited by Toby Segaran and Jeff Hammerbacher, 73–84. Beijing, China, and Sebastopol, CA: O'Reilly Media, 2009.

Hankinson, R. J. "Galen's Concept of Scientific Progress." *Aufstieg und Niedergang der Römischen Welt* 2, no. 2 (1994): 1775–89.

Hardtwig, Wolfgang. *Deutsche Geschichtskultur im 19. und 20. Jahrhundert*. Munich: Oldenbourg Verlag, 2013.

Harland, W. B., et al., eds. *The Fossil Record*. London: Geological Society, 1967.

Harnack, Adolf. *Geschichte der Königlich Preussischen Akademie der Wissenschaften zu Berlin*. 3 vols. Berlin: Reichsdruckerei, 1900.

Harrison, George W. M. "Problems with the Genre of Problems: Plutarch's Literary Innovations." *Classical Philology* 95 (2000): 193–99.

Harry, Debra. "Indigenous People and Gene Disputes." *Chicago-Kent Law Review* 84, no. 1 (2009): 147–96.

Hatch, Warren L. *Selective Guide to Climatic Data Sources*. Washington, DC: US Department of Commerce, 1983.

Hayden, Cori. *When Nature Goes Public: The Making and Unmaking of Bioprospecting in Mexico*. Princeton: Princeton University Press, 2003.

Head, Randolph. "Mirroring Governance: Archives, Inventories and Political Knowledge in Early Modern Switzerland and Europe." *Archival Science* 7 (2008): 317–29.

Heberden, William. *Commentaries on the History and Cure of Diseases*. Boston: Wells and Lilly, 1818.

Hecht, Alan. "Meeting the Challenge of Climate Services in the 1980s." *Bulletin of the American Meteorological Society* 65 (1984): 365–66.

Heesen, Anke te. *The World in a Box: The Story of an Eighteenth Century Picture Encyclopedia*. Chicago: University of Chicago Press, 2002.

Heesen, Anke te, and Emma Spary, eds. *Sammeln als Wissen: Das Sammeln und seine wissenschaftsgeschichtliche Bedeutung*. Göttingen: Wallstein Verlag, 2002.

Heilbron, John L. *The Sun in the Church: Cathedrals as Solar Observatories*. Cambridge, MA: Harvard University Press, 1999.

Heitjan, Isabel. "Zur Erstausgabe der Beobachtungen Tycho Brahes." *Libri* 14, no. 3 (1964): 189–226.

Henderson, Gabriel. "Governing the Hazard of Climate: The Development of the National Climate Program Act, 1977–1981." *Historical Studies in the National Sciences* 46, no. 2 (April 2016): 207–42. doi: 10.1525/hsns.2016.46.2.207

Herder, Johann G. *Ideen zur Philosophie der Geschichte der Menschheit*. In *Herders Sämtliche Werke*, vol. 14 (of 33), edited by Bernhard Suphan. Berlin: Weidmann, 1883.

Herner, Saul. *Deep Subject Indexing by Manual Permutation Methods*. Washington, DC: Herner and Co., 1963.

Herodotus. *Histories*. Translated by Alfred Denis Godley. Cambridge, MA: Harvard University Press, [1920] 1981–95.

Herzog, Johann J., Philip Schaff, Albert Hauck, Charles C. Sherman, George W. Gilmore, and Samuel M. Jackson. *The New Schaff-Herzog Encyclopedia of Religious Knowledge: Embracing Biblical, Historical, Doctrinal, and Practical Theology and Biblical, Theological, and Ecclesiastical Biography from the Earliest Times to the Present Day*. New York: Funk and Wagnalls, 1909.

Herzog, Johann J., Philip Schaff, Samuel M. Jackson, and David S. Schaff. *A Religious Encyclopaedia: Or Dictionary of Biblical, Historical, Doctrinal, and Practical Theology: Based on the Real-Encyklopädie of Herzog, Plitt, and Hauck*. New York: Funk & Wagnalls, 1891.

Hess, Volker, and J. Andrew Mendelsohn. "Case and Series: Medical Knowledge and Paper Technology, 1600–1900." *History of Science* 48 (2010): 287–314.

Hess, Volker, and J. Andrew Mendelsohn, eds. *Paper Technology: Ein Forschungsinstrument der frühneuzeitlichen Wissenschaft*, special issue of *NTM Zeitschrift für Geschichte der Wissenschaften, Technik und Medizin/Journal of the History of Science, Technology and Medicine* 21, 2013.

Hess, Volker, and J. Andrew Mendelsohn. "Sauvages' Paperwork: How Disease Classification Arose from Scholarly Note-Taking." *Early Science and Medicine* 19 (2014): 471–503.

Hett, Walter S. Introduction. *Aristotle. Problems.* 2 vols, translated by Walter S. Hett. London: Heinemann; Cambridge, MA: Harvard University Press, 1957.

Hevelius, Johannes. *Annus climactericus, sive rerum uranicarum observationum annus quadragesimus nonus.* Gdańsk: Sumptibus auctoris, typis Dav.-Frid. Rhetii, 1685.

Hevelius, Johannes. *Cometographia, totam naturam cometarum . . . . Accesit, omnium cometarum, à mundo condito hucusquè ab historicis, philosophis, & astronomis annotatorum, historia.* Gdańsk: Auctoris typis, & sumptibus, imprimebat Simon Reiniger, 1668.

Hevelius, Johannes. *Machinae coelestis pars posterior; rerum uranicarum observationes.* Gdańsk: In aedibus auctoris, eiusą; typis, & sumptibus: imprimebat Simon Reiniger, 1679.

Hevelius, Johannes. *Prodromus cometicus, historia, cometae anno 1664 exorti.* Gdańsk: Auctoris typis, et sumptibus, Imprimebat Simon Reiniger, 1665.

Hicks, R. D. *Diogenes Laertius' Lives of Eminent Philosophers.* 2 vols. Cambridge, MA: Harvard University Press, [1925] 1972.

Hilgartner, Stephen, and Sherry I. Brand-Rauf. "Data Access, Ownership, and Control: Toward Empirical Studies of Access Practices." *Knowledge: Creation, Diffusion, Utilization* 15, no. 4 (1994): 355–72.

Hillier Parry, Caleb. *An Inquiry into the Symptoms and Causes of the Syncope Anginosa Commonly Called Angina Pectoris; Illustrated by Dissections.* London: Cadell and Davies, 1799.

Hine, Harry M. *An Edition with Commentary of Seneca, Natural Questions,* Book Two. New York: Arno Press, 1981.

Hippocrates. *Hippocratic Corpus Aphorisms.* Translated by F. Adams. In the Internet Classics Archive, http://classics.mit.edu/Hippocrates/aphorisms.html. Accessed June 4, 2013.

Hjørland, Birger. *Information Seeking and Subject Representation: An Activity-theoretical Approach to Information Science.* Westport, CT: Greenwood, 1997.

Hoang, Pierre. *Catalogue des éclipses de soleil et de lune relatées dans les documents chinois et collationnées avec le canon de Th. Ritter v. Oppolzer.* Shanghai: Mission Catholique, 1925.

Holberton, J. H. "Case of Slow Pulse with Fainting Fits . . ." and discussion, Royal Medical and Surgical Society. *Lancet* 35, no. 916 (1841): 892.

Holtz, Bärbel. "Preußens Kulturstaatlichkeit im langen 19. Jahrhundert im Fokus seines Kultusministeriums." In *Kulturstaat und Bürgergesellschaft: Preußen, Deutschland und Europa im 19. und frühen 20. Jahrhundert,* ed. Wolfgang Neugebauer and Bärbel Holtz, 55–77. Berlin: Akademie Verlag, 2010.

Hooke, Robert. *The Posthumous Works of Robert Hooke, Containing His Cutlerian Lectures, and Other Discourses, Read at the Meetings of the Illustrious Royal Society.* London: printed by Sam, 1705.

Hope, Richard. *The Book of Diogenes Laertius, Its Spirit and Its Method.* New York: Columbia University Press, 1930.

Horning, Rob. "Know Your Product." *New Inquiry,* July 29, 2015. http://thenewinquiry.com/blogs/marginal-utility/know-your-product/.

Horstmann, Anja, and Vanina Kopp. *Archiv—Macht—Wissen: Organisation und Konstruktion von Wissen und Wirklichkeiten in Archiven.* Frankfurt am Main: Campus-Verlag, 2010.

Houlden, Michael A., and F. Richard Stephenson. *A Supplement to the Tuckerman Tables.* Philadelphia: American Philosophical Society, 1986.

Hsia, Florence C. "Chinese Astronomy for the Early Modern European Reader." *Early Science and Medicine* 13, no. 5 (2008): 417–50.

Hsia, Florence C. *Sojourners in a Strange Land: Jesuits and their Scientific Missions in Late Imperial China*. Chicago: University of Chicago Press, 2009.

Hubble Second Decade Committee. *The Hubble Data Archive: Towards the Ultimate Union Archive of Astronomy*. Baltimore: Space Telescope Science Institute, 2000.

Huber, Peter J., and Salvo de Meis. *Babylonian Eclipse Observations from 750 BC to 1 BC*. Milan: Mimesis; IsIAO, 2004.

Hübner, Emil. *Über mechanische Copien von Inschriften*. Berlin: Weidmannsche Buchhandlung, 1881.

Huchard, Henri. *Maladies du coeur et des vaisseaux*. Paris: Doin, 1889.

Hughes, Agatha C. *Systems, Experts, and Computers: The Systems Approach in Management and Engineering, World War II and After*. Cambridge, MA: MIT Press, 2000.

Hughes Bennett, John. "Case II. Case of Hypertrophy of the Spleen and Liver in which Death Took Place from Suppuration of the Blood." *Edinburgh Medical and Surgical Journal* 64 (1845): 413–23.

Hunger, Hermann, and Rudolf Dvořák. *Ephemeriden von Sonne, Mond und hellen Planeten von –1000 bis –601*. Vienna: Verlag der Österreichischen Akademie der Wissenschaften, 1981.

Hunger, Hermann, and David Pingree. *Astral Sciences in Mesopotamia*. Boston: Brill, 1999.

Hunt, Lynn, Margaret C. Jacob, and Wijnand Mijnhardt. *The Book That Changed Europe: Bernard and Picart's Religious Ceremonies of the World*. New York: Belknap Press, 2010.

Hunter, Kathryn M. *Doctors' Stories: The Narrative Structure of Medical Knowledge*. Princeton: Princeton University Press, 1993.

Hunter, Kathryn M. "'There Was This One Guy': The Uses of Anecdotes in Medicine." *Perspectives in Biology and Medicine* 29, no. 4 (1986): 619–30.

Hunter, Michael. *Science and the Shape of Orthodoxy*. Suffolk, UK: Boydell Press, 1995.

Hurwitz, Brian. "Form and Representation in Clinical Case Reports." *Literature and Medicine* 25, no. 2 (2006): 216–40.

Hutchinson, Jonathan. "Series Illustrating the Connexion between Bronzed Skin and Disease of the Supra-Renal Capsules." *Medical Times & Gazette* 2 (1855): 593–94.

Illich, Ivan. *In the Vineyard of the Text: A Commentary to Hugh's* Didascalicon. Chicago: University of Chicago Press, 1996.

Institut de France–Académie des Sciences. *Congrès astrophotographique international tenu à l'Observatoire de Paris pour le levé de la Carte du Ciel*. Paris: Gauthier-Villars, 1887.

Inwood, Brad, and Jaap Mansfeld, eds. *Assent and Argument: Studies in Cicero's Academic Books: Proceedings of the 7th Symposium Hellenisticum* (Utrecht, August 21–25, 1995). Leiden: Brill, 1997.

Inwood, M. J. "Problematic Problems." Review of Aristote, *Problèmes*, I, sections I through X by Pierre Louis. *Classical Review*, n.s. 42, no. 2 (1992): 285–86.

James, William. *Principles of Psychology*. 2 vols. New York: Holt, 1890.

Janković, Vladimir. "Climates as Commodities: Jean-Pierre Purry and the Modelling of the Best Climate on Earth." *Studies in History and Philosophy of Modern Physics* 41 (2010): 201–7.

Janković, Vladimir. "Working with Weather: Atmospheric Resources, Climate Variability, and the Ascent of Industrial Meteorology 1950–2010." *History of Meteorology* 7 (2015): 98–111.

Jardine, Nicholas, James A. Secord, and Emma C. Spary, eds. *Cultures of Natural History*. Cambridge: Cambridge University Press, 1996.

Jenne, Roy L. *Data Sets for Meteorological Research*. NCAR Technical Note IA-111.Boulder, CO: National Center for Atmospheric Research, 1975.

Jenne, Roy L., and Dennis H. Joseph, *Techniques for the Processing, Storage, and Exchange of Data*. Technical Note NCAR-TN/STR-93. Boulder, CO: National Center for Atmospheric Research, 1974.

Jewson, Nicholas D. "The Disappearance of the Sick-Man from Medical Cosmology." *Sociology* 10 (1974): 369–85.

Jimerson, Randall C. "Embracing the Power of Archives." *American Archivist* 69, no. 1 (2006): 19–32.

Johns, Adrian. *The Nature of the Book: Print and Knowledge in the Making*. Chicago: University of Chicago Press, 1998.

Johnson, Kristin. *Ordering Life: Karl Jordan and the Naturalist Tradition*. Baltimore: Johns Hopkins University Press, 2012.

Jones, Alexander. "Ptolemy's Ancient Planetary Observations." *Annals of Science* 63, no. 3 (2006): 255–90.

Jones, Derek. "The Scientific Value of the Carte du Ciel." *Astronomy & Geophysics* 41, no. 5 (2000): 16–20.

Josephus, Flavius. *Judean Antiquities 1–4*. Translated by Louis H. Feldman. In *Flavius Josephus: Translation and Commentary*, vol. 3. Edited by Steve Mason. Leiden: Brill, 2000.

Kahlert, Thorsten. "Theodor Mommsen, informelle Netzwerke und die Entstehung des Corpus Inscriptionum Latinarum um 1850." In *Geschichtsforschung in Deutschland und Österreich im 19. Jahrhundert: Ideen–Akteure–Institutionen*, edited by Christine Ottner and Klaus Ries, 180–97. Stuttgart: Franz Steiner Verlag, 2014.

Kahn, Charles. "Writing Philosophy: Prose and Poetry from Thales to Plato." In *Written Texts and the Rise of Literate Culture in Ancient Greece*, edited by Harvey Yunis, 139–61. Cambridge: Cambridge University Press, 2003.

Kahn, Patricia, and David Hazledine. "NAR's New Requirement for Data Submission to the EMBL Data Library: Information for Authors." *Nucleic Acids Research* 16, no. 10 (May 25, 1988): i–iv.

Kates, Robert W. "The Interaction of Climate and Society." In *Climate Impact Assessment: Studies of the Interaction of Climate and Society*, ICSU/Scope no. 27, edited by R. W. Kates, J. H. Ausubel, and M. Berberian, 3–36, New York; John Wiley, 1985.

Katz, Richard W. "Assessing Impact of Climatic Change on Food Production." *Climatic Change* 1 (1977): 85–96.

Kaufmann, Doris. "Dreams and Self-Consciousness: Mapping the Mind in the Late Eighteenth and Early Nineteenth Centuries." In *Biographies of Scientific Objects*, edited by Lorraine Daston, 67–85. Chicago: University of Chicago Press, 2000.

Kaufmann, Claudia, and Walter Leimgruber. *Was Akten bewirken können: Integrations- und Ausschlussprozesse eines Verwaltungsvorgangs*. Zürich: Seismo Verlag, 2008.

Kelley, Donald R. *Faces of History: Historical Inquiry from Herodotus to Herder*. New Haven: Yale University Press, 1998.

Kelley, Donald R. *Fortunes of History: Historical Inquiry from Herder to Huizinga*. New Haven: Yale University Press, 2003.

Kelty, Christopher M. *Two Bits*. Durham: Duke University Press, 2008.

Kepler, Johannes. *Ad Vitellionem paralipomena*. Frankfurt: Apud Claudium Marnium & Hæredes Joannis Aubrii, 1604.

Kidwell, Susan M., and J. John Sepkoski, Jr. "The Nature of the Fossil Record." In *Evolution: Investigating the Evidence*, edited by Judy Scotchmoor and Dale A Springer, 61–76. Pittsburgh: Paleontological Society, 1999.

Kimball, Ralph. "The Database Market Splits." http://www.kimballgroup.com/1995/09/01/the-database-market-splits/. Accessed June 18, 2013,

Kirk, Geoffrey S. "The Sources for Presocratic Philosophy." In *The Presocratic Philosophers*, 2nd ed., edited by Geoffrey S. Kirk, John E. Raven, and Malcolm Schofield, 1–7. Cambridge: Cambridge University Press, 1983.

Kirk, Geoffrey S., John E. Raven, and Malcolm Schofield. *The Presocratic Philosophers: A Critical History with a Selection of Texts*. 2nd ed. Cambridge: Cambridge University Press, 1983.

Kleinberg, Jon M. "Authoritative Sources in a Hyperlinked Environment." *Journal of the Association of Computing Machinery* 46, no. 5 (September 1999): 604–32. doi:10.1145/324133.324140.

Klonk, Charlotte. *Spaces of Experience: Art Gallery Interiors 1800–2000*. New Haven: Yale University Press, 2009.

Kocka, Jürgen, ed. *Die Königliche Preussische Akademie der Wissenschaften*. Berlin: Akademie Verlag, 1999.

Kögler, Ignatius. *Observationes eclipsium, variorumque caelestium congressuum habitae in Sinis*. Lucca: Typis Salvatoris, & Jo. Dominici Marescandoli, 1745.

Kohavi, Ronny. "Origin of 'Diapers and Beer,'" July 6, 2000. http://www.kdnuggets.com/news/2000/n14/8i.html.

Kohler, Robert E. *Lords of the Fly: Drosophila Genetics and the Experimental Life*. Chicago: University of Chicago Press, 1994.

Kohler, Robert E. "Paul Errington, Aldo Leopold, and Wildlife Ecology: Residential Science." *Historical Studies in the Natural Sciences* 41, no. 2 (May 2011): 216–54.

Kontler, László. "The Uses of Knowledge and the Symbolic Map of the Enlightened Monarchy of the Habsburgs: Maximilian Hell as Imperial and Royal Astronomer (1755–1792)." In *Negotiating Knowledge in Early Modern Empires: A Decentered View*, edited by László Kontler et al., 79–105. New York: Palgrave Macmillan, 2014.

Kowal, Emma E. "Orphan DNA: Indigenous Samples, Ethical Biovalue, and Postcolonial Science." *Social Studies of Science* 43, no. 4 (2013): 577–97.

Krajewski, Markus. *Paper Machines: About Cards and Catalogs, 1548–1929*. Translated by Peter Krapp. Cambridge, MA: MIT Press, [2002] 2011.

Kramer, Adam D. I., Jamie E. Guillory, and Jeffrey T. Hancock. "Experimental Evidence of Massive-scale Emotional Contagion through Social Networks." *Proceedings of the National Academy of Sciences* 111, no. 24 (June 17, 2014): 8788–90.

Kremer, Richard L. "Bernard Walther's Astronomical Observations." *Journal for the History of Astronomy* 11 (1980): 174–91.

Kremer, Richard L. "Text to Trophy: Shifting Representations of Regiomontanus's Library." In *Lost Libraries: The Destruction of Great Book Collections since Antiquity*, edited by James Raven, 75–90. Houndmills, Basingstoke, Hampshire: Palgrave Macmillan, 2004.

Kremer, Richard L. "The Use of Bernard Walther's Astronomical Observations: Theory and Observation in Early Modern Astronomy." *Journal for the History of Astronomy* 12 (1981): 124–32.

Kremer, Richard L. "Walther's Solar Observations: A Reply to R. R. Newton." *Quarterly Journal of the Royal Astronomical Society* 24 (1983): 36–47.

Kronk, Gary W. *Cometography: A Catalog of Comets*. Vol. 1. Cambridge: Cambridge University Press, 1999.

Kucera, Henry, and W. Nelson Francis. *Computational Analysis of Present-Day American English*. Providence: Brown University Press, 1967.

Kurke, Leslie. *Aesopic Conversations: Popular Tradition, Cultural Dialogue, and the Invention of Greek Prose*. Princeton: Princeton University Press, 2010.

Kurland, Leonard T., and Craig A. Molgaard. "The Patient Record in Epidemiology." *Scientific American* 245, no. 4 (1981): 46–55.

Kurtz, Michael J., and Guenther Eichhorn. "The Historical Literature of Astronomy, via ADS." *Library and Information Services in Astronomy III (LISA III): Proceedings of a Conference Held in Puerto de la Cruz, Tenerife, Spain, April 21–24, 1998*, edited by Uta Grothkopf et al, 293. San Francisco: Astronomical Society of the Pacific, 1998.

Kushner, David. "The Controversy Surrounding the Secular Acceleration of the Moon's Mean Motion." *Archive for History of Exact Sciences* 39, no. 4 (1989): 291–316.

Lacaille, Nicolas-Louis de. "Extrait de quelques observations astronomiques, faites au Collège Mazarin pendant l'année 1743." *Histoire de l'Académie royale des sciences: année [1743]: avec les mémoires de mathématique & de physique . . . tirés des registres de cette académie* (1746): 159–90.

La Condamine, Charles-Marie de. *Mesure des trois premiers degrés du méridien dans l'hémisphere austral: tirée des observations de Mrs. de l'Académie royale des sciences, envoyés par le roi sous l'équateur*. Paris: De l'Imprimerie royale, 1751.

Lalande, Joseph-Jérôme. *Histoire céleste française, contenant les observations faites par plusieurs astronomes français*. Paris: De l'Imprimerie de la république, 1801.

Lamy, Jérôme, ed. *La Carte du Ciel: Histoire et actualité d'un projet scientifique international*. Paris: Observatoire de Paris, 2008.

Lamy, Jérôme. "La Carte du Ciel et l'ajustement des pratiques (fin XIXe–debut XXe siècle)." In *La Carte du Ciel: Histoire et actualité d'un projet scientifique international*, edited by Jérôme Lamy, 45–67. Paris: Observatoire de Paris, 2008.

Langville, Amy N., and Carl D. Meyer. *Google's PageRank and Beyond: The Science of Search Engine Rankings*. Princeton: Princeton University Press, 2006.

Lankford, John. "Amateurs and Astrophysics: A Neglected Aspect in the Development of a Scientific Specialty." *Social Studies of Science* 11 (1981): 275–303.

Lankford, John. "The Impact of Photography on Astronomy." In *Astrophysics and Twentieth-Century Astronomy to 1950*, edited by Owen Gingerich, 16–39. Cambridge: Cambridge University Press, 1984.

Lansbergen, Philips van. *Tabulae motuum coelestium perpetuae; ex omnium temporum obser-vationibus constructae, temporumque omnium observationibus consentientes. Item novae et genuinae motuum coelestium theoricae. & astronomicarum observationum thesaurus*. Middelburg: Apud Zachariam Romanum, 1632.

Larfeld, Wilhelm. *Handbuch der griechischen Epigraphik*. 2 vols. Leipzig: O. R. Reisland, 1907.

Laval, Antoine Francois. *Voyage de la Louisiane, fait par ordre du roy en l'année mil sept cent vingt: dans lequel sont traitées diverses matieres de physique, astronomie, géographie & marine*. Paris: Chez Jean Mariette, 1728.

La Vopa, Anthony J. *Grace, Talent, and Merit: Poor Students, Clerical Careers, and Profes-sional Ideology in Eighteenth-Century Germany*. Cambridge: Cambridge University Press, 1988.

Laycock, S., et al. "Digital Access to a Sky Century at Harvard: Initial Photometry and Astrometry." *Astronomical Journal* 140, no. 4 (2010): 1062–77.

Leese, John A, Arthur L. Booth, and Frederic A. Godshall. *Archiving and Climatological Applications of Meteorological Satellite Data*. ESSA Tech. Report NESC 53. Suitland, MD: NESS, 1970.

Legaspi, Michael C. *The Death of Scripture and the Rise of Biblical Studies*. Oxford: Oxford University Press, 2010.

Lemov, Rebecca. "Filing the Total Human Experience: Anthropological Archives at Mid-Twentieth Century." In *Social Knowledge in the Making*, edited by Charles Camic, Neil Gross, and Michèle Lamont, 119–50. Chicago: University of Chicago Press, 2011.

Le Monnier, Pierre-Charles. *Histoire celeste, ou Recueil de toutes les observations astronomiques faites par ordre du roy*. Paris: Chez Briasson, 1741.

Le Monnier, Pierre-Charles. *Observations de la lune, du soleil, et des étoiles fixes: pour servir a la physique celeste et aux usages de la navigation*. Paris: De l'Imprimerie royale, 1751–73.

Le Monnier, Pierre-Charles. "Recherches sur la hauteur du pole de Paris [14 June 1738]." *Histoire de l'Académie royale des sciences: année [1738]: avec les memoires de mathematique & de physique . . . tirés des registres de cette académie* (1740): 209–25.

Leonelli, Sabina. "Packaging Small Facts for Re-use: Databases in Model Organism Biology." In *How Well Do Facts Travel?*, edited by Peter Howlett and Mary S. Morgan, 325–48. Cambridge: Cambridge University Press, 2011.

Leonelli, Sabina, and Rachel A. Ankeny. "Re-Thinking Organisms: The Impact of Databases on Model Organism Biology." *Studies in History and Philosophy of Science Part C: Studies in History and Philosophy of Biological and Biomedical Sciences* 43, no. 1 (March 2012): 29–36. doi:10.1016/j.shpsc.2011.10.003.

Leopold, J. H. "Christiaan Huygens and His Instrument Makers." In *Studies on Christiaan Huygens*, edited by H. J. M. Bos et al., 221–33. Lisse: Swets & Zeitlinger, 1980.

Leroux des Tillets, Jean-Jacques. *Commission de l'Instruction publique. Académie de Paris: Faculté de Médecine—Clinique interne: Société d'Instruction médical: règlement*. Paris: Migneret, 1818.

Lévi-Strauss, Claude. *The Savage Mind*. Chicago: University of Chicago Press, 1966.

Lewin, Richard. "Proposal to Sequence the Human Genome Stirs Debate." *Science* 232, no. 4758 (June 27, 1986): 1598–1600.

"Librarians Stop Abortion Stop-Listing." *American Libraries* 39, no. 5 (May 2008): 23.

Lieberman, Bruce S., and Roger Kaesler. *Prehistoric Life: Evolution and the Fossil Record*. Oxford: Blackwell Publishing, 2010.

Lieutaud, Joseph, and Antoine Portal. *Historia anatomico-medico*. Paris: Vincent, 1767.

Light, Jennifer S. "When Computers Were Women." *Technology and Culture* 40, no. 3 (July 1999): 455–83.

Linacre, Edward. *Climate Data and Resources: A Reference and Guide*. London: Routledge, 1992.

Lipphardt, Veronika. "The Jewish Community of Rome: An Isolated Population? Sampling Procedures and Bio-Historical Narratives in Genetic Analysis in the 1950s." *BioSocieties* 5, no. 3 (September 2010): 306–29.

Lloyd, G. E. R. *Aristotelian Explorations*. Cambridge: Cambridge University Press, 1996.

Lorace, Loretta. *Becoming Bucky Fuller*. Cambridge: MIT Press, 2009.

Louis, Pierre. *Aristote: Problème, texte établi et traduit*. Paris: Belles Lettres, 1991–94.

Love, Heather. "Close but Not Deep: Literary Ethics and the Descriptive Turn." *New Literary History* 41, no. 2 (Spring 2010): 371–91.

Lovejoy, Arthur O. *The Great Chain of Being: A Study in the History of an Idea*. Baltimore: Johns Hopkins University Press, 1948.

Luhn, Hans Peter. "The Automatic Creation of Literature Abstracts." *IBM Journal of Research and Development* 2, no. 2 (April 1958): 159–65.

Luhn, Hans Peter. "A Business Intelligence System." *IBM Journal of Research and Development* 2, no. 4 (October 1958): 314–19.

Luhn, Hans Peter. "Keyword-in-Context Index for Technical Literature." *American Documentation* 11, no. 4 (1960): 288–95.

Luhn, Hans Peter. "A New Method of Recording and Searching Information." *American Documentation* 4, no. 1 (1953): 14–16.

Luhn, Hans Peter. "Selective Dissemination of New Scientific Information with the Aid of Electronic Processing Equipment." *American Documentation* 12, no. 2 (1961): 131–38.

Luhn, Hans Peter. "A Statistical Approach to Mechanized Encoding and Searching of Literary Information." *IBM Journal of Research and Development* 1, no. 4 (October 1957): 309–17.

Luhn, Hans Peter, and P. C. Janaske, eds. *Automation and Scientific Communication: Short Papers Contributed to the Theme Sessions of the 26th Annual Meeting of the American Documentation Institute at Chicago, Pick-Congress Hotel, October 6–11, 1963 (American Documentation Institute, 1963).* American Documentation Institute, 1963.

Lukashova, M. V., and L. I. Rumyantseva. "Canon of Solar Eclipses from 1000 to 2050 for Russia." *Solar System Research* 32, no. 2 (1998): 166–70.

Luther, Frederic. *Microfilm: A History, 1839–1900.* Annapolis: National Microfilm Association, 1959.

Lyell, Charles. *Principles of Geology.* London: John Murray, 1830.

Lyell, Charles. *Principles of Geology.* 2nd ed. London: John Murray, 1853.

Macaulay, Thomas B. "History." In *The Works of Lord Macaulay: Essays and Biographies*, vol. 1 (vol. 7 of collected works). London: Longmans, 1896.

Mach, Ernst. *Die Geschichte und die Wurzel des Satzes von der Erhaltung der Arbeit.* 2nd ed. Leipzig: Johann Ambrosius Barth, [1872] 1909.

Maddox, John. "Making Authors Toe the Line." *Nature* 342 (1989): 855.

Mahoney, Michael S. "Christiaan Huygens, the Measurement of Time and Longitude at Sea." In *Studies on Christiaan Huygens*, edited by H. J. M. Bos et al., 234–70. Lisse: Swets & Zeitlinger, 1980.

Mallarmé, Stéphane. *Divagations.* Paris: Bibliothéque-Charpentier, 1897.

Manetti, Daniela. "The Role of Doxography." In *Ancient Histories of Medicine: Essays in Medical Doxography and Historiography in Classical Antiquity*, edited by Philip J. Van der Eijk, 95–141. Leiden: Brill, 1999.

Manfredi, Eustachio. *De gnomone meridiano bononiensi ad divi petronii: deque observationibus astronomicis eo instrumento ab ejus constructione.* Bologna: Ex Typographia Laelii a Vulpe, 1736.

Manoff, Marlene. "Theories of the Archive from Across the Disciplines." *Libraries and the Academy* 4 (2004): 9–25.

Manovich, Lev. "The Database as Symbolic Form." *Convergence* 5, no. 2 (June 1999): 80–99.

Manovich, Lev. "Database as Symbolic Form." In *The Database Aesthetics: Art in the Age of Information Overflow*, edited by Victoria Vesna, 39–60. Minneapolis: University of Minnesota Press, 2007.

Manovich, Lev. "The Database as Symbolic Form." In *The Language of New Media*, 218–43. Cambridge, MA: MIT Press, 2001.

Mansfeld, Jaap. "Aristotle, Plato, and the Preplatonic Doxography and Chronography." In *Storiografia e dossografia nella filosofia antica*, edited by G. Cambiano, 1–59. Turin:

Tirrenia, 1986. Reprinted in Jaap Mansfeld, *Studies in the Historiography of Greek Philosophy* (Assen: Van Gorcum, 1990).

Mansfeld, Jaap. "Deconstructing Doxography." *Philologus* 146 (2002): 277–86.

Mansfeld, Jaap. "Doxography of Ancient Philosophy." In *The Stanford Encyclopedia of Philosophy* (Spring 2004 ed.), edited by Edward N. Zalta. http://plato.stanford.edu/archives/spr2004/entries/doxography-ancient/.

Mansfeld, Jaap. "Doxography of Ancient Philosophy." In *The Stanford Encyclopedia of Philosophy* (Summer 2012 ed.), edited by Edward N. Zalta. http://plato.stanford.edu/archives/sum2012/entries/doxography-ancient/.

Mansfeld, Jaap. "Physikai doxai and Problemata physica from Aristotle to Aëtius (and Beyond)." In *Theophrastus: his Psychological, Doxographical and Scientific Writings*, edited by W. W. Fortenbaugh and D. Gutas, 63–111. New Brunswick: Rutgers University Studies in Classical Humanities (RUSCH) 5, 1992.

Mansfeld, Jaap. *Prolegomena: Questions to Be Settled before the Study of an Author, or a Text.* Leiden: Brill, 1994.

Mansfeld, Jaap. "Sources." In *The Cambridge History of Hellenistic Philosophy*, edited by Keimpe Algra, Jonathan Barnes, Jaap Mansfeld, and Malcolm Schofield, 3–30. Cambridge: Cambridge University Press, 1999.

Mansfeld, Jaap, and D. T. Runia. *Aëtiana: The Method and Intellectual Context of a Doxographer.* Leiden: Brill, 1997.

Mantell, Gideon Algernon. *The Fossils of the South Downs; or, Illustrations of the Geology of Sussex.* London: Lupton Relfe, 1822.

Manuel, Frank E. *Isaac Newton: Historian.* New York: Belknap Press, 1963.

Marchand, Suzanne. *Down from Olympus: Archaeology and Philhellenism in Germany, 1750–1970.* Princeton: Princeton University Press, 1996.

Marchand, Suzanne. *German Orientalism in the Age of Empire: Religion, Race, and Scholarship.* New York: Cambridge University Press, 2009.

Marchand, Suzanne. "Where Does History Begin? J. G. Herder and the Problem of Near Eastern Chronology in the Age of Enlightenment." *Eighteenth-Century Studies* 47, no. 2 (2014): 157–75.

Markow, Therese Ann, P. W. Hedrick, K. Zuerlein, J. Danilovs, J. Martin, T. Vyvial, and C. Armstrong. "HLA Polymorphism in the Havasupai: Evidence for Balancing Selection." *American Journal of Human Genetics* 53, no. 4 (October 1993): 943–52.

Markow, Therese Ann, and Irving I. Gottesman. "Behavioral Phenodeviance: A Lerneresque Conjecture." *Genetica* 89, no. 1 (1993): 297–305.

Markow, Therese Ann, and J. F. Martin. "Inbreeding and Developmental Stability in a Small Human Population." *Annals of Human Biology* 20, no. 4 (1993): 389–94.

Markow, Therese Ann, and Kevin Wandler. "Fluctuating Dermatoglyphic Asymmetry and the Genetics of Liability to Schizophrenia." *Psychiatry Research* 19, no. 4 (December 1986): 323–28.

Maskelyne, Nevil. *Astronomical observations made at the Royal Observatory at Greenwich, in the years 1765, 1766, 1767, 1768, and 1769. . . . Published by the President and Council of the Royal Society, at the public expence, in obedience to His Majesty's Command.* London: Printed by W. and J. Richardson, 1774.

Maslow, Abraham. "A Theory of Human Motivation." *Psychological Review* 50, no. 4 (1943): 370–96.

Maupertuis, Pierre-Louis Moreau de. *La figure de la terre, déterminée par les observations de Messieurs de Maupertuis, Clairaut, Camus, Le Monnier . . . & de M. l'Abbé Out-*

hier... *accompagnés de M. Celsius... faites par ordre du roy au cercle polaire.* Paris: De l'Imprimerie royale, 1738.

Mayer, Christian. *Solis et lunae eclipseos observatio astronomica... facta Schwezingae in specula nova electorali... comparata pluribus Europae celebrioribus observationibus.* Mannheim: Ex Typographejo Electorali-Aulico, 1766.

Mayer-Schönberger, Viktor. *Delete: The Virtue of Forgetting in the Digital Age.* Princeton: Princeton University Press, 2011.

Mayer-Schönberger, Viktor, and Kenneth Cukier. *Big Data: A Revolution That Will Transform How We Live, Work, and Think.* Boston: Houghton Mifflin Harcourt, 2013.

Mayhew, Robert. *Aristotle: Problems.* Cambridge, MA: Harvard University Press, 2011.

Mayo, Herbert. "An Account of Some Cases of Slowness of the Pulse." *London Medical Gazette* 22 (1838): 232–38.

McAllister, James. "Climate Controversies and the Demand for Access to Empirical Data." *Philosophy of Science* 79 (2012): 871–80.

McCarthy, John. "Phenomenal Data Mining: From Data to Phenomena." *ACM SIGKDD Explorations Newsletter* 1, no. 2 (2000): 24–29.

McCarty, Willard. "Beyond the Word: Modelling Literary Context." Canadian Symposium on Text Analysis, Fredricton, NB, 2006.

McClellan, James E. *Specialist Control: The Publications Committee of the Académie royale des sciences (Paris), 1700-1793.* Philadelphia: American Philosophical Society, 2003.

McCray, W. Patrick. "Amateur Scientists, the International Geophysical Year, and the Ambitions of Fred Whipple." *Isis* 97, no. 4 (December 2006): 634–58.

McGann, Jerome. "Information Technology and the Troubled Humanities." In *The Digitial Humanities: A Reader,* edited by Melissa Terras, Julianne Nyhan, and Edward Vanhoutte, 49–66. Burlington, VT: Ashgate, 2013.

McKenzie, Kenneth. "Means and End in Making a Concordance, with Special Reference to Dante and Petrarch." In *Twenty-Fifth Annual Report of the Dante Society*, 18–46. Boston: Ginn & Co, 1907.

McKeon, Robert M. "Les débuts de l'astronomie de précision." *Physis* 13 (1971): 225–88; 14 (1972): 221–42.

Meinel, Christoph. "Enzyklopädie der Welt und Verzettelung des Wissens: Aporien der Empirie bei Joachim Jungius." In *Enzyklopädien der Frühen Neuzeit: Beiträge zu ihrer Forschung,* edited by Franz M. Eybl et al., 162–87. Tübingen: Max Niemeyer, 1995.

Meiners, Christoph. *Versuch über die Religionsgeschichte der älteste Völker besonders der Egyptier.* Göttingen: Johann Christian Dieterich, 1775.

Mejer, Jørgen. "Ancient Philosophy and the Doxographical Tradition." In *A Companion to Ancient Philosophy,* edited by Mary Louise Gill and Pierre Pellegrin, 20–33. Oxford: Blackwell, 2006.

Mejer, Jørgen. *Diogenes Laertius and His Hellenistic Background.* Wiesbaden: Steiner, 1978.

Mello, Michelle M., and Leslie E. Wolf. "The Havasupai Indian Tribe Case: Lessons for Research Involving Stored Biologic Samples." *New England Journal of Medicine* 363, no. 3 (2010): 204–7.

Mendelsohn, J. Andrew. "The World on a Page: Making a General Observation in the Eighteenth Century." In *Histories of Scientific Observation,* edited by Lorraine Daston and Elizabeth Lunbeck, 396–420. Chicago: University of Chicago Press, 2011.

Mendelsohn, J. Andrew, and Volker Hess. "Case and Series: Medical Knowledge and Paper Technology, 1600-1900." *History of Science* 48 (2010): 287–314.

Mercator, Gerhard. *Chronologia. Hoc est, Temporum demonstratio exactissima ab initio mundi usque ad Annum Domini M. D. LXVIII, ex eclipsibus et obseruationibus astronomicis omnium temporum*. Cologne: Apud hæredes Arnoldi Birckmanni, 1569.

Mercator, Gerhard, ed. *Tabulae geographicae Cl: Ptolemei ad mentem autoris restitutae & emendate*. [Cologne]: [typis Godefridi Kempensis], 1578.

Middlesex Hospital. "Cases of Bronzed Skin, etc.," communicated, with Remarks, by S. W. Sibley. *Medical Times & Gazette* 1 (1856): 188–89.

Miller, Peter N. "Mapping Peiresc's Mediterranean: Geography and Astronomy, 1610–36." In *Communicating Observations in Early Modern letters (1500–1675): Epistolography and Epistemology in the Age of the Scientific Revolution*, edited by Dirk van Miert, 135–60. London: Warburg Institute, 2013.

Mitman, Gregg, and Kelley Wilder, eds. *Documenting the World: Film, Photography, and the Scientific Record*. Chicago: University of Chicago Press, 2016.

Momigliano, Arnaldo. "Ancient History and the Antiquarian." *Journal of the Warburg and Courtauld Institutes* 13 (1950): 285–315.

Momigliano, Arnaldo. *The Classical Foundations of Modern Historiography*. Berkeley: University of California Press, 1990.

Momigliano, Arnaldo. "Gibbon's Contribution to Historical Method." In Arnaldo Momigliano, *Contributo alla Storia degli Studi Classici*, 195–211. Rome: Edizione di Storia e Letteratura, 1955.

Momigliano, Arnaldo. "The Place of Ancient History in Modern Historiography." In *Settimo Contributo alla Storia degli Studi Classici e del Mondo Antico*, edited by Arnaldo Momigliano, 13–36. Rome: Edizioni di Storia e Letteratura, 1984.

Mommsen, Adelheid. *Theodor Mommsen im Kreise der Seinen*. 2nd ed. Berlin: Verlag E. Ebering, 1937.

Mommsen, Theodor. *Reden und Aufsätze*. 2nd ed. Berlin: Weidmannsche Buchhandlung, 1905.

Mommsen, Theodor. *Tagebuch der französisch-italienischen Reise 1844/1845*, edited by Gerold and Brigitte Walser. Bern: Verlag Herbert Lang, 1976.

Mommsen, Theodor. *Ueber Plan und Ausführung eines Corpus Inscriptionum Latinarum*. Berlin: A. W. Schade, 1847.

Moore, Lara Jennifer. *Restoring Order: The École des Chartes and the Organization of Archives and Libraries in France, 1820–1870*. Duluth, MN: Litwin Books, 2001.

Moran, Bruce T. "Wilhelm IV of Hesse-Kassel: Informal Communication and the Aristocratic Context of Discovery." In *Scientific Discovery: Case Studies*, edited by Thomas Nickles, 67–96. Dordrecht: D. Reidel Pub. Co, 1980.

Moretti, Franco. *Distant Reading*. London: Verso, 2013.

Morgagni, Giovanni Battista. *The Seats and Causes of Diseases Investigated by Anatomy*. 3 vols. Translated by Benjamin Alexander. London: Printed for A. Millar, T. Cadell, and Johnson and Payne, 1769.

Morrell, Jack. *John Phillips and the Business of Victorian Science*. Aldershot, UK: Ashgate, 2005.

Morrison, Leslie V., M. R. Lukac, and F. Richardson Stephenson. "Catalogue of Observations of Occultations of Stars by the Moon for the Years 1623 to 1942 and Solar Eclipses for the Years 1621 to 1806." *Royal Greenwich Observatory Bulletins* 186 (1981): 5–7, 36–54.

Morrison, Leslie V., and F. Richard Stephenson. "Historical Values of the Earth's Clock Error ΔT and the Calculation of Eclipses." *Journal for the History of Astronomy* 35, no. 3 (2004): 327–36.

Mosley, Adam. *Bearing the Heavens: Tycho Brahe and the Astronomical Community of the Late Sixteenth Century.* Cambridge: Cambridge University Press, 2007.

Mosley, Adam. "Past Portents Predict: Cometary Historiae and Catalogues in the Sixteenth and Seventeenth Centuries." In *Celestial Novelties on the Eve of the Scientific Revolution: 1540–1630,* edited by Dario Tessicini and Patrick Boner, 1–32. Florence: L. S. Olschki, 2013.

Mosley, Adam. "Reading the Heavens: Observation and Interpretation of Astronomical Phenomena in Learned Letters circa 1600." In *Communicating Observations in Early Modern Letters (1500–1675): Epistolography and Epistemology in the Age of the Scientific Revolution,* edited by Dirk van Miert, 115–34. London: Warburg Institute, 2013.

Most, Glenn W. *Doubting Thomas.* Cambridge, MA: Harvard University Press, 2007.

Most, Glenn W. "*Quellenforschung.*" In *The Making of the Humanities,* edited by Rens Bod, Jaap Maat, and Thijs Weststeijn, 207–17. Amsterdam: Amsterdam University Press, 2014.

Mouchez, Ernest B. *La Photographie astronomique à l'Observatoire de Paris et la Carte du Ciel.* Paris: Gauthier-Villars, 1887.

Muhlack, Ulrich. *Geschichtswissenschaft im Humanismus und in der Aufklärung.* Munich: C. H. Beck, 1991.

Muhlack, Ulrich. "Herodotus and Thucydides in the View of Nineteenth-Century German Historians." In *The Western Time of Ancient History,* edited by Alexandra Lianeri, 179–209. Cambridge: Cambridge University Press, 2011.

Müller-Wille, Staffan. "Ein Anfang ohne Ende: Das Archiv der Naturgeschichte und die Geburt der Biologie." In *Macht des Wissens: Die Entstehung der modernen Wissensgesellschaft,* edited by R. van Dülmen and S. Rauschenbach, 587–605. Cologne: Böhlau Verlag, 2004.

Müller-Wille, Staffan. "Claude Lévi-Strauss on Race, History, and Genetics." *BioSocieties* 5, no. 3 (2010): 330–47.

National Academy of Sciences. *Report of the Climate Board Ad Hoc Panel on Climate Impacts to the National Climate Program Office regarding Social Science Climate Impact Research.* Washington, DC: National Academy of Sciences, 1981.

National Academy of Sciences. *Report of the Climate Data Management Workshop.* Washington, DC: US Department of Commerce, 1980.

National Academy of Sciences. *A Strategy for the National Climate Program: Report of the Workshop to Review the Preliminary National Climate Program Plan, Woods Hole, MA, July 16–21, 1979.* Washington, DC: National Academy of Sciences, 1980.

National Academy of Sciences. *Toward a U.S. Climate Program Plan: Report of the Workshop to Review the U.S. Climate Program Plans.* Washington, DC: National Academy of Sciences, 1978.

National Academy of Sciences. *Understanding Climatic Change: A Program for Action.* Washington, DC: National Academy of Sciences, 1975.

National Climate Program: Early Achievements and Future Directions. Report of the Woods Hole Workshop July 15–19, 1985.

National Climate Program Act, 95-367 (Sept. 17, 1978). www.epw.senate.gov/ncpa.pdf, accessed 1 July 2014.

National Research Council. *Meeting the Challenge of Climate.* Washington, DC: National Academies Press, 1982.

Nelson, Theodor H. *Literary Machines.* Swarthmore, PA: Theodor H. Nelson, 1987.

Neugebauer, Otto. *Astronomical Cuneiform Texts: Babylonian Ephemerides of the Seleucid*

Period for the Motion of the Sun, the Moon, and the Planets. London: Published for the Institute for Advanced Study by Lund Humphries, 1955.

Neugebauer, Otto. *A History of Ancient Mathematical Astronomy*. Berlin: Springer-Verlag, 1975.

Neugebauer, Paul V., and Richard Hiller. "Spezieller Kanon der Mondfinsternisse für Vorderasien und Ägypten von 3450 bis 1 v. Chr." *Astronomische Abhandlungen* 9, no. 2 (1934).

Neugebauer, Paul V., and Richard Hiller. "Spezieller Kanon der Sonnenfinsternisse für Vorderasien und Ägypten für die Zeit von 900 v. Chr. bis 4200 v. Chr." *Astronomische Abhandlungen* 8, no. 4 (1931).

Newhall, X. X., E. M. Standish, and J. G. Williams. "DE 102: A Numerically Integrated Ephemeris of the Moon and Planets Spanning Forty-four Centuries." *Astronomy and Astrophysics* 125, no. 1 (1983): 150–67.

Newton, Robert R. "An Analysis of the Solar Observations of Regiomontanus and Walther." *Quarterly Journal of the Royal Astronomical Society* 23 (1982): 67–93.

Newton, Robert R. *Ancient Astronomical Observations and the Accelerations of the Earth and Moon*. Baltimore: Johns Hopkins University Press, 1970.

Newton, Robert R. *Ancient Planetary Observations and the Validity of Ephemeris Time*. Baltimore: Johns Hopkins University Press, 1976.

Newton, Robert R. *A Canon of Lunar Eclipses for the Years –1500 to –1000*. Laurel, MD: Johns Hopkins University, Applied Physics Laboratory, 1977.

Newton, Robert R. *The Crime of Claudius Ptolemy*. Baltimore: Johns Hopkins University Press, 1977.

Newton, Robert R. *Medieval Chronicles and the Rotation of the Earth*. Baltimore: Johns Hopkins University Press, 1972.

Newton, Robert R. *The Moon's Acceleration and Its Physical Origins*. 2 vols. Baltimore: Johns Hopkins University Press, 1979–84.

Nielsen, Michael A. *Reinventing Discovery: The New Era of Networked Science*. Princeton: Princeton University Press, 2012.

Nierenberg, Nicolas, Walter R. Tschinkel, and Victoria J. Tschinkel. "Early Climate Change Consensus at the National Academy: The Origins and Making of *Changing Climate*." *Historical Studies in the Natural Sciences* 40 (2010): 318–49.

Noel, François. *Observationes mathematicæ, et physicæ in India et China*. Prague: typis Universit: Carolo Fernandeæ, in Collegio Soc. Jesu ad S. Clementem, per Joachimum Joannem Kamenicky Factorem, 1710.

Nolan, Richard L. "Information Technology Management since 1960." In *Nation Transformed by Information: How Information Has Shaped the United States from Colonial Times to the Present*, edited by Alfred Dupont Chandler and James W. Cortada, 217–56. New York: Oxford University Press, 2000.

Nyhan, Julianne, Andrew Flynn, and Anne Welsh. "Oral History and the Hidden Histories Project: Towards Histories of Computing in the Humanities." *Literary and Linguistic Computing*. First published online (30 July 2013) at http://dsh.oxfordjournals.org/content/early/2015/01/11/llc.fqt044.

Oertzen, Christine von. "Science in the Cradle: Milicent Shinn and Her Home-Based Network of Baby Observers, 1890–1910." *Centaurus* 55, no. 2 (2013): 175–95.

"Of the Judgement of Some of the English Astronomers." *Philosophical transactions* 9 (February 12, 1665/6): 150–51.

Ogilvie, Brian W. *The Science of Describing: Natural History in Renaissance Europe*. Chicago: University of Chicago Press, 2006.

Ogle, John. "Fibrinous Deposit Infiltrated, and in Masses within the Substance of the Walls of the Heart: Tendency to the Formation of an Aneurismal Pouch: Peculiarities in the Pulse." *Transactions of the Pathological Society of London* 8 (1857): 118–21.

O'Hara, Kieran, et al. "Memories for Life: A Review of the Science and Technology." *Journal of the Royal Society Interface* 3, no. 8 (2006): 351–65.

O'Hora, Nathy P. "Astrographic Catalogues of British Observatories." In *Mapping the Sky: Past Heritage and Future Directions*, edited by Suzanne Débarat, J. A. Eddy, H. K. Eichhorn, and A. R. Upgren, 135–38. Dordrecht: Kluwer, 1988.

Olmsted, J. W. "The 'Application' of Telescopes to Astronomical Instruments, 1667–1669: A Study in Historical Method." *Isis* 40, no. 3 (1949): 213–25.

Ophuijsen, Johannes M. van. "Where Have All the Topics Gone?" In *Peripatetic Rhetoric after Aristotle*, edited by W. W. Fortenbaugh and D. Mirhady, 131–73. New Brunswick, NJ: Transaction Publishers, 1994.

Oppolzer, Theodor. *Canon der Finsternisse*. Vienna: aus der Kaiserlich-Königlichen Hof- und Staatsdruckerei in Commission bei K. Gerold, 1887.

Oppolzer, Theodor. *Canon of Eclipses: Canon der Finsternisse*. Translated by Owen Gingerich. New York: Dover Publications, 1962.

Opsomer, Jan. "Zetematic Structures in Plutarch." Paper given in a 28 Workshop on Ancient Scientific, Technical and Medical Writing, at Topoi (Berlin), March 15, 2013.

Osborne, Thomas. "The Ordinariness of the Archive." *History of the Human Sciences* 12 (1999): 51–64.

Ossendrijver, Mathieu. *Babylonian Mathematical Astronomy*. Dordrecht: Springer, 2012.

Ossendrijver, Mathieu. "Science in Action: Networks in Babylonian Astronomy." In *Babylon: Wissenskultur in Orient und Okzident*, edited by Eva Christiane Cancik-Kirschbaum, Margarete van Ess, and Joachim Marzahn, 213–21. Berlin: Walter de Gruyter, 2011.

Ottner, Christine, and Klaus Ries, eds. *Geschichtsforschung in Deutschland und Österreich im 19. Jahrhundert: Ideen—Akteure—Institutionen*. Stuttgart: Franz Steiner Verlag, 2014.

Outhier, Réginald. *Journal d'un voyage au nord, en 1736. & 1737*. Paris: Chez Piget . . . Durand, 1744.

Palmer, Carole L., Nicholas M. Weber, and Melissa H. Cragin. "The Analytic Potential of Scientific Data: Understanding Re-Use Value." ASIST 2011, October 9–13, New Orleans, LA. http://www.asis.org/asist2011/proceedings/submissions/174_FINAL _SUBMISSION.pdf.

Pankenier, David W. "On the Reliability of Han Dynasty Solar Eclipse Records." *Journal of Astronomical History and Heritage* 15, no. 3 (2012): 200–212.

Parham, P., K. L. Arnett, E. J. Adams, A.-M. Little, K. Tees, L. D. Barber, S. G. E. Marsh, T. Ohta, T. Markow, and M. L. Petzl-Erler. "Episodic Evolution and Turnover of HLA-B in the Indigenous Human Populations of the Americas." *Tissue Antigens* 50, no. 3 (1997): 219–32.

Parkins, Phyllis V. "Approaches to Vocabulary Management in Permuted Title Indexing of Biological Abstracts." In *Automation and Scientific Communication: Proceedings of the ADI Annual Meeting*, 27–28. Washington, DC: ADI, 1963.

Parkins, Phyllis V. "The Computerization of Biological Abstracts." In *Biological Abstracts/ BIOSIS*, 123–39. New York: Springer, 1976.

Parks, Cara. "Books of Forgetting: Why We Can't Stop Writing about What We Can't Remember." *New Republic,* May 12, 2014.

Parry, Bronwyn. *Trading the Genome: Investigating the Commodification of Bio-Information.* New York: Columbia University Press, 2004.

Pasquale, Frank. "The Algorithmic Self." *Hedgehog Review* 17, no. 1 (Spring 2015). http://www.iasc-culture.org/THR/THR_article_2015_Spring_Pasquale.php.

Pasquale, Frank. *The Black Box Society: The Secret Algorithms That Control Money and Information.* Cambridge, MA: Harvard University Press, 2015.

Patzer, A. *Der Sophist Hippias als Philosophiehistoriker.* Freiburg: Alber, 1986.

Pearson, Lionel, and F. H. Sandbach. *Plutarch Moralia XI.* Cambridge, MA: Harvard University Press, 1965.

Pedersen, Olaf. *A Survey of the Almagest.* Edited by Alexander Jones. New York: Springer, 2011.

Pennebaker, J. W. *The Secret Life of Pronouns: What Our Words Say about Us.* New York: Bloomsbury Press, 2011.

Perec, Georges. *An Attempt at Exhausting a Place in Paris.* Translated by Marc Lowenthal. Cambridge, MA: Wakefield Press, 2010.

Perec, Georges. *La Disparition.* Paris: Gallimard, 1989.

Perec, Georges. *Espèces d'espaces.* Paris: Editions Galilée, 1974.

Perec, Georges. *L'infra-ordinaire.* Paris: Seuil, 1989.

Perec, Georges. *Species of Spaces and Other Pieces.* London: Penguin Books, 1997.

Perutz, Max F. "Refinement of Hemoglobin and Myoglobin." *Acta Crystallographica Section A,* supplement S (1975): 31.

Petau, Denis. *Opus de doctrina temporum,* 4 vols. Paris: Sumptibus Sebastiani Cramoisy, 1627.

Peters, Francis E. *Greek Philosophical Terms: A Historical Lexicon.* New York: New York University Press, 1967.

Peurbach, Georg von. "Canones pro compositione et usu gnomonis geometrici." In *Scripta clarissimi mathematici M. Ioannis Regiomontani,* edited by Johann Schöner, 61r–78v. Nuremberg: Apud Ioannem Montanum & Vlricum Neuber, 1544.

Peurbach, Georg von. *Quadratu[m] geometricu[m] praeclarissimi mathematici Georgii Burbachii.* In *Scripta clarissimi mathematici M. Ioannis Regiomontani,* edited by Johann Schöner. Nuremberg: Impressum . . . per Ioannem Stuchs, 1516.

Pfeiffer, Rudolph. *History of Classical Scholarship from the Beginning to the End of the Hellenistic Age.* Oxford: Clarendon Press, [1968] 1978.

Phillips, John. *Illustrations of the Geology of Yorkshire; or a Description of the Strata and Organic Remains of the Yorkshire.* York: Thomas Wilson and Sons, 1829.

Phillips, John. *Life on the Earth: Its Origin and Succession.* London: Macmillan and Co., 1860.

Pickstone, John V. *Ways of Knowing: A New History of Science, Technology, and Medicine.* Chicago: University of Chicago Press, 2001.

Picolet, Guy, ed. *Jean Picard et les débuts de l'astronomie de précision au XVIIe siècle.* Paris: Editions du CNRS, 1987.

Pinches, Theophilus G., and J. N. Strassmaier. *Late Babylonian Astronomical and Related Texts,* edited by A. J. Sachs and J. Schaumberger. Providence: Brown University Press, 1955.

Pine, Jason. "Meth Labs, Alchemical Ontology, and Homespun Worlds." Harvard University Anthropology Department Colloquium, March 2013.

Pingré, Alexandre-Guy. *Annales célestes du dix-septième siècle*, edited by Guillaume Bigourdan. Paris: Gauthier-Villars, 1901.

Pingré, Alexandre-Guy. *Cométographie, ou, traité historique et théorique des comètes.* 2 vols. Paris: De l'Imprimerie royale, 1783.

Pingré, Alexandre-Guy. "Chronologie des éclipses de soleil & de lune qui ont été visibles sur terre, depuis le pôle boréal jusque vers l'equateur, durant les dix siècles qui ont précédé l'ère chrétienne, par M. Pingré." *Histoire de l'Académie royale des inscriptions et belles-lettres [HAIBL 1776-1779]* 42 (1786): 78–150.

Pingré, Alexandre-Guy. "Discours préliminaire sur la chronologie des éclipses" and "Chronologie des eclipses." In Maur-François Dantine et al., *L'art de vérifier les dates de faits historiques, des chartes, des chroniques, et autres anciens monumens depuis la naissance de notre seigneur. . . . Ouvrage nécessaire à ceux qui veulent avoir un parfaite conoissance de l'histoire. Par des religieux benedictins de la congrégation de S. Maur,* 39–89. 2nd ed. Paris: Chez G. Desprez, 1770, 39–89.

Pitt, Joseph C. "The Epistemology of the Very Small." In *Discovering the Nanoscale,* edited by Davis Baird, Alfred Nordmann, and Joachim Schummer, 157–63. Amsterdam: IOS Press, 2004.

Plante, Raymond L., et al. "Building Archives in the Virtual Observatory Era." *Software and Cyberinfrastructure for Astronomy: 27–30 June 2010, San Diego, California, United States,* edited by Nicole M. Radziwill and Alan Bridger. Bellingham, WA: SPIE, 2010, 77400K-1-12.

Pliny. *Historia naturalis. Natural History.* Vol. 1. Translated by H. Rackham. Cambridge, MA: Harvard University Press, 1942.

Plutarch. *Moralia.* XIII, Part I. Translated by Harold Cherniss. Cambridge, MA: Loeb Classical Library, [1976] 2000.

Plutarch. *Plutarchi Moralia,* edited by J. Mau. Leipzig: Teubner, 1971.

Plutarch. *Plutarch's Morals.* Translated from the Greek by several hands, corrected and revised by William W. Goodwin. Boston: Little, Brown, and Company; Cambridge: Press of John Wilson and Son, 1874.

Poincaré, Henri. *La science et l'hypothèse.* Paris: Flammarion, [1902] 1968.

Pomata, Gianna. "Observation Rising: Birth of an Epistemic Genre, 1500–1650." In *Histories of Scientific Observation,* edited by Lorraine Daston and Elizabeth Lunbeck, 45–80. Chicago: University of Chicago Press, 2011.

Pomata, Gianna. "Sharing Cases: The *Observationes* in Early Modern Medicine." *Early Science and Medicine* 15 (2010): 193–236.

Pomian, Krzysztof. *Collectors and Curiosities: Paris and Venice, 1500–1800.* Cambridge, UK: Polity Press, 1990.

Pompe, Hedwig, and Leander Scholz. *Archivprozesse: Die Kommunikation der Aufbewahrung.* Cologne: Dumont, 2002.

Power, Dan. "Origins of Beer & Diapers." *DSS News* 3, no. 23 (2002). http://www.dssresources.com/newsletters/66.php.

Price, Derek J. de Solla. *Little Science, Big Science.* New York: Columbia University Press, 1963.

Pritzl, Kurt. "Endoxa as Appearances." *Ancient Philosophy* 14 (1994): 41–51.

Protein Data Bank. *Newsletter* 3 (1976): 2.

Protein Data Bank. Policies & References. http://www.rcsb.org/pdb/static.do?p=general_information/about_pdb/policies_references.html. Accessed July 10, 2014.

[Pseudo-Plutarch]. *The Complete Works of Plutarch: Essays and Miscellanies.* Vol. 3. New

York: Crowell, 1909. http://ebooks.adelaide.edu.au/p/plutarch/nature/. Accessed June 23, 2014.

Ptolemy, Claudius. *Ptolemy's Almagest*. Translated by G. J. Toomer. London: Duckworth, 1984.

Radin, Joanna. "Latent Life: Concepts and Practices of Human Tissue Preservation in the International Biological Program." *Social Studies of Science* 43, no. 4 (August 1, 2013): 484–508.

Ranke, Leopold von. *Aus Werk und Nachlass*, vol. 4, *Vorlesungs-Einleitungen*, edited by Volker Dotterweich and Walter Peter Fuchs. Munich: R. Oldenbourg Verlag, 1975.

Ranke, Leopold von. *Zur Kritik neuerer Geschichtsschreiber*. Leipzig: G. Reimer, 1824.

Rasmussen, Anne. "L'Internationale scientifique, 1890–1914." Thesis, Ecole des Hautes Etudes en Sciences Sociales, Paris, 1995.

Raup, David M. *The Nemesis Affair: A Story of the Death of Dinosaurs and the Ways of Science*. New York: W. W. Norton & Co., 1986.

Reardon, Jenny. *Race to the Finish: Identity and Governance in an Age of Genomics*. Princeton: Princeton University Press, 2005.

Rebenich, Stefan. "'Unser Werk lobt keinen Meister': Theodor Mommsen und die Wissenschaft von Altertum." In *Theodor Mommsen: Gelehrter, Politiker und Literat*, edited by Josef Wiesehöfer, 185–205. Stuttgart: Franz Steiner Verlag, 2005.

Rebenich, Stefan, and Gisa Franke, eds. *Theodor Mommsen und Friedrich Althoff: Briefwechsel 1882–1903*. Munich: Oldenbourg Verlag, 2012.

Regiomontanus, Johannes. *De cometae magnitudine, longitudineq[ue] ac de loco eius vero, problemata XVI*. Nuremberg: apud Fridericum Peypus, 1531.

Regiomontanus, Johannes, and Georg von Peurbach. *Epytoma Joa[n]nis de Mo[n]te Regio in almagestu[m] Ptolomei*. [Venice]: impressionis . . . Johannis ha[m]man, 1496.

Reiner, Erica. "Babylonian Celestial Divination." In *Ancient Astronomy and Celestial Divination*, edited by N. M. Swerdlow, 21–37. Cambridge, MA: MIT Press, 1999.

Reinhold, Erasmus. *Prutenicae tabulae coelestium motuum*. Tübingen: Per Ulricum Morhardum, 1551.

Reiser, Stanley J. *Medicine and the Reign of Technology*. Cambridge: Cambridge University Press, 1978.

Rennell, James. *The Geographical System of Herodotus Examined; And Explained by a Comparison with Those of Other Ancient Authors, and with Modern Geography*. London: W. Bulmer and Co., 1800.

Report of the Climate Board Ad Hoc Panel on Climate Impacts to the National Climate Program Office Regarding Social Science Climate Impact Research. Washington, DC: National Academy of Sciences, 1981.

Report of the Climate Data Management Workshop. Washington, DC: US Department of Commerce, 1980.

Revue Germanique Internationale. *La fabrique internationale de la science: Les congrès scientifiques de 1865 à 1945*. Paris: CNRS Editions, 2010.

Reynaud Montlosier, François-Dominique de. *Essai sur la théorie des volcans d'auvergne*. Paris: Imprimerie de Landriot et Rousset, 1789 [1802].

Rheticus, Georg Joachim. *Ad clarissimum virum D. Ioannem Schonerum, de libris revolutionum eruditissimi viri, & mathematici excellentissimi reverendi D. Doctoris Nicolai Copernici Torunnaei, Canonici Varmiensis, per quendam iuvenem, mathematicae studiosum narratio prima*. [Gdańsk]: [Excusum . . . per Franciscum Rhodum], [1540].

Rheticus, Georg Joachim. *Georgii Joachimi Rhetici Narratio prima*, edited and translated by Henri Hugonnard-Roche and Jean-Pierre Verdet. Wrocław: Ossolineum, 1982.

Riccioli, Giovanni Battista. *Almagestum novum astronomiam veterem novamque complectens*. Bologna: Ex Typographia Hæredis Victorij Benatij, 1651.

Riccioli, Giovanni Battista. *Astronomiae reformatae tomi duo*. Bologna: Ex Typographia Hæredis Victorij Benatij, 1665.

Riccioli, Giovanni Battista. *Geographiae et hydrographiae reformatae*. Venice: Typis Ioannis La Noù, 1672.

Rimmer, Matthew. "The Genographic Project: Traditional Knowledge and Population Genetics." *Australian Indigenous Law Review* 11, no. 2 (2007): 33–54.

Robinson, Peter J. "Use of the National Environmental Data Referral Service." *Bulletin of the American Meteorological Society* 65 (1984): 1310–15.

Robson, Eleanor. "Astronomical Diaries and Related Texts from Babylonia." *Journal of the Royal Asiatic Society* 17 (2007): 61–63.

Rochberg, Francesca. "Babylonian Horoscopy: The Texts and Their Relations." In *Ancient Astronomy and Celestial Divination*, edited by N. M. Swerdlow, 39–59. Cambridge, MA: MIT Press, 1999.

Rochberg, Francesca. *The Heavenly Writing: Divination, Horoscopy, and Astronomy in Mesopotamian Culture*. Cambridge: Cambridge University Press, 2004.

Ropelewski, C. F., M. C. Predoehl, and M. Platto, eds. *The Interim Climate Data Inventory: A Quick Reference to Selected Climate Data*. Washington, DC: US Department of Commerce, 1980.

Rose, Herbert J. *The Roman Questions of Plutarch: A New Translation with Introductory Essay and a Running Commentary*. Oxford: Clarendon Press, 1924.

Rosenberg, Charles E. "The Tyranny of Diagnosis: Specific Entities and Individual Experience." *Milbank Quarterly* 80, no. 2 (2002): 237–60.

Rosenberg, Daniel. "An Eighteenth-Century Time Machine: The *Encyclopedia* of Diderot and d'Alembert." *Postmodernism and the Enlightenment*, edited by Daniel Gordon, 45–66. New York: Routledge, 2001.

Rosenberg, Daniel. "Early Modern Information Overload." *Journal of the History of Ideas* 64, no. 1 (January 1, 2003): 1–9.

Rosenberg, Daniel."Electronic Memory." In *Histories of the Future*, edited by Daniel Rosenberg and Susan Harding, 123–52. Durham: Duke University Press, 2005.

Ross, Andrew. "Is Global Culture Warming Up?" *Social Text* 28 (1991): 3–30.

Rothacker, Ernst. *Logik und Systematik der Geisteswissenschaften*. Bonn: H. Bouvier u. Co. Verlag, 1947.

Roy, Deb. "The Birth of a Word." TED talk. http://www.ted.com/talks/deb_roy_the_birth _of_a_word?language=en.

Rubin, Paul. "Indian Givers:" *Phoenix New Times News,* May 27, 2004.

Rudwick, Martin J. S. *Bursting the Limits of Time: The Reconstruction of Geohistory in the Age of Revolution*. Chicago: University of Chicago Press, 2005.

Runia, David T. "The Placita Ascribed to Doctors in Aëtius' Doxography on Physics." In *Ancient Histories of Medicine: Essays in Medical Doxography and Historiography in Classical Antiquity*, edited by Philip J. van der Eijk, 189–250. Leiden: Brill, 1999.

Runia, David T. "What Is Doxography." In *Ancient Histories of Medicine: Essays in Medical Doxography and Historiography in Classical Antiquity*, edited by Philip J. van der Eijk, 33–55. Leiden: Brill, 1999.

Ryghaug, Marianne, and Tomas Moe Skjølsvold. "The Global Warming of Climate Science: Climategate and the Construction of Scientific Facts." *International Studies in the Philosophy of Science* 24, no. 3 (2010): 287–307.

Sachs, Abraham J. "Babylonian Observational Astronomy." *Philosophical Transactions of the Royal Society of London. Series A, Mathematical and Physical Sciences* 276, no. 1257 (1974): 43–50.

Sachs, Abraham J. "A Classification of the Babylonian Astronomical Tablets of the Seleucid Period." *Journal of Cuneiform Studies* 2, no. 4 (1948): 271–90.

Sachs, Abraham J., and Hermann Hunger. *Astronomical Diaries and Related Texts from Babylonia*. Vienna: Verlag der Österreichischen Akademie der Wissenschaften, 1988–.

"Sacred Electronics." *Time*, December 31, 1956, 40.

Sadler, Donald H. "Prediction of Eclipses." *Nature* 211, no. 5054 (1966): 1119–21.

Said, Said S., F. Richard Stephenson, and Wafiq Rada. "Records of Solar Eclipses in Arabic Chronicles." *Bulletin of the School of Oriental and African Studies* 52, no. 1 (1989): 38–64.

Salton, Gerard. "Historical Note: The Past Thirty Years in Information Retrieval." *Journal of the American Society for Information Science* 38, no. 5 (1987): 375.

Samimian-Darash, Limor, and Paul Rabinow. *Modes of Uncertainty*. Chicago: University of Chicago Press, 2015.

Sattler, Hubert. *Die Basedow'sche Krankheit*. Leipzig: W. Engelmann, 1909–10.

Scaliger, Joseph Justus. *Thesaurus temporum*. Amsterdam: Apud Joannem Janssonium, 1658.

Schaffer, Simon. "Babbage's Intelligence: Calculating Engines and the Factory System." *Critical Inquiry* 21 (1994): 203–27.

Scheiner, Julius. *Die Photographie der Gestirne*. Leipzig: Wilhelm Engelmann, 1897.

Schenk, Dietmar. *Aufheben, was nicht vergessen werden darf. Archive vom alten Europa bis zur digitalen Welt*. Stuttgart: Franz Steiner Verlag, 2013.

Schierbaum, Martin, ed. *Enzyklopädistik 1550–1650: Typen und Transformationen von Wissensspeichern und Medialisierung des Wissens*. Berlin: LIT Verlag, 2009.

Schiller, Friedrich. "Über naive und sentimentalische Dichtung." First published in the journal *Die Horen*, edited by Friedrich Schiller. Tübingen: Cotta'sche Verlagsbuchhandlung, 1795–96.

Schindewolf, Otto H. *Basic Questions in Paleontology: Geologic Time, Organic Evolution, and Biological Systematics*. Translated by Judith Schaefer. Chicago: University of Chicago Press, 1993.

Schmidt, Manfred G. "Spiegelbilder römischer Lebenswelt: Inschrift-Clichés aus dem Archiv des Corpus Inscriptionum Latinarum ausgewählt und kommentiert." In *150 Jahre Corpus Inscriptionum Latinarum*, 3–32. Berlin: Walter de Gruyter, 2003.

Schneider, Stephen H. "Climate Change and the World Predicament: A Case Study for Interdisciplinary Research." *Climatic Change* 1 (1977): 21–43.

Schöner, Johann, ed. *Scripta clarissimi mathematici M. Ioannis Regiomontani*. Nuremberg: Apud I. Montanum & V. Neuber, 1544.

Schopf, Thomas J. M. "Evolving Paleontological Views on Deterministic and Stochastic Approaches." *Paleobiology* 5, no. 3 (1979): 337–52.

Schramm, Helmar, Ludger Schwarte, and Jan Lazardig, eds. *Instruments in Art and Science: On the Architectonics of Cultural Boundaries in the 17th Century*. Berlin: De Gruyter, 2008.

Schroeder-Gudehaus, Brigitte. "Die Akademie auf internationalem Parkett: Die Programmatik der internationalen Zusammenarbeit wissenschaftlicher Akademien und ihr

Scheitern im Ersten Weltkrieg." In *Die Königliche Preussische Akademie der Wissen-schaften*, edited by Jürgen Kocka, 175–95. Berlin: Akademie Verlag, 1999.

Schüll, Natasha Dow. "Data for Life: Wearable Technology and the Design of Self-care." *BioSocieties* (7 March 2016): 1–17. doi:10.1057/biosoc.2015.47

Schütrumpf, Eckart E. "Hermann Diels." In *Classical Scholarship: A Biographical Ency-clopedia*, edited by Ward W. Briggs and William C. M. Calder III, 52–60. New York: Garland, 1990.

Scott, Ridley, director. *Blade Runner* (film). Warner/Ladd Company, 1982.

Segal, David. "This Man is Not A Cyborg. Yet." *New York Times*, June 1, 2013. http://www.nytimes.com/2013/06/02/business/dmitry-itskov-and-the-avatar-quest.html?_r=0.

Selective Guide to Climatic Data Sources: Superintendent of Documents, U.S. Government Print-ing Office, Washington DC. Asheville, NC: National Climatic Center, 1969.

Senebier, Jean. *L'art d'observer*. 2 vols. Geneva: Chez Cl. Philibert & Bart. Chirol, 1775.

Sepkoski, David. "The Database before the Computer?" *Osiris* 32, *Data Histories*. (forth-coming).

Sepkoski, David. *Rereading the Fossil Record: The Growth of Paleobiology as an Evolutionary Discipline*. Chicago: University of Chicago Press, 2012.

Sepkoski, David. "Towards 'A Natural History of Data': Evolving Practices and Episte-mologies of Data in Paleontology, 1800–2000." *Journal of the History of Biology* 46, no. 3 (August 2013): 401–44. doi:10.1007/s10739-012-9336-6.

Sepkoski, J. John, Jr. *A Compendium of Fossil Marine Families*. Milwaukee: Milwaukee Public Museum, 1982.

Sewell, Robert, Śaṅkara Bālakṛṣṇa Dīkshita, and Robert Gustav Schram. *The Indian Calendar: With Tables for the Conversion of Hindu and Muhammadan into A.D. Dates, and Vice Versa: with Tables of Eclipses Visible in India by Dr. Robert Schram of Vienna*. London: Swan Sonnenschein & Co., Ltd., 1896.

Shapin, Steven. "The House of Experiment in 17th-Century England." *Isis* 79, no. 298 (September 1988): 373–404.

Shapin, Steven. *A Social History of Truth: Civility and Science in Seventeenth-Century En-gland*. Chicago: University of Chicago Press, 1994.

Sharples, R. "Pseudo-Alexander or Pseudo-Aristotle: Medical Puzzles and Physical Problems." In *Aristotle's* Problemata *in Different Times and Tongues*, edited by Pieter De Leemans and Michèle Goyens, 21–31. Leuven: Leuven University Press, 2006.

Sheynin, Oscar. "The Treatment of Observations in Early Astronomy." *Archives for History of Exact Sciences* 46 (1993): 153–92.

Siegert, Bernhard. "Perpetual Doomsday." In *Europa: Kultur der Sekretäre*, edited by Bernhard Siegert and Joseph Vogl, 63–78. Zürich: diaphanes, 2003.

Silberschatz, Avi, Michael Stonebraker, and Jeffrey D. Ullman. "Database Systems: Achievements and Opportunities." *ACM Sigmod Record* 19, no. 4 (1990): 6–22.

Silberschatz, Avi, and Alexander Tuzhilin. "What Makes Patterns Interesting in Knowl-edge Discovery Systems." *IEEE Transactions on Knowledge and Data Engineering* 8, no. 6 (1996): 970–74. doi:http://doi.ieeecomputersociety.org/10.1109/69.553165.

Sinclair, John McHardy. *Corpus, Concordance, Collocation*. Oxford: Oxford University Press, 1991.

Slawski, Bill. "New Google Approach to Indexing and Stopwords." *SEO by the Sea*, January 17, 2008. http://www.seobythesea.com/2008/01/new-google-approach-to-indexing-and-stopwords/.

Slotsky, Alice Louise. *The Bourse of Babylon: Market Quotations in the Astronomical Diaries of Babylonia*. Bethesda, MD: CDL Press, 1997.

Small, Jocelyn Penny. *Wax Tablets of the Mind: Cognitive Studies of Memory and Literacy in Classical Antiquity*. London: Routledge, 1997.

Smith, Merritt Roe, and Leo Marx, eds. *Does Technology Drive History?: The Dilemma of Technological Determinism*. Cambridge: MIT Press, 1994.

Smith, Robin. *Aristotle Topics*. Books 1 and 8, with excerpts from related texts. Translated and with a commentary by Robin Smith. Oxford: Clarendon, 1997.

Smith, V. Kerry. "Economic Impact Analysis and Climate Change: An Overview and Proposed Research Agenda." *Final Report to the National Climate Program Office*, NOAA. Washington, DC: Department of Commerce, 1980.

Smith, William. *Strata Identified by Organized Fossils*. London: W. Arding, 1816.

Smith, William. *Stratigraphical System of Organized Fossils with Reference to the Specimens of the Original Geological Collection in the British Museum*. London: E. Williams, 1817.

Snell, Willebrord. *Coeli & siderum in eo errantium observationes Hassiacae*. Leiden: Apud Iustum Colsterum, 1618.

Social Science Climate Impact Research. Washington, DC: National Academy of Sciences, 1981.

Soll, Jacob. *The Information Master: Jean-Baptiste Colbert's Secret State Intelligence System*. Ann Arbor: University of Michigan Press, 2009.

Sommer, Marianne. "History in the Gene: Negotiations between Molecular and Organismal Anthropology." *Journal of the History of Biology* 41, no. 3 (2008): 473–528.

Souciet, Etienne, ed. *Observations mathématiques, astronomiques, geographiques, chronologiques, et physiques, tirées des anciens livres chinois, ou faites nouvellement aux Indes et a la Chine, par les peres de la compagnie de Jesus*. Paris: Chez Rollin, 1729–32.

Sparck Jones, Karen. "Comments on Grand Challenge Document (GCD): 'Memories for Life': Managing Information over a Human Lifetime." 2004. http://citeseerx.ist .psu.edu/showciting;jsessionid=37C6DDAC15CCC7D9CAF09696A6434D7D?cid= 977594. Accessed November 21, 2014.

Sparck Jones, Karen. "Four Notes on Memories for Life," 2004. Comments on Grand Challenge Document (GCD): Memories for Life: Managing Information over a Human Lifetime. http://www.nesc.ac.uk/esi/events/Grand_Challenges/gcconf04 /submissions/28.pdf. Accessed November 21, 2014.

Spek, R. J. van der. "New Evidence from the Babylonian Astronomical Diaries Concerning Seleucid and Arsacid History." *Archiv für Orientforschung* 44/45 (1997/1998): 167–75.

Spens, Thomas. "History of a Case in Which There Took Place a Remarkable Slowness of the Pulse." *Medical and Philosophical Commentaries* 7 (1792): 458–65.

Stammen, Theo, and Wolfgang E. J. Weber, eds. *Wissenssicherung, Wissensordnung und Wissensverarbeitung: Das Europäische Modell der Enzyklopädien*. Berlin: Akademie Verlag, 2004.

Stancari, Vittorio Francesco. *Schedae mathematicae post ejus obitum collectae ejusdem Observationes astronomicae*, edited by Eustachio Manfredi. Bologna: Typis Jo: Petri Barbiroli, 1713.

Stanley, Matthew. "Predicting the Past: Ancient Eclipses and Airy, Newcomb, and Huxley on the Authority of Science." *Isis* 103, no. 2 (2012): 254–77.

Steedman, Carolyn. *Dust: The Archive and Cultural History*. Manchester: Manchester University Press, 2001.

Steele, John M. *Ancient Astronomical Observations and the Study of the Moon's Motion (1691–1757).* New York: Springer, 2012.

Steele, John M. "Applied Historical Astronomy: An Historical Perspective." *Journal for the History of Astronomy* 35, no. 3 (2004): 337–55.

Steele, John M. *Observations and Predictions of Eclipse Times by Early Astronomers.* Dordrecht: Kluwer Academic Publishers, 2000.

Steele, John M. "A Re-analysis of the Eclipse Observations in Ptolemy's *Almagest.*" *Centaurus* 42 (2000): 89–108.

Steere, William Campbell, Hazel A. Philson, and Phyllis V. Parkins. *Biological Abstracts/BIOSIS: The First Fifty Years, the Evolution of a Major Science Information Service.* New York: Plenum Press, 1976.

Stephenson, F. Richard. *Historical Eclipses and Earth's Rotation.* Cambridge: Cambridge University Press, 1997.

Stephenson, F. Richard, and David H. Clark. *Applications of Early Astronomical Records.* New York: Oxford University Press, 1978.

Stephenson, F. Richard, and David A. Green. *Historical Supernovae and Their Remnants.* Oxford: Clarendon Press, 2002.

Stephenson, F. Richard, and M. A. Houlden. "The Accuracy of Tuckerman's Solar and Planetary Tables." *Journal for the History of Astronomy* 12 (1981): 133–38.

Stephenson, F. Richard, and Leslie V. Morrison. "Long-term Changes in the Rotation of the Earth: 700 B.C. to A.D. 1980." *Philosophical Transactions of the Royal Society of London. Series A, Mathematical and Physical Sciences* 313, no. 1524 (1984): 47–70.

Stephenson, F. Richard, and Said S. Said. "Records of Lunar Eclipses in Medieval Arabic Chronicles." *Bulletin of the School of Oriental and African Studies* 60, no. 1 (1997): 1–34.

Stevens, Hallam. *Life Out of Sequence: A Data-Driven History of Bioinformatics.* Chicago: University of Chicago Press, 2013.

Stevens, Mary Elizabeth. *Automatic Indexing: A State-of-the-Art Report.* National Bureau of Standards Monograph 91. Washington, DC: U.S. Government Printing Office, 1965.

Stevens, Mary Elizabeth, Vincent E. Giuliano, and Laurence B. Heilprin. *Statistical Association Methods for Mechanized Documentation.* National Bureau of Standards Miscellaneous Publications 269. Washington, DC: U.S. Government Printing Office, 1965.

Stevenson, Lloyd G. "Exemplary Disease: The Typhoid Pattern." *Journal of the History of Medicine and Allied Sciences* 37 (1982): 159–81.

Stigler, Stephen M. *The History of Statistics: The Measurement of Uncertainty before 1900.* Cambridge, MA: Belknap Press of Harvard University Press, 1986.

Stokes, William. "Observations on Some Cases of Permanently Slow Pulse." *Dublin Quarterly Journal of Medical Science* 2 (1846): 73–85.

Stolberg, Michael. "Formen und Funktionen medizinischer Fallberichte in der frühen Neuzeit (1500–1800)." In *Fallstudien: Theorie—Geschichte—Methode,* edited by Johannes Süßmann, Susanne Scholz, and Gisela Engel, 81–95. Berlin: trafo, 2007.

Stoler, Ann Laura. *Along the Archival Grain: Epistemic Anxieties and Colonial Common Sense.* Princeton: Princeton University Press, 2009.

Stone, Linda, and Paul F. Lurquin. *A Genetic and Cultural Odyssey: The Life and Work of L. Luca Cavalli-Sforza.* New York: Columbia University Press, 2005.

Storm, Dunlop, and M. Michèle Gerbaldi, eds. *Stargazers: The Contribution of Amateurs to Astronomy.* Berlin: Springer-Verlag, 1988.

Strasser, Bruno J. "Collecting, Comparing, and Computing Sequences: The Making of

Margaret O. Dayhoff's *Atlas of Protein Sequence and Structure, 1954-1965.*" *Journal of the History of Biology* 43, no. 4 (2010): 623–60.

Strasser, Bruno J. "Collecting Nature: Practices, Styles, and Narratives." *Osiris*, 2nd ser., 27 (2012): 303–40.

Strasser, Bruno J. "Collections." In *Eine Naturgeschichte für das 21. Jahrhundert: hommage à, zu Ehren von, in honor of Hans-Jörg Rheinberger*, edited by Safia Azzouni et al., 25–27. Berlin: Max-Planck-Institut für Wissenschaftsgeschichte, 2011.

Strasser, Bruno J. "Data-Driven Sciences: From Wonder Cabinets to Electronic Databases." *Studies in History and Philosophy of Science Part C: Studies in History and Philosophy of Biological and Biomedical Sciences* 43, no. 1 (March 2012): 85–87. doi:10.1016/j.shpsc.2011.10.009.

Strasser, Bruno J. "The Experimenter's Museum: GenBank, Natural History, and the Moral Economies of Biomedicine." *Isis* 102, no. 1 (2011): 60–96.

Streete, Thomas. *Astronomia Carolina: A New Theory of the Coelestial Motions: Composed According to the Best Observations.* London: Printed for Lodowick Lloyd, 1661.

Strong, James. *The Exhaustive Concordance of the Bible: Showing Every Word of the Text of the Common English Version of the Canonical Books, and Every Occurrence.* London: Hodder & Stoughton, 1890.

Sturdy, Steve. "Knowing Cases: Biomedicine in Edinburgh, 1887–1920." *Social Studies of Science* 37 (2007): 659–89.

Sturtevant, Alfred H., and Theodosius Dobzhansky. "Inversions in the Third Chromosome of Wild Races of Drosophila Pseudoobscura, and Their Use in the Study of the History of the Species." *Proceedings of the National Academy of Sciences of the United States of America* 22, no. 7 (July 1936): 448–50.

Sussman, Joel L. "Protein Data Bank Deposits." *Science News Letter* 282, no. 5396 (December 11, 1998): 1993.

Süßmann, Johannes, Susanne Scholz, and Gisela Engel, eds. *Fallstudien: Theorie— Geschichte—Methode.* Berlin: trafo, 2007.

Swerdlow, N. M., ed. *Ancient Astronomy and Celestial Divination.* Cambridge, MA: MIT Press, 1999.

Swerdlow, Noel M. "Astronomy in the Renaissance." In *Astronomy before the Telescope*, edited by Christopher Walker, 187–230. London: British Museum Press, 1996.

Swerdlow, Noel M. *The Babylonian Theory of the Planets.* Princeton: Princeton University Press, 1998.

Swerdlow, Noel M. "Tycho, Longomontanus, and Kepler" on Ptolemy's Solar Observations and Theory, Precession of the Equinoxes, and Obliquity of the Ecliptic." In *Ptolemy in Perspective: Use and Criticism of His Work from Antiquity to the Nineteenth Century*, edited by Alexander Jones, 151–202. Dordrecht: Springer, 2010.

Swerdlow, Noel M., and Otto Neugebauer. *Mathematical Astronomy in Copernicus's De revolutionibus.* New York: Springer-Verlag, 1984.

System Development Corporation and J. Citron. *A Permutation Index to the Preprints of the International Conference on Scientific Information.* Santa Monica: System Development Corporation, 1959.

Szalay, András S., and R. J. Brunner. "Astronomical Archives of the Future: A Virtual Observatory." *Future Generation Computer Systems* 16, no. 1 (1999): 63–72.

Taavitsainen, Irma. "Changing Conventions of Writing: The Dynamics of Genres, Text Types, and Text Traditions." *European Journal of English Studies* 5, no. 2 (2001): 139–50.

Talboys Wheeler, James. *The Life and Travels of Herodotus in the Fifth Century before Christ: An Imaginary Biography Founded on Fact, Illustrative of the History, Manners, Religion, Literature, Art and Social Condition of the Greeks, Egyptians, Persians, Babylonians, Hebrews, Scythians and Other Ancient Nations in the Days of Pericles and Nehemiah*. Vol. 1. London: Longman, 1855.

Tallbear, Kim. "Narratives of Race and Indigeneity." *Journal of Law, Medicine and Ethics* 35, no. 3 (2007): 412–24.

Tasman, Paul. *Indexing the Dead Sea Scrolls by Electronic Literary Data Processing Methods*. White Plains, NY: IBM, 1958.

Tasman, Paul. "Literary Data Processing." *IBM Journal of Research and Development* 1, no. 3 (1957): 249–56.

Taub, Liba. *Ancient Meteorology*. London: Routledge, 2003.

Taub, Liba. "Cosmology and Meteorology." In *Cambridge Companion to Epicureanism*, edited by James Warren, 105–24. Cambridge University Press, 2009.

Taub, Liba. "On the Variety of 'Genres' of Greek Mathematical Writing: Thinking about Mathematical Texts and Modes of Mathematical Discourse." In *Writing Science*, edited by Markus Asper, 333–36. Berlin: De Gruyter, 2013.

Taub, Liba. "Presenting a 'Life' as a Guide to Living: Ancient Accounts of the Life of Pythagoras." In *The History and Poetics of Scientific Biography*, edited by Thomas Söderqvist, 17–36. Aldershot: Ashgate, 2007.

Taub, Liba. "'Problematising' the *Problems*: The *Problemata* in Relation to Other Question-and-Answer Texts." In *The Aristotelian* Problemata Physica: *Philosophical and Scientific Investigations*, edited by Robert Mayhew, 413–36. Leiden: Brill, 2015.

Terrall, Mary. *Catching Nature in the Act: Réaumur and the Practice of Natural History in the Eighteenth Century*. Chicago: University of Chicago Press, 2014.

Terrall, Mary. *The Man Who Flattened the Earth: Maupertuis and the Sciences in the Enlightenment*. Chicago: University of Chicago Press, 2002.

Terras, Melissa. "For Ada Lovelace Day: Father Busa's Female Punch Card Operatives." Melissa Terras Blog, http://melissaterras.blogspot.com/2013/10/for-ada-lovelace-day-father-busas.html. Accessed April 10, 2014.

Thelen, Shawn, Sandra Mottner, and Barry Berman. "Data Mining: On the Trail to Marketing Gold." *Business Horizons* 47 (2004): 26. doi:10.1016/j.bushor.2004.09.005.

Theroux, Marcel. *Strange Bodies*. New York: FSG, 2013.

Thies, David. "Stop Words Are Dead! Did I Miss Another Memo?" *SEO Fast Start*, January 9, 2008. http://www.seofaststart.com/stop-words-are-dead/.

Thomas, David Hurst. *Skull Wars: Kennewick Man, Archaeology, and the Battle for Native American Identity*. New York: Basic Books, 2000.

Thomson Reuters. "Thomson Reuters launches data citation index for discovering global data sets" (2 April 2013). http://thomsonreuters.com/content/press_room/science/730914. Accessed November 17, 2014.

Tieleman, Teun L. *Galen and Chrysippus on the Soul*. Leiden: Brill, 1996.

Tihon, Anne, and Raymond Mercier, eds. *Ptolemaiou procheiroi kanones*. Louvain-la-Neuve: Université catholique de Louvain, Institut orientaliste, 2011.

Tomezsko, George. *Fully Occupied Years: The Rise and Fall of a Company Called BIOSIS*. Philadelphia: Xlibris, 2006.

Trousseau, Armand. "Du goître exophthalmique," lessons in the Clinique de la faculté de médecine de Paris: Hôtel-Dieu, with case reports by attending physicians. *Union médicale* 8 (1860): 434–39.

Tsvetkov, M., et al., eds. *Proceedings of the International Workshop on Virtual Observatory: Plate Content Digitization, Archive Mining [and] Image Sequence Processing*. [Sofia], Bulgaria: Heron Press, 2005.

Tuckerman, Bryant. *Planetary, Lunar, and Solar Positions, 601 B.C. to A.D. 1 at Five-day and Ten-day Intervals*. Philadelphia: American Philosophical Society, 1962.

Tuckerman, Bryant. *Planetary, Lunar, and Solar Positions, A.D. 2 to A.D. 1649 at Five-day and Ten-day Intervals*. Philadelphia: American Philosophical Society, 1964.

Tully, Françoise Le Guet, Jean Davoigneau, Jérôme Lamy, Jérôme de La Noë, Jean-Michel Rousseau, and Hamid Sadsouad. "Les traces matérielles de la Carte du Ciel: Le cas des observatoires d'Alger et Bordeaux." In *La Carte du Ciel: Histoire et actualité d'un projet scientifique international*, edited by Jérôme Lamy, 213–35. Paris: Observatoire de Paris, 2008.

Tully, Françoise Le Guet, Jérôme de La Noë, and Hamid Sadsaoud. "L'Opération de la Carte du Ciel dans le contexte institutionnel et technique de l'astronomie française à la fin du XIXe siècle." In *La Carte du Ciel: Histoire et actualité d'un projet scientifique international*, edited by Jérôme Lamy, 69–107. Paris: Observatoire de Paris, 2008.

Turner, Herbert H. *The Great Star Map*. New York: E. P. Dutton, 1912.

Uhlir, Paul F. *Board on Research Data and Information, Policy and Global Affairs, National Research Council. For Attribution—Developing Data Attribution and Citation Practices and Standards*. Washington, DC: National Academy Press, 2012.

Uhlir, Paul F. "Information Gulags, Intellectual Straightjackets, and Memory Holes: Three Principles to Guide the Preservation of Scientific Data." *Data Science Journal* 10 (2010): 1–5.

Ullman, Jeffrey D. "CS 345 Data Mining Lecture Notes" (unpublished, 2000). https://web .archive.org/web/20030623052236/http://www-db.stanford.edu/~ullman/mining /overview.pdf .

Ullman, Jeffrey D. "The MIDAS Data-Mining Project at Stanford." In *Database Engineering and Applications, 1999. IDEAS'99. International Symposium Proceedings*, 1999, 460–64. http://ieeexplore.ieee.org/xpls/abs_all.jsp?arnumber=787298.

UNESCO. *The Race Concept: Results of an Inquiry*. Paris: [UNESCO], 1952.

Van Helden, Albert. *Measuring the Universe: Cosmic Dimensions from Aristarchus to Halley*. Chicago: University of Chicago Press, 1985.

Varlejs, Jana. "The Technical Report and Its Impact on Post–World War II Information Systems: Two Case Studies." In *The History and Heritage of Scientific and Technological Information Systems: Proceedings of the 2002 Conference of the American Society for Information Science and Technology*, edited by W. Boyd Rayward and Mary Ellen Bowden, 89–99. Medford, NJ: Information Today, for ASIS&T and Chemical Heritage Foundation, 2004.

Vegetti, Mario. "Tradition and Truth: Forms of Philosophical-scientific Historiography in Galen's *De placitis*." In *Ancient Histories of Medicine: Essays in Medical Doxography and Historiography in Classical Antiquity*, edited by Philip J. van der Eijk, 333–57. Leiden: Brill, 1999.

Veilleux, Mary P. "Permuted Title Word Indexing Procedures for a Man/Machine System." In *Machine Indexing*, 1962, 77–111. Washington, DC: American University.

Venezky, Richard L. *A Microfiche Concordance to Old English: The High-Frequency Words*, edited by Sharon Butler. Newark, DE: University of Delaware, 1983.

Venezky, Richard L. *A Microfiche Concordance to Old English: The High-Frequency Words*, edited by Sharon Butler. Toronto: Published for the Dictionary of Old English

Project, Centre for Medieval Studies, University of Toronto by the Pontifical Institute of Mediaeval Studies, 1985.

Venezky, Richard L. "Computer-Aided Humanities Research at the University of Wisconsin." *Computers and the Humanities* 3, no. 3 (January 1, 1969): 129–38.

Vesna, Victoria. "Seeing the World in a Grain of Sand: The Database Aesthetics of Everything." In *Database Aesthetics: Art in the Age of Information Overflow*, edited by Victoria Vesna, 1–39. Minneapolis: University of Minnesota Press, 2007.

Vidal, Fernando, and Nélia Dias, eds. *Endangerment, Biodiversity and Culture*. London: Routledge, 2016.

Vidal, Fernando. *The Sciences of the Soul: The Early Modern Origins of Psychology*. Chicago: University of Chicago Press, 2011.

Virchow, Rudolf. "Weisses Blut." *Neue Notizen aus dem Gebiete der Natur- und Heilkunde* 36 (1845): cols. 150–56.

Vismann, Cornelia. *Akten: Medientechnik und Recht*. Frankfurt am Main: Fischer Taschenbuch Verlag, 2000.

Voltaire. "Le Pyrrhonisme de l'Histoire" (1768). In *Oeuvres Complètes de Voltaire*, vol. 14. Stuttgart: Frères Hartmann à la Haye, 1829.

Wailoo, Keith, Alondra Nelson, and Catherine Lee, eds. *Genetics and the Unsettled Past: The Collision of DNA, Race, and History*. New Brunswick, NJ: Rutgers University Press, 2012.

Walford, Antonia. "Data Moves: Taking Amazonian Climate Science Seriously." *Cambridge Anthropology* 30, no. 2 (2012): 101–17

Walker, Richard T. "A Method for the Rapid and Accurate Deposition of Nucleic Acid Sequence Data in an Acceptably-Annotated Form." In *Biomolecular Data, a Resource in Transition*, edited by Rita Colwell, 45–51. Oxford: Oxford University Press, 1989.

Wallhauser, Andrew. "Hodgkin's Disease." *Archives of Pathology* 16 (1933): 522–62, 672–712.

Wallis, John. *The Correspondence of John Wallis*, edited by Philip Beeley and Christoph J. Scriba. Oxford: Oxford University Press, 2003.

Watson, James D. *The Double Helix: A Personal Account of the Discovery of the Structure of DNA: Text, Commentary, Reviews, Original Papers*. New York: Touchstone, [1968] 2001.

Weart, Spencer. *The Discovery of Global Warming*. Cambridge, MA: Harvard University Press, 2008.

Weber, Max. "Wissenschaft als Beruf." In Max Weber, *Gesammelte Aufsätze zur Wissenschaftslehre*, 3rd ed., edited by Johannes Winckelmann, 582–613. Tübingen: J. C. B. Mohr, 1968.

Webster Prentiss, Daniel. "Report of Three Cases of Remarkably Slow Pulse, to which Is Appended Brief Abstracts of Ninety-Three Cases of Slow Pulse Found Recorded in Medical Journals in the Library of the Surgeon-General's Office, Washington DC." *Transactions of the Association of American Physicians* 4 (1889): 120–59.

Weimer, Théo. *Brève histoire de la Carte du Ciel en France*. Paris: Observatoire de Paris, 1987.

Weimer, Théo. "Naissance et développement de la Carte du Ciel." In *Mapping the Sky: Past Heritage and Future Directions*, edited by Suzanne Débarat, J. A. Eddy, H. K. Eichhorn, and A. R. Upgren, 29–32. Dordrecht: Kluwer, 1988.

Weinberg, Bella Haas. "Predecessors of Scientific Indexing Structures in the Domain of Religion." In *History and Heritage of Scientific and Technological Information Systems: Proceedings of the 2002 Conference of the American Society for Information Science and Technology*, edited by W. Boyd Rayward and Mary Ellen Bowden, 126–36. Medford, NJ: Information Today, 2004.

Weinberger, David. *Too Big to Know: Rethinking Knowledge Now That the Facts Aren't the*

Facts, Experts Are Everywhere, and the Smartest Person in the Room Is the Room. New York: Basic Books, 2011.

Weinrich, Harald. "The Textual Function of the French Article." In *Literary Style: A Symposium*, edited by Seymour Chatman, 221–40. New York: Oxford University Press, 1971.

"Weisses Blut und Milztumoren." *Schmidts Jahrbücher der in- und ausländischen Gesammten Medicin* 57 (1848): 181–88.

Weld, Kristin. *Paper Cadavers: The Archives of Dictatorship in Guatemala.* Durham: Duke University Press, 2014.

Wells, Spencer. *Deep Ancestry: Inside the Genographic Project.* Washington, DC: National Geographic, 2006.

Wells, Spencer. *Genographic Project Ethical Framework.* n.p, n.d.

Wendelin, Godefroy. *Eclipses lunares ab anno M. D. LXXIII. ad M. DC. XLIII. observatae quibus Tabulae Atlanticae superstruuntur earumque idea proponitur.* Antwerp: Apud Hieronymum Verdussium, 1644.

Werlich, Egon A. *A Text Grammar of English.* Heidelberg: Quelle & Meyer, 1976.

Werrett, Simon. "The Astronomical Capital of the World: Pulkova Observatory in the Russia of Tsar Nicholas." In *The Heavens on Earth: Observatories and Astronomy in Nineteenth-century Science and Culture*, edited by David Aubin, Charlotte Bigg, and H. Otto Sibum, 33–57. Durham: Duke University Press, 2010.

White, Graeme L. "The Carte du Ciel—The Australian Connection." In *Mapping the Sky: Past Heritage and Future Directions*, edited by Suzanne Débarat, J. A. Eddy, H. K. Eichhorn, and A. R. Upgren, 45–51. Dordrecht: Kluwer, 1988.

Wickert, Lothar, ed. *Theodor Mommsen–Otto Jahn: Briefwechsel 1842–1868.* Frankfurt am Main: Vittorio Klostermann, 1962.

Widmalm, Sven. "A Commerce of Letters: Astronomical Communication in the Eighteenth Century." *Science Studies* 5, no. 2 (1992): 43–58.

Wiesehöfer, Josef, ed. *Theodor Mommsen: Gelehrter, Politiker und Literat.* Stuttgart: Franz Steiner Verlag, 2005.

Wilkinson, Alec. "Remember This?" *New Yorker*, May 28, 2007. 38–44.

Wilks, Samuel. "Cases of Enlargement of the Lymphatic Glands and Spleen (or, Hodgkin's Disease) with Remarks." *Guy's Hospital Reports*, 3rd ser., 11 (1865): 56–67.

Williamson, George S. *The Longing for Myth in Germany: Religion and Aesthetic Culture in Germany from Romanticism to Nietzsche.* Chicago: University of Chicago Press, 2004.

Williamson, Kirsty, and Graeme Johanson, eds. *Research, Information, Systems, and Contexts.* Prahran, Australia: Tilde, 2013.

Wilson, Curtis. *The Hill-Brown Theory of the Moon's Motion: Its Coming-to-be and Short-lived Ascendancy (1877–1984).* New York: Springer, 2010.

Wilson, Curtis. "Predictive Astronomy in the Century after Kepler." In *Planetary Astronomy from the Renaissance to the Rise of Astrophysics.* Part A: *Tycho Brahe to Newton*, edited by René Taton and Curtis Wilson, vol. 2A, 161–206. Cambridge: Cambridge University Press, 1989.

Winchester, Simon. *The Map That Changed the World: William Smith and the Birth of Modern Geology.* New York: HarperCollins, 2001.

Wing, Vincent. *Astronomia Britannica. . . . Cui accesit observationum astronomicarum synopsis compendiaria, ex quâ astronomiae britannicae certitudo affatim elucescit.* London: Typis Johannis Macock, 1669.

Winslett, Marianne, and Rakesh Agrawal. "Rakesh Agrawal Speaks Out on Where the Data Mining Field Is Going, Where It Came From, How to Choose Problems and

Open Up New Fields, Our Responsibilities to Society as Technologists, What Industry Owes Academia, and More," 2003. http://www.sigmod.org/publications/interview/pdf/D15.rakesh-final-final.pdf.

Winterhalter, Albert G. *The International Astrophotographical Congress and a Visit to Certain European Observatories and Other Insitutions: Report to the Superintendent.* Washington, DC: Government Printing Office, 1889.

Witte, Barthold C. *Der preussische Tacitus: Aufstieg, Ruhm und Ende des Historikers Barthold Georg Niebuhr, 1776–1831.* Düsseldorf: Droste Verlag, 1979.

Witten, Ian H., Alistair Moffat, and Timothy C. Bell. *Managing Gigabytes: Compressing and Indexing Documents and Images.* 2nd ed. San Francisco: Morgan Kaufmann Publishers, 1999.

Wlodawer, Alexander. "Deposition of Macromolecular Coordinates Resulting From Crystallographic and NMR Studies." *Nature Structural Biology* 4, no. 3 (March 1997): 173–74.

Wolf, Gary, and Ernesto Ramirez. *Quantified Self-Public Health Report* (2014). http://quantifiedself.com/symposium/Symposium-2014/QSPublicHealth2014_Report.pdf.

Wood, Robert. *An Essay on the Original Genius and Writings of Homer* (1769). London: H. Hughs, 1775.

Worthington, W. C., Sr. "Remarkable Slowness of the Pulse." *Lancet* 36, no. 926 (1841): 336–37.

Wright, Alex. *Cataloging the World: Paul Otlet and the Birth of the Information Age.* Oxford: Oxford University Press, 2014.

Wyatt, Sally. "Technological Determinism Is Dead; Long Live Technological Determinism." In *Handbook of STS Studies*, edited by Edward J. Hackett, Olga Amsterdamska, Michael Lynch, and Judy Wajcman, 165–80. Cambridge, MA: MIT Press, 2008.

Xu, Zhentao, Yaoting Jing, and David W. Pankenier. *East Asian Archaeoastronomy: Historical Records of Astronomical Observations of China, Japan and Korea.* Amsterdam: Gordon & Breach, 2000.

Yatsuhashi, Akira. "In the Bird Cage of the Muses: Archiving, Erudition, and Empire in Ptolemaic Egypt." Ph.D. thesis, Duke University, 2010.

Yunis, Harvey, ed. *Written Texts and the Rise of Literate Culture in Ancient Greece.* Cambridge: Cambridge University Press, 2003.

Zedelmaier, Helmut. *Bibliotheca Universalis und Bibliotheca Selecta: Das Problem der Ordnung des gelehrten Wissens in der Frühen Neuzeit.* Cologne: Böhlau, 1992.

Zhang, Tian, Raghu Ramakrishnan, and Miron Livny. "BIRCH: An Efficient Data Clustering Method for Very Large Databases." In *Proceedings of the 1996 ACM SIGMOD International Conference on Management of Data, Montreal, Quebec, Canada, June 4–6, 1996*, edited by H. V. Jagadish and Inderpal Singh Mumick, 103–14. Montreal: ACM Press, 1996.

Zhmud, Leonid. "The Historiographical Project of the Lyceum: The Peripatetic History of Science, Philosophy and Medicine." *Antike Naturwissenschaft und ihre Rezeption* 13 (2003): 109–26.

Zhmud, Leonid. *The Origin of the History of Science in Classical Antiquity.* Translated by Alexander Chernoglazov. Berlin: De Gruyter, 2006.

Zhmud, Leonid. "Revising Doxography: Hermann Diels and His Critics." *Philologus* 145 (2001): 219–43.

Zinner, Ernst. *Regiomontanus, His Life and Work* [1938]. Translated by Ezra Brown. Amsterdam: North-Holland, 1990.

Zuckerkandl, Emile, and Linus Pauling. "Molecules as Documents of Evolutionary History." *Journal of Theoretical Biology* 8, no. 2 (1965): 357–66.

Page numbers in italics indicate illustrations.

astronomical (celestial scientific) records: applied historical astronomy and, 17, 18, 22; celestial histories from, 19, 29–30, 34–37, 47nn78–80, 50nn104–5, 50n110, 51n113; and collection, data, 26–28, 37–38, 113; collectivity/collective knowledge and, 24, 32, 170; data deluge and, 37; during Middle Ages, 17, 21, 24–25; modern period of, 5, 20, 21–22, 25–26, 27–28, 37–38, 41n28, 45n58; opportunism and, 32; overview and description of, 4–10, 12, 17–18; ownership of, 8, 33–34, 49n97; print publication/transcriptions of, 25–26, 29–30, 31–34, 44n50, 45n55, 45nn57–58, 48n89, 49n93, 49nn97–98; science and, 37–38; time/timescales and, 7, 22, 329–30, 331; uncanonized archive of, 28–31, 35, 47nn78–80, 50n105. See also canonical archive of astronomical records

astrophotography, 169, 171, 180n49, 180n53
Astrophysics Data System (ADS), NASA's, 17, 37
authorship, data, 185, 190–91, 196–97, 198, 199
autopsia epigraphy, 86, 90, 103
Auzout, Adrian, 31

Babylonian archive, 19–21, 22–24, 30, 34, 39n10, 39n13, 39n16, 45n58
Bacon, Francis, 144
Bailly, Jean-Sylvain, 19
Bainbridge, John, 42n36
Bartholin, Erasmus, 29–30, 47n79
Basedow'sche Krankheit (Graves' disease), 92–93, 99
BASIC (Biological Abstracts' Subject in Context), 300–301, 302
Battānī, al-, 21, 43n48
Bayle, Pierre, 142–43
Behm, Johannes, 23
Bell, Gordon: as lifelogger, 247–48, 255, 263, 265n1, 265n6, 267n29, 269n53; MyLifeBits by, 248, 255–56, 264, 265, 267n32
Bergman, Torbern Olof, 89
Berman, Helen, 191
Bernal, Martin, 148
Bessel, Friedrich, 170
Best, Stephen, 305
BIBCON, 276
Bible concordance(s), 284, 285–87, *292–93*, 297
bibliographies, 92–93, 101, 102
big data: data deluge and, 12, 37, 185, *186*, 187, 188, 193, 199; data mining of, 96, 311–12, 314, 316, 323–24; data sharing and, 185,

186–87; punch card use and, 283–84; in writing and electronic archives, 291
Big Science, 160, 161, 176, 177n6. *See also* science
biocolonialism, and human genetic material, 204, 208–10. *See also* neocolonialism, and human genetic material; postcolonialism, and human genetic material
biographical texts: archiving scientific ideas in Antiquity and, 117–18, 132n34, 133n36; doxographical texts or *doxai* as analogous to, 118, 120, 132nn33–34, 132n36; overview and description of, 132n34
Biological Abstracts' Subject in Context (BASIC), 300–301, 302
Birdsell, Joseph, 207
Blade Runner (film), 264
Bliss, Nathaniel, 34
Blumenbach, Johann Friedrich, 89
Boas, Franz, 205
Böckh, August, 147, 163, 164, 165
Bodin, Jean, 143
Borges, Jorge Luis, 304, 310n75
Borghesi of San Marino, Count Bartolomeo, 164, 168, 178n19
Bormann, Eugen, 167
Borri, Cristoforo, 27
Bossuet, Jacques-Bénigne, 137–38, 140–41, 155n10
Bouguer, Pierre, 35–36
Boulliau, Ismail, 26, 27, 29
Bowker, Geoffrey C., 3, 305
Bradley, James, 33, 34, 50n106
Brahe, Tycho, and astronomical records: and collection, data, 27; Hesse-Kassel observations, 26, 45n57; ownership of observations, 8; print publication/transcriptions, 26, 29, 30, 32, 45n57, 48n89; Tuckerman tables, 21; uncanonized archive, 26, 30, 47n78; Uraniborg observatory, 87
Brin, Sergey, 320, 321–24
Brongniart, Alexandre, 57, 60, *61*, 62, 65, 68
Bronn, H. G., 68, *69*, 71, 72, 75
Bünting, Heinrich, 28
Burnett, William, 89, 91, 92, 93–94
Burtin, François-Xavier, 89
Busa, Roberto, and topics: Aquinas scholarship and concordance, 9, 274–75, 284, 287, *288*, 289, *289*, *290*, 291, 294–95, *294*; concordances, 284, 287, *288*, 289, *289*, *290*, 291, 294, 309n66; digital humanities, 289, 309n62; linguistic structure

Busa, Roberto (*continued*)
   analysis, 273–75, 283; stop lists, 284;
   textual analysis/textuality, 295
Bush, Vannevar, 251, 256, 267n31
Buttmann, Philipp, 145

Calvisius, Seth, 23
canonical archive of astronomical records:
   Babylonian archive as, 19–21, 22–24, 30,
   34, 39n10, 39n13, 39n16, 45n58; eclipse
   canons and, 10, 21–23, 24, 27, 28, 29,
   35, 41n31, 330; Egyptian archive as, 19,
   22–24, 39n10, 45n58; in Greece, ancient,
   19, 22–23, 30, 42n36; Mesopotamian
   archive as, 4, 5, 19, 21, 329–30; Ptolemy
   and, 23–25, 36–37, 42n36, 43n48, 44n49,
   45n58; and Rome, ancient, 18–20, 22–23,
   39n13, 45n58; royal canon as, 23–25,
   42nn36–37, 43n41, 45n58; uncanonized
   archive versus, 28–31, 35, 47nn78–80,
   50n105. *See also* astronomical (celestial
   scientific) records
Carte du Ciel: as archive for future research,
   159–62, 171, 172–73, 176; archives as data
   and, 173; astrophotography and, 169,
   171, 180n49, 180n53; digital/electronic
   database of, 11; geopolitics of archival
   projects for future research and, 162,
   169–71; industrialism and, 173–74,
   175, 180n65; positivism and, 97; print
   publication/transcriptions and, 171–73,
   172, 180n53; women's role in, 174, 175,
   180n65; workers'/students' role in, 174,
   180n65
cases/case reports, in library medical case
   research, 87–88, 106nn7–8
Cassini family of astronomers: Cassini,
   Giovanni Domenico, 35; Cassini,
   Jacques, 35, 50n110; Cassini, Jean-
   Dominique, 36; Cassini de Thury, César-
   François, 33, 51n113; print publication/
   transcriptions of, 32–33
Cavalli-Sforza, L. Luca, 208–9, 210, 217
ceaseless curation/curation, in self-archiving
   practices, 256, 264, 267n31
celestial histories, 19, 29–30, 34–37, 47nn78–
   80, 50nn104–5, 50n110, 51n113. *See also*
   astronomical (celestial scientific) records
celestial scientific (astronomical) records.
   See astronomical (celestial scientific)
   records; canonical archive of astronomical records
Centre de données astronomiques de

Strasbourg's Set of Identifications,
   Measurements, and Bibliography for
   Astronomical Data (SIMBAD), 37
Church, George, 262
church archives (ecclesiastical archives), 4–5,
   164–65, 175
Cicero, 119, 120
CIG (*Corpus Inscriptionum Graecarum*),
   147, 163. See also *Corpus Inscriptionum
   Graecarum* (*CIG*)
CIL (*Corpus Inscriptionum Latinarum*). See
   Corpus Inscriptionum Graecarum (*CIG*);
   Corpus Inscriptionum Latinarum (*CIL*)
citation, data, 196, 197–98, 199
Climate Research Board, NRC's, 227–28,
   229, 231, 236. *See also* United States (US)
   National Research Council (NRC)
climatological data archiving: of climate
   change, 224, 226, *236*, 237, 239; climate
   defined and, 225; climatic futures and,
   223, 224, 227; climatology and, 224, 227;
   and collection, data, 223, 224, 225–26,
   231; of data, climate, 223–26, 234, *235*,
   239; data hygiene and, 236; economics/
   socioeconomics of, 223–24, 226–31, 233,
   238; empiricism in, 224, 237, 240; and
   flow/workflow, data/information, 235–
   38; for future research, 236, 239, 240–41;
   knowledge production/epistemic practices and, 230–31, 233, 238; and management, data, 223–24, 226, 227, 228–31,
   232, 233–35, *235*; open-endedness of, 7,
   8, 236; overview of, 7, 8, 9, 12, 128–29,
   223–26; and policy, climate/environ-
   mental, 224, 226, 228, 234, *235*, 238; and
   politics, environmental, 223–24, 237–39;
   and products, climate data, 223, 224, 227,
   229, 230, 233–36, *235*; and transformation, data, 224, 231, 234, 238; warehouse
   model for, 224–25. *See also* United States
   (US) National Climate Program
climatology, 224, 227. *See also* climatological
   data archiving
clinical medicine, and library medical case
   research, 89–90, 92–93, 94, 99, 100,
   102–3, 104
collection, data (acquisition of data): archival
   projects for future research and, 165–66,
   165, *167*, 168, 171, 173, 178n26; archive₄ or
   numerical data of fossils, 68, 71–72, 74–
   75, 82n40; archiving scientific ideas in
   Antiquity, 114–15; astronomical records,
   26–28, 37–38, 113; climatological data ar-

Dobzhansky, Theodosius, 205–7, 208, 214
Dodson, Michael "Mick," 209–10, 216
Domaszewski, Alfred von, 167
d'Orbigny, Alcide, 65, 68
doxographical texts (*doxai*): as archive/
　archiving opinions, 10, 115, 120–21, 124,
　128–29; biographical texts as analogous
　to, 118, 120, 132nn33–34, 132n36; as
　data, 124, 133n50; *endoxa* or reputable
　opinions and, 122, 123, 124, 126, 127,
　133n49; ethical philosophy as, 118,
　119, 131n21, 132n33; as genre critique,
　119–20, 121, 132nn29–31, 132nn33–34,
　132n37; on medicinal subjects, 118;
　on natural philosophy, 117, 118, 119,
　121; *placita* texts and, 118, 119, 122; and
　reconfiguration, data/information, 121,
　132n37; and retrieval, data/information,
　120–21; as second natures, 128–29; and
　storage, data, 114, 115, 121; as term of
　use, 118, 131n20; as third nature, 129;
　transmission of data/information in,
　121; users of, 10, 115, 121–26, 131nn20–
　23, 133n47, 133nn49–50, 134n54–55,
　134n61, 134n64, 135nn65–66. *See also*
　archiving scientific ideas in Antiquity
dreams, nightly, 249–50, 260, *264*, 265
Dressel, Heinrich, 167
Dreyer, John L. E., 29
Droysen, Johann G., 150, 152–53
Duncker, Max, 149
Dunn, L. C., 205
Dunthorne, Richard, 21, 35
Dymaxion Chronofile, 253–54, 258, 267n26

earth as archive: antiquarianism and, 57–60,
　61–62, 81n21; archive₀ or the earth and,
　56, 62, 65, *66–67*, 68, *69*, 70–71, *70*; and
　contingency, historical, 57; first nature
　and, 56; narrative approach to, 57, 59,
　60–61, *61*, 70–72, *70*, *72*
ecclesiastical archives (church archives), 4–5,
　164–65, 175
eclipse canons, 10, 21–23, 24, 27, 28, 29, 35,
　41n31, 330. *See also* canonical archive of
　astronomical records
economics/socioeconomics: of climatological
　data archiving, 223–24, 226–31, 233,
　238; moral economies in data sharing
　and, 185, 190–91, 193, 196, 198. *See also*
　power relations
Edwards, Paul, 284, 309n64
Egyptian archive, 19, 22–24, 39n10, 45n58

Eichhorn, J. G., 145
Eijk, Philip J. van der, 120, 131n20
Eldredge, Niles, 53–54
electronic archives, and writing. *See* writing,
　and electronic archives
electronic database of fossil archive (archive₅),
　74–75, 77–78, *78*, 79. *See also* digital/
　electronic database(s)
EMBL (European Molecular Biology Labora-
　tory) data library, 195–96
empiricism: in climatological data archiving,
　224, 237, 240; in data mining, 324; in
　library medical case research, 92, 96,
　97, 101, 102; in self-archiving practices,
　249–50
encyclopedic compilations, 97–98, 103
endoxa (reputable opinions), 122, 123, 124,
　126, 127, 133n49. *See also* doxographical
　texts (*doxai*)
Engelbart, Douglas C., 299, 309n58
environmental (climate) policy, 224, 226,
　228, 234, *235*, 238. *See also* climatological
　data archiving
epigraphy, 160, 161, 163, 164, 165–66, 174,
　179n33
epilepsy, in library medical case research,
　88–89, 90, 91–92, 93–94, 97, 102
epistemic practices (knowledge production).
　See knowledge production (epistemic
　practices)
Eskildsen, Kaspar Risbjerg, 139–40, 153–54
ethical philosophy, as doxographical texts/
　doxai, 118, 119, 131n21, 132n33
ethics, scientific, 211, 212, 215–16, 218
European Molecular Biology Laboratory
　(EMBL) data library, 195–96
evidence-based medicine, 104
experimental life sciences: archival projects
　for future research and, 188; data sharing
　and, 187–89, 194; GenBank and, 187, 191,
　193–94, 195–96; Protein Data Bank and,
　2, 8, 9, 191–93

Fabian, Johannes, 203
Fabroni, Angelo, 139
Facebook, 258, 259, 260, 272
families' role, in archival projects, 174–75
fantasies about archives: archives/archive
　practices and, 3, 11–12, 13n10; self-
　archiving practices and, 251–52, 257,
　258, 267n19; in writing and electronic
　archives, 12, 294–95
Fayyad, Usama, 311, 312, 319

human history (human sciences) (*continued*) and, 2, 13, 60; and contingency, historical, 54–55; fossil records as compared with, 80n10; narrative approach to, 71, *73*; spindle diagrams of, 71, *73. See also* science

Hunger, Hermann, 20

IBM: Aquinas scholarship and concordance, 9, 287, *288*, 294; BASIC (Biological Abstracts' Subject in Context) program, 300; data mining group at, 312, 313, 317–18, 320–21; Genographic Project funding by, 214; KWIC (Keyword-in-Context) indexing system, 295; stop lists, 282, *282*; women in writing and electronic archives, 9, 302

Illuminator of the Path (*Me'ir Nativ*), 285, 307n30

immortality, 3, 12, 329, 330. *See also* archival projects, for future research

indexes: computers and, 276, 280, 302; Data Citation Index, 198; fantasies about archives and, 11; index reading and, 297, 305; KWIC indexing system and, 282–83, *282*, *283*, 295–304, *296*, *298*; on magnetic tape, 301, 309n66; overview and history of, 284, 331; punch cards usage and, 302; and retrieval, data/information, 321

indigenous genetics, and human genetic material, 203–4, 208, 209–14, 215–16, 218

industrialism, and archival projects for future research, 160, 163, 166, 168, 173–74, 175

information (data). *See* big data; collection, data (acquisition of data); collections; data (information); data mining; data sharing; digital/electronic database(s); storage, data; transformation, data

infraordinary, in concordance(s), 280, 283

InnerNet, 257–61, *261*, 268nn38–39

interestingness, rules/value of, 102, 317, 326n25

Internet. *See* World Wide Web (Web), and data mining

Itskov, Dmitry, 256, 257, 263

Jablonski, P. E., 141

Jahn, Otto, 164

James, William, 275–76, 283, 306n7

Jenner, Edward, 99

Joachim, Johann, 59

Joint Declaration of Data Citation Principles in 2013, 197–98

Jones, William, 142

Josephus, Flavius, 18–19

KDD (knowledge discovery in databases), 311, 314–15, 326n14

Kellermann, Olaus, 164, 178n19

Kelley, Donald, 139

Kepler, Johannes, 21, 26

keyword(s): stop lists and, 283; in writing and electronic archives, 295–96, 297–98, *298*, *299*, 300, 303–4

Keyword-in-Context (KWIC) indexing system, 282–83, *282*, *283*, 295–304, *296*, *298*

Kimball, Ralph, 313

Klebs, Elimar, 167

Klumpke, Dorothea, 174

knowing: library medical case research and, 90, 95, 96, 97, 99, 100, 102 (*see also* unknown(s), and library medical case research); self-archiving practices and, 249, 259, 261–62, 266n8

knowledge discovery in databases (KDD), 311, 314–15, 326n14

knowledge production (epistemic practices): archives and, 55, 56; climatological data archiving, 230–31, 233, 238; in data mining, 312, 313–14, 315, 326n14; data sharing and, 187–88, 189, 190, 195; fossil records and, 55, 56, 75; human genetic material and, 204, 208, 216; paleontology and, 55, 56, 187; in scientific archives, 187

Kosuri, Sriram, 262

KWIC (Keyword-in-Context) indexing system, 282–83, *282*, *283*, 295–304, *296*, *298*

Lacaille, Nicolas-Louis de, 36

La Condamine, Charles-Marie de, 35–36

La Hire, Philippe de, 35

Lalande, Joseph-Jérôme, 21, 36

Landsberg, Helmut, 230

Lansbergen, Philips van, 27

La Vopa, Anthony J., 147

Leibniz, Gottfried Wilhelm, 13n10, 168

Le Mercier, François, 31

Le Monnier, Pierre-Charles, 34–35, 36, 50nn104–5

Lepoix, Charles, 90

leukemia, 93, 102

Lévi-Strauss, Claude, 264

Lhwyd, Edward, 58–59

Melanchthon, Philip, 25, 28
memory (remembering) practices. *See* remembering (memory) practices
Mesopotamian archive, 4, 5, 19, 21, 329–30
metaphors, archival, 7, 10, 11, 76–77
Meyers, Edgar, 191
Michaelis, J. D., 145
Middle Ages: archives defined during, 187; astronomical records during, 17, 21, 24–25; historicism or historical truth during, 151, 152, 159; human history, 58; and indexes, history of, 284; library medicine during, 103; and organization, scholarly, 166, 168; royal canonical archives, 5; scientific archives during, 6
Miller, Peter N., 145
modern period: of astronomical records, 5, 20, 21–22, 25–26, 27–28, 37–38, 41n28, 45n58; in library medical case research, 96
Momigliano, Arnaldo, 59, 139, 145
Mommsen, Theodor, and topics: Big Science as term of use, 160; *CIL*, 162–69, *165*; collection, data, 164, 165–66, *165*, *167*, 168, 171, 173, 178n26; geopolitics of archival projects for future research, 162–63, 164, 168; historical skepticisms, 148; industrialism, 160, 166, 168; modern and ancient historians' methodological divide, 149, 153; organization, scholarly, 161, 163, 164–65, 166–68, 169; workers'/students' role in data collection, 167, 174, 179n33
moral economy(ies), 185, 190–91, 193, 196, 198
Moretti, Franco, 275, 305
Morgagni, Giovanni Battista, 89, 92, 93–94, 97, 102
Moritz, Landgraf, of Hesse, 26, 45n57
Mosso, Ugolino, 250
Mouchez, Ernest, 169, 170–71
Muhlack, Ulrich, 139
Müller, Johannes, 139
Müller, K. O., 148
Muratori, Lodovico, 164
MyLifeBits, 248, 255–56, 264, 265, 267n32

narrative approach: in cases in library medical case research, 87–88; databases and, 75; earth as archive and, 57, 59, 60–61, *61*, 70–72, *70*, *72*; in human history/human sciences, 71; in paleontology, 54–55, 60–61, *61*, 70–72, *70*

NASA (National Aeronautics and Space Administration), 10, 17, 23, 37, 330
National Center for Atmospheric Research, 236
National Climatic Data Center (NCDC), 235, 238. *See also* National Oceanic and Atmospheric Administration (NOAA)
National Environmental Satellite, Data, and Information Service (NESDIS), 235, 238
National Institutes of Health (NIH), 193, 194–95
National Oceanic and Atmospheric Administration (NOAA), 226, 229, 230, 233, 235, 238
National Science Foundation (NSF), 199, 300
Native Americans, 210–13, 214, 215, 218
naturalist sciences, and data sharing, 187, 188–89, 190, 194
natural philosophy, and archiving scientific ideas in Antiquity, 10, 116, 117, 118, 121, 124, 125, 127
NCDC (National Climatic Data Center), 235, 238. *See also* National Oceanic and Atmospheric Administration (NOAA)
neocolonialism, and human genetic material, 204, 210–14. *See also* biocolonialism, and human genetic material; postcolonialism, and human genetic material
NESDIS (National Environmental Satellite, Data, and Information Service), 235, 238
Neugebauer, Otto, 9, 20–21
Newton, Isaac, 33, 59, 141
Niebuhr, Barthold Georg, 146, 147, 149
Nielson, Michael, 186–87, 198
NIH (National Institutes of Health), 193, 194–95
NOAA (National Oceanic and Atmospheric Administration), 226, 229, 230, 233, 235, 238
nomothetic paleontology, 54, 56, 76, 77. *See also* paleontology and paleontological archives
NRC (United States [US] National Research Council), 223–24, 227–28, 229, 231, 234, 236, 242n15
NSA (United States [US] National Security Agency), 272, 324, 331
NSF (National Science Foundation), 199, 300
numerical data of fossils (archive$_4$), 68, 71–72, 74–75, 82n40

objectification of self (self-commodification), 252–53, 255, 263

Ptolemy, Claudius (*continued*)
36–37, 42n36, 43n48, 44n49, 45n58; collection, data, 27, 36, 113–14; collection of opinions, 123–24; uncanonized archive of astronomical records, 28–29
PubMed, 105
pulse, and Stokes-Adams disease, 88–89, 90, 91–92, 93–94, 97
punch cards usage, stop lists and, 283–84, *289*, *290*
punctuated equilibria model, 53–54, 79

Quantified Self movement, 256–57, 330. *See also* self-archiving practice(s)
Quellenkritik (source criticism), 144–45, 147, 148, 149, 150, 151–52, 153, 154

Rabinow, Paul, 262
RAFI (Rural Advancement Foundation International), 209
Ranke, Leopold von, and topics: father of history as discipline, 139, 153, 154; historicism or historical truth, 139, 151; history/historians' identification with archives, 2, 79; modern and ancient historians' methodological divide, 138, 139, 151–54, 157n46; modern historians' reliance on archives, 138, 152–53; narrative approach, 153; *Quellenkritik* or source criticism, 150, 151–52, 153, 154; universal history, 138, 153, 154; the Vatican's prototypical archive, 4
Raynal, Abbé, 140
reconfiguration, data/information: archiving scientific ideas in Antiquity and, 115–16, 121, 132n37; and metaphors, archival, 10; and retrieval, information, 114, 129n5; of scientific archives, 6, 10
Regiomontanus, Johannes, 25–26, 43n48, 45n55, 47n80
Reinhold, Erasmus, 28
Reiser, Stanley J., 91
remembering (memory) practices: in archives, 3, 10; in library medical case research, 87, 92–93; in scientific archives, 3; self-archiving practices and, 248, 252, 257, 261, 262, 263–64, 265n5
reputable opinions (*endoxa*), 122, 123, 124, 126, 127, 133n49. *See also* doxographical texts (*doxai*)
retrieval, data/information: archival practices, 2, 5; archiving scientific ideas in Antiquity and, 113, 114, 115, 129n9;

doxographical texts or *doxai* and, 120–21; indexes and, 321; in libraries, 1; in library medical case research, 92, 101, 105; and reconfiguration, information, 114, 129n5; scientific archives/practices and, 7–8, 9–10; writing in electronic archiving practices and, 321, 327n37. *See also* indexes
retrieval, information, access to collections and, 8, 13n14
review articles, 101–3
Reynaud Montlosier, François-Dominique de, 60
Riccioli, Giovanni Battista, 28–29
Richards, Frederic, 191
Roberts, Isaac, 171
Robertson, William, 137–38, 142
Rokitansky, Carl, 102
Rollin, Charles, 137–38, 140, 148
Rome, ancient: antiquarian approach to human history and, 58, 59; archives of, 5, 18–20, 22–23, 39n13, 45n58. *See also* archiving scientific ideas in Antiquity; doxographical texts (*doxai*); Greece, ancient
Room with a View, A (Forster), 281
Rosenberg, Charles, 100
Ross, Andrew, 224
Roy, Deb, 248, 265n4
royal canonical archive, 23–25, 42nn36–37, 43n41, 45n58
Rudwick, Martin J. S., 57, 59
Runia, David T., 117, 118, 124, 131n20, 134n54
Rural Advancement Foundation International (RAFI), 209
Rymer, Thomas, 151

Sachs, Abraham J., 20, 21, 39n13
Sadler, Donald H., 41n28
Samimian-Darash, Limor, 262
Savigny, Friedrich Carl von, 164
Scaliger, Joseph Justus, 42n36, 164
Schickard, Wilhelm, 47n78
Schiller, Friedrich, 144
Schindewolf, Otto H., 79
Schlosser, F. C., 150
Schlözer, A. L., 140
Schmidt, Johannes, 179n33
Schöner, Johann, 25–26, 44n50, 45n55
Schopf, Thomas J. M., 76–77
Schüll, Natasha, 250, 268n39
science: astronomical records, 37–38; Big

Vidal, Fernando, 266n8
Villemain, Abel-François, 161, 164
Virchow, Rudolf, 102–3
Voltaire, 140, 141–42, 143, 146, 151
Voss, Johannes, 145, 147

Wallis, John, 47n79
Walther, Bernard, 25–26, 30, 43n48, 45n55, 47n80
warehouse model, for climatological data archiving, 224–25
Watson, James, 190–91
Web (World Wide Web), and data mining, 11, 311–12, 320, 321–24
Weber, Max, 176
Weinberger, David, 186–87
Weinrich, Harald, 275, 283
Weld, Kristen, 250
Wells, Spencer, 214–15, 216–17
Welser, Marcus, 164
Wendelin, Godefroy, 26, 27
Werner, Johannes, 47n80
Wheeler, James Talboys, 149
Wilhelm IV of Hesse-Kassel, 26, 30, 45n58
Wolf, Friedrich A., 145, 146
women: in archival projects used for future research, 174, 175, 180n65; and authorship, data, 190–91; in writing and electronic archives, 9, *289*, *290*, 302
Woodward, John, 58–59, 65, 68
Wordle, 273, *274*, 305
Wordle word clouds, 273, *274*, 305
workers'/students' role, in archival projects for future research, 167, 174, 179n33, 180n65

workflow/flow, data/information. *See* flow/workflow, data/information
World Climate Program, 227
World Meteorological Organization, 225, 239
World Wide Web (Web), and data mining, 311–12, 320, 321–24
writing, and electronic archives: archives and, 271–72, 305; BASIC and, 300–301, 302; big data and, 291; computers and, 276, 280, 281, 282; digital humanities and, 287, 289, 300–302, *303*, 304–5, 309n62; distant reading and, 275, 305; fantasies about archives and, 12, 294–95; index reading and, 297, 305 (*see also* indexes); keywords and, 295–96, 297–98, *298*, *299*, 300, 303–4; KWIC indexing system and, 282–83, *282*, *283*, 295–304, *296*, *298*; linguistic structure analysis, 273–76, 281–82, 283–84, 304, 306n7, 310n75, 324; magnetic tape and, 301, 309n66; overview and description of, 4, 12, 271–72, 305; and retrieval, data/information, 129n5, 321; textual analysis/textuality and, 291, 295–300, *296*, *298*, *299*, 309n58; women's role in, 9, 302. *See also* concordance(s); stop list(s); stop word(s)
Wunderlich, Carl, 87

Zacut, Abraham, 30, 47n80
Zhmud, Leonid, 123, 124–25, 132n30, 132n37
Zuckerkandl, Emile, 207

CPSIA information can be obtained
at www.ICGtesting.com
Printed in the USA
LVHW03s0955140718
583472LV00005B/15/P

9 780226 432366